i-mode: A Primer

i-mode: A Primer

Nik Frengle

M&T Books
An imprint of Hungry Minds, Inc.

Best-Selling Books • Digital Downloads • e-Books • Answer Networks •
e-Newsletters • Branded Web Sites • e-Learning

New York, NY • Cleveland, OH • Indianapolis, IN

i-mode: A Primer

Published by
M&T Books
An imprint of Hungry Minds, Inc.
909 Third Avenue
New York, NY 10022
www.hungryminds.com

Copyright © 2002 Hungry Minds, Inc. All rights reserved. No part of this book, including interior design, cover design, and icons, may be reproduced or transmitted in any form, by any means (electronic, photocopying, recording, or otherwise) without the prior written permission of the publisher.

Library of Congress Control Number: 2001092895

ISBN: 0-7645-4884-0

Printed in the United States of America

10 9 8 7 6 5 4 3 2 1

1B/RZ/RS/QR/IN

Distributed in the United States by Hungry Minds, Inc.

Distributed by CDG Books Canada Inc. for Canada; by Transworld Publishers Limited in the United Kingdom; by IDG Norge Books for Norway; by IDG Sweden Books for Sweden; by IDG Books Australia Publishing Corporation Pty. Ltd. for Australia and New Zealand; by TransQuest Publishers Pte Ltd. for Singapore, Malaysia, Thailand, Indonesia, and Hong Kong; by Gotop Information Inc. for Taiwan; by ICG Muse, Inc. for Japan; by Intersoft for South Africa; by Eyrolles for France; by International Thomson Publishing for Germany, Austria, and Switzerland; by Distribuidora Cuspide for Argentina; by LR International for Brazil; by Galileo Libros for Chile; by Ediciones ZETA S.C.R. Ltda. for Peru; by WS Computer Publishing Corporation, Inc., for the Philippines; by Contemporanea de Ediciones for Venezuela; by Express Computer Distributors for the Caribbean and West Indies; by Micronesia Media Distributor, Inc. for Micronesia; by Chips Computadoras S.A. de C.V. for Mexico; by Editorial Norma de Panama S.A. for Panama; by American Bookshops for Finland.

For general information on Hungry Minds' products and services please contact our Customer Care department within the U.S. at 800-762-2974, outside the U.S. at 317-572-3993 or fax 317-572-4002.

For sales inquiries and reseller information, including discounts, premium and bulk quantity sales, and foreign-language translations, please contact our Customer Care department at 800-434-3422, fax 317-572-4002 or write to Hungry Minds, Inc., Attn: Customer Care Department, 10475 Crosspoint Boulevard, Indianapolis, IN 46256.

For information on licensing foreign or domestic rights, please contact our Sub-Rights Customer Care department at 212-884-5000.

For information on using Hungry Minds' products and services in the classroom or for ordering examination copies, please contact our Educational Sales department at 800-434-2086 or fax 317-572-4005.

For press review copies, author interviews, or other publicity information, please contact our Public Relations department at 317-572-3168 or fax 317-572-4168.

For authorization to photocopy items for corporate, personal, or educational use, please contact Copyright Clearance Center, 222 Rosewood Drive, Danvers, MA 01923, or fax 978-750-4470.

LIMIT OF LIABILITY/DISCLAIMER OF WARRANTY: THE PUBLISHER AND AUTHOR HAVE USED THEIR BEST EFFORTS IN PREPARING THIS BOOK. THE PUBLISHER AND AUTHOR MAKE NO REPRESENTATIONS OR WARRANTIES WITH RESPECT TO THE ACCURACY OR COMPLETENESS OF THE CONTENTS OF THIS BOOK AND SPECIFICALLY DISCLAIM ANY IMPLIED WARRANTIES OF MERCHANTABILITY OR FITNESS FOR A PARTICULAR PURPOSE. THERE ARE NO WARRANTIES WHICH EXTEND BEYOND THE DESCRIPTIONS CONTAINED IN THIS PARAGRAPH. NO WARRANTY MAY BE CREATED OR EXTENDED BY SALES REPRESENTATIVES OR WRITTEN SALES MATERIALS. THE ACCURACY AND COMPLETENESS OF THE INFORMATION PROVIDED HEREIN AND THE OPINIONS STATED HEREIN ARE NOT GUARANTEED OR WARRANTED TO PRODUCE ANY PARTICULAR RESULTS, AND THE ADVICE AND STRATEGIES CONTAINED HEREIN MAY NOT BE SUITABLE FOR EVERY INDIVIDUAL. NEITHER THE PUBLISHER NOR AUTHOR SHALL BE LIABLE FOR ANY LOSS OF PROFIT OR ANY OTHER COMMERCIAL DAMAGES, INCLUDING BUT NOT LIMITED TO SPECIAL, INCIDENTAL, CONSEQUENTIAL, OR OTHER DAMAGES.

Trademarks: Hungry Minds, the Hungry Minds logo, M&T Books, the M&T Books logo, and Professional Mindware are trademarks or registered trademarks of Hungry Minds, Inc. in the United States and other countries and may not be used without written permission. All other trademarks are the property of their respective owners. Hungry Minds, Inc., is not associated with any product or vendor mentioned in this book.

 is a trademark of Hungry Minds, Inc.

 is a trademark of Hungry Minds, Inc.

About the Author

 Nik Frengle was born in Portland, Oregon, but now makes his home in Tochigi, Japan with his son Christopher and wife Meiko. He is a Mobile Industry Analyst with IntaDev, Incorporated, in Tokyo, Japan. He has been writing since he was six and programming computers since he was ten. He got his first mobile phone in 1994, and it weighed over a pound. Besides technical writing, he enjoys writing short stories and mystery novels.

Credits

SENIOR ACQUISITIONS EDITOR
Sharon Cox

SENIOR PROJECT EDITOR
Jodi Jensen

DEVELOPMENT EDITOR
Marla Reece-Hall

TECHNICAL EDITOR
Jean-Baptiste Minchelli

COPY EDITORS
Susan Christophersen
Nancy Sixsmith

EDITORIAL MANAGER
Mary Beth Wakefield

SENIOR VICE PRESIDENT, TECHNICAL PUBLISHING
Richard Swadley

VICE PRESIDENT AND PUBLISHER
Joseph B. Wikert

SENIOR PERMISSIONS EDITOR
Carmen Krikorian

MEDIA DEVELOPMENT SPECIALIST
Travis Silvers

PROJECT COORDINATOR
Nancee Reeves

GRAPHICS AND PRODUCTION SPECIALISTS
Karl Brandt, Sean Decker,
Brian Drumm, Gabriele McCann,
Jackie Nicholas, Barry Offringa,
Jacque Schneider, Betty Schutte,
Jeremey Unger, Erin Zeltner

QUALITY CONTROL TECHNICIANS
Laura Albert, John Greenough,
Carl Pierce

PROOFREADING AND INDEXING
TECHBOOKS Production Services

COVER IMAGE
© Noma/Images.com

SPECIAL HELP
Sara Shlaer

Foreword

The term *i-mode* has been on the lips of mobile Internet watchers for almost three years. During this time, the concept has evolved way beyond NTT DoCoMo's original black and white cell phones to include Java applications, location-based services, and – with a view to the future – will mature into a range of 3G services under the FOMA brand.

As i-mode has grown, rival mobile and Internet players have learned to emulate some of its successes. Fixated with technology differences between WML and cHTML for the better part of last year, Europe's mobile community began only recently to talk of i-mode's success in terms of business models. Countless carriers, including Vodafone, Deutsche Telekom's T-Motion portal and Norway's Telenor have now put in place commercial strategies that borrow, for the main part, from i-mode models.

In the realm of cell phone technology, international players are now learning valuable lessons. Rival carriers have long admired NTT DoCoMo's ability to have handset vendors sculpt their phones around its definition of the i-mode vision. Slowly, but surely, international carriers have also begun to seek out new sources to help them develop cell phones to their own specifications. Following this aspect of the i-mode model should help them offer customers a considerably more fulfilling mobile data experience than in the past.

Now that more carriers are beginning to catch up with i-mode, content providers and other players can start taking advantage of mobile data with far greater effectiveness. The lessons learned from Japan will provide the backbone for these companies as they move into the wireless space.

From an international perspective, however, the lessons don't stop there. Nik Frengle's research brings into superb relief some very finely tuned lessons from Japan. Although, undeniably, there have been many winners in the i-mode market, the number of unreported failures is even greater. Business models, technology issues, and marketing methodology have all played their part in helping i-mode players generate new business. By focusing on all three issues, *i-mode: A Primer* provides a key insight into the best and worst of i-mode practices.

i-mode's relevance doesn't end there. Although the mood for 3G has generally quieted outside Japan, significant investment in 3G services will soon be on the agenda for all players in the mobile market. If DoCoMo can maintain the head start it has with i-mode into its FOMA offering, the rest of the world will continue to look toward Japan both for inspiration and to avoid some of the pitfalls that affected many novices to mobile Internet the first time around.

Danny Williams
Editorial Director, Mobile Internet
Baskerville, part of Informa Telecoms Group
www.telecoms.com/mi

Preface

NTT DoCoMo of Japan's i-mode wireless Web-enabled mobile phones is the single most successful mobile data system in the world today. With nearly 27 million subscribers, i-mode users represent about one in five Japanese people. i-mode has changed behavior in Japanese society at large by making access to Internet e-mail and online information ubiquitous. Although the full social implications of this always-on, always-connected Internet won't be known for some time, the business and technical implications can and must be understood by those in positions to implement mobile Internet applications in other parts of the world, including the U.S.

Using a packet-switched network, a variant of HTML, TCP/IP, and standard e-mail and Web protocols, i-mode is a system designed to take advantage of the ready availability and low cost of the infrastructure already used for the wired Internet. In taking advantage of this infrastructure, i-mode has unleashed the potential of content providers by providing a reasonable cost of entry and attractive potential rewards based on NTT DoCoMo's billing system. In much the same way that the wired Internet gave anyone with an ISP account a platform to share something with the world, i-mode allows entry to just about anyone who can use a text editor. Certainly, as on the wired Internet, much of this material is amateurish and shallow. By using open standards, however, to give everyone an equal shot, NTT DoCoMo has helped the outstanding to stand out.

Who Should Read This Book

This book is targeted at both those who have a simple interest in the technology and business sides of i-mode, and those who wish to develop new services and applications for i-mode. As I wrote this book, there were no other books about i-mode currently available in English. As a moderator on a mailing list about i-mode, I run into questions every day from those interested in i-mode who don't know where to go for answers. This book is an attempt to answer many of those questions, both of a technical and a business nature.

It's also a clear and straightforward discussion of how to develop applications for i-mode. cHTML and Java applications are both covered at a comprehensive beginning level, and a user should be able to take away a strong basic understanding of the techniques and considerations that go into developing i-mode applications.

Hardware and Software Requirements

Because this book is not strictly a development book, you can get quite a lot out of it without even turning on a computer. To use the database applications on your desktop, however, you need Windows 95 or later, or you need to already have PHP compiled into an Apache server and have mySQL installed. The applications on the enclosed CD-ROM require Windows; the version depends on the application. These applications include emulators, an IDE, sound editing software, a text editor, and others.

If you don't plan to use any of these specific applications but only wish to view the examples, which are also on the CD-ROM, you should be fine with a computer capable of opening CD-ROMs in the ISO 9660 format.

How This Book Is Organized

Part I, "Definition and History of i-mode," looks at the software and hardware of i-mode, as well as the company behind it. This part looks at the software that actually runs i-mode, as well as the content currently available for i-mode. Network protocols, network architecture, and network behavior are all discussed. You get a brief look at the history and people behind i-mode, and I look ahead at what the future holds.

Part II, "The i-mode Environment," takes a broad look at how i-mode is being used, how it fits into Japanese society, and how it addresses the needs of its users. I also discuss the requirements for becoming an official NTT DoCoMo Information Provider.

Part III, "Developing i-mode Applications," covers developing cHTML and Java content for i-mode. I talk about how to develop within the memory, processing power, network speed, and size constraints of a mobile phone handset. I show you each cHTML tag and talk about how to reproduce these on a handset. Then I show you how to develop a mobile address book using cHTML, PHP, and mySQL. The rest of the chapter takes a look at the most basic classes available in DoCoMo's DoJa API, shows you how to develop an addition to your previously developed address book, introduces you to a simple but enjoyable game, and talks about sound formats and how to convert sounds for use with i-mode.

Part IV, "Case Studies of i-mode: Implementations and Services," looks at three companies that provide official i-mode content. I walk you through each of their business models and discuss the success or failure of each model up to now. Finally, we analyze the lessons we can learn from each of these companies.

The appendixes give you information about what's included on the CD-ROM that accompanies this book, along with providing a list of i-mode sites, the API for i-mode Java, a quick cHTML reference, and a list of graphical emoji characters available on the i-mode handset.

Companion CD-ROM

The accompanying CD-ROM includes source code examples from the book, a number of useful applications, and an electronic, searchable copy of the book. Check out Appendix A for more details about the contents of the CD.

Navigating This Book

If you are reading this book because of an overall interest in i-mode, I would suggest starting at the beginning and reading straight through. In Parts I and II, you should get a clear picture of exactly what i-mode is, who is using it, who is developing for it, and the company that created it. In Part III, you get exposed to the tools you need to develop content for i-mode. First is cHTML, followed by sound formats, PHP, and Java. For each of these tools, a sample application is developed from the ground up. Although a familiarity with HTML, server-side scripting, and Java are helpful in these chapters, they are by no means required to benefit from the lessons taught. I have taken some of the pain out of the process of learning a new language by providing sample applications that are useful enough to be interesting, yet easy enough to be grasped by nearly anyone. After you have learned to code your killer app, Part IV provides case studies of three companies that develop i-mode content to help you learn how to sell your application.

If your main interest is in developing for i-mode, you should probably start in Part III, which takes you directly into developing i-mode Web pages using a combination of cHTML, PHP, mySQL, and Java applications using NTT DoCoMo's API. This is not an advanced book, so the examples are fairly basic. No prior experience with i-mode is assumed. Yet it's a useful introduction even for those who make their living writing HTML pages or coding Java applications. I hope that this book points you in the right direction so that you can apply your previous knowledge of those fields to this new platform.

If your primary interest is in the business aspects of i-mode, I suggest that you read Parts II and IV first. These two sections cover the business environment of i-mode and give specific examples of companies in the business of content production for i-mode. I provide general background information, specific examples of implementation, and a short guide on how to become an official content provider with NTT DoCoMo.

Conventions

Here are some conventions that I use throughout the book to help you use it more efficiently:

- *Italics* indicate a new term that I'm defining, represent placeholder text (especially when used in code), or add emphasis.

- A special `monofont` typeface is used throughout the book to indicate code, a filename or pathname, or an Internet address.

Icons Used in This Book

Icons appear in the text to indicate important or especially helpful items. Here's a list of the icons and their functions:

Notes provide additional or critical information and technical data on the current topic.

The Tip icon points you to useful techniques and helpful hints..

Cross-Reference icons point you to someplace where you can find more information on a particular topic.

The Caution icon is your warning of a potential problem or pitfall.

Further Information

All of the code examples can be found on my web site at `http://book.eimode.com/`. If you don't want to install Apache, PHP, or mySQL, you can still sample the applications built on them by going to this site. You won't be able to put your own content on the site to test, but for simply viewing examples, this is a reasonable alternative to installing a Web server on your own computer. This site also includes a links area, as well as a forum for asking questions and getting answers to common questions about i-mode.

I welcome your feedback. You can reach me at my home on the Internet: Eimode, at `http://book.eimode.com` (the sample code is also available here).

Acknowledgments

It takes so many people to create a book of this sort, and authors really only know about the people we deal with directly. So I would first like to acknowledge, not by name, because I don't know any of your names, the people in the background who helped to make this book a reality. I know and appreciate what you have done: Thank you to the graphics people who created or re-created my graphics, the proofreaders who dealt with my poor spelling, the legal people who dealt with my slow trickle of permission letters, and all the rest who have worked behind the scenes but whose contributions have been so important to the completion of this project.

I would like to thank Sharon Cox, Senior Acquisitions Editor at Hungry Minds, for asking me to do this project in the first place and for trusting and believing in me. Without Sharon, this book would never have happened, and I would not have had the chance to write a book that badly needed to be written.

To my agent, Carol Susan Ross, I can only say thank you for all your support, good advice, and tough-as-nails approach to the publishing business. You have taught me so many lessons about the business of writing, and I hope to learn more from you in the future.

To Chandani Thapa, Project Editor at Hungry Minds in New York, I thank you for your firm and cheerful approach to the project, and I'm sorry that we couldn't see it through to the end together.

To Jodi Jensen, Senior Project Editor at Hungry Minds in Indianapolis, thank you for your much-needed help in bringing the book to completion and in getting it into the shape it needed to be for publication.

I would like to thank my son, Christopher. Although he hasn't understood what it is papa is doing at his computer all the time, he has foregone the play that we both really wanted. I promise that if I do another book, I will adjust the schedule to have more time to play.

To Meiko, my wife, I would just like to say that I love you and thank you for your support. I know that it has been hard to be married to me during the time I have been working on this book, and I am sorry for what you have had to put up with. I won't promise that this will be the last book, but I will try harder in the future to make sure that it's the last one I write without days off.

Last, and far from least, I would like to thank Marla Reece-Hall, of Reece-Hall Editorial Services. You are very, very good at what you do, and it is with the utmost humility and gratitude that I say that without you, this book would not be the book it is, and it would almost certainly not be one that I am as proud of. You taught me more about writing in our short and intense six weeks than I had learned in the previous sixteen years. You are a wonderful editor and an amazing person, and you have all my prayers and good wishes for a happy and healthy future.

Contents at a Glance

Foreword................................vii

Preface..................................ix

Acknowledgments..........................xiii

Part I	Definition and History of i-mode
Chapter 1	What Is i-mode?........................... 3
Chapter 2	Getting to Know DoCoMo 23
Chapter 3	i-mode Software 41
Chapter 4	i-mode Hardware 57

Part II	The i-mode Environment
Chapter 5	The Direction of Mobile Network Development ... 71
Chapter 6	The Audience 81
Chapter 7	The Developers — Getting a Piece of the i-mode Pie 89

Part III	Developing i-mode Applications
Chapter 8	Discovering the Lost Joy of Coding Small 105
Chapter 9	cHTML, the Language Used for Creating i-mode Pages 123
Chapter 10	Playing Sounds in i-mode 157
Chapter 11	Programming in cHTML: A Tutorial 163
Chapter 12	i-Appli: The i-mode Version of Java 211
Chapter 13	Programming I-applis: A Tutorial 237
Chapter 14	Creating an i-Appli Game 265

Part IV	Case Studies of i-mode: Implementations and Services
Chapter 15	Case Study 1: Walkerplus.com 291
Chapter 16	Case Study 2: Index Corporation 295
Chapter 17	Case Study 3: Nikkei 307

Appendix A: What's On the CD-ROM? 313

Appendix B: A Complete List of Official
             i-mode Sites 317

Appendix C: i-mode Java API 385

Appendix D: cHTML and X-HTML Basic Tags . . . 439

Appendix E: Emoji Symbol Codes 449

Index. 457

Hungry Minds, Inc. End-User License Agreement. . . 486

Contents

Foreword vii

Preface ix

Acknowledgments xiii

Part 1	Definition and History of i-mode
Chapter 1	**What Is i-mode?** 3
	Exploring Handsets 4
	Using the handset for Internet browsing 8
	Using the handset for e-mailing 9
	Using the handset for phoning 11
	Exploring i-mode Services 13
	A brief description of packet networks 13
	Specific services and an exploration
	of the revenue model 14
	i-Appli, the i-mode version of Java 15
Chapter 2	**Getting to Know DoCoMo** 23
	The Little Spin-off that Could 23
	A brief chronology 24
	Key players 27
	The Path Toward i-mode 29
	Strictly business as usual 29
	Catering to a new market 30
	Leveraging the existing user base 30
	Going with what works 31
	Strategies and Growth 33
	Updating the infrastructure 34
	Knowing the market 36
	Expanding into the global market 37
Chapter 3	**i-mode Software** 41
	The Software Architecture of i-mode 41
	All that power in the palm of your hand—
	the system structure 41
	The layers between the Access browser and the RTOS 43

	The i-mode Browser and the cHTML Standard 45
	XHTML in brief 47
	The differences between cHTML and i-mode HTML 50
	Running midlets in the browser 51
	Speaking in i-mode: Language issues and globalization 52
	Taking a Look at Content Providers 53
Chapter 4	**i-mode Hardware** **57**
	NTT's Packet-Switched Standard 57
	Can a "proprietary" network compete with WAP? 57
	A technical definition of the network architecture for i-mode networks 60
	The i-mode server 61
	Packet-Switching Using DoCoMo's PDC-P Standard 64

Part II	**The i-mode Environment**
Chapter 5	**The Direction of Mobile Network Development** ... **71**
	The Revolt Against the Fixed Line: If a Home Phone Costs This Much, Let's Go Mobile! 71
	Search For the Killer App: A Build-Up Before the 3G Storm 73
	Wireless: A David versus Goliath Struggle 76
	L-mode, and the i-moding of the Fixed-line Internet 77
	Other networked appliance plans 78
Chapter 6	**The Audience** **81**
	Who Uses i-mode and for What? 81
	Young people – the standard-bearers of i-mode 81
	How businesses have implemented i-mode-based solutions 83
	How Does i-mode Fit the Needs of Its Audience? 84
	Messaging, the little app that made i-mode 84
	Customization and the need for uniqueness 85
Chapter 7	**The Developers – Getting a Piece of the i-mode Pie** **89**
	A Further Exploration of the Revenue Model 89
	New entrepreneurs – a little bit of revenue multiplied many times equals a fortune 91
	Developers' relationship with DoCoMo 92
	NTT DoCoMo Standard for Inclusion on i-mode Menu 92
	Statement of principles about contents 92
	i-mode menu site principles of morality 93
	Content requirement 1: It realizes a value for the i-mode user ... 95
	Content requirement 2: Miscellaneous 96
	Benign Protector or Malignant Monopolist? 98
	Preparing a Partnership Proposal 100

Part III Developing i-mode Applications

Chapter 8 Discovering the Lost Joy of Coding Small 105
Working with i-mode's Memory, Storage,
and Screen Limitations 105
   Coding compactly 106
   Trimming graphics by color 107
   Bringing graphics down to screen size 108
Connection Speed, Limitations, and Considerations 110
   A server-side model: Getting around i-mode limitations
    to provide rich content 111
   Using server-side Perl, PHP, ASP, JSP, and other scripting
    languages to bring richness to i-mode 111
   Using SQL databases to host dynamic content 112
Using Emulators and Editors 117
   Finding an emulator 118
   Coding with editors 119
   Working with the language of i-mode 120

Chapter 9 cHTML, the Language Used for Creating
i-mode Pages 123
File Formats Used in cHTML 123
Using cHTML Tags in i-mode 124
   &XXX; 124
   <a></a> 126
   <base> 129
   <blink></blink> 129
   <blockquote></blockquote> 130
   <body></body> 131

 131
    133
   <dir><li></dir> 134
   <dl><dt><dd></dl> 134
    135
   <font></font> 136
   <form></form> 137
   <input> 137
   <select></select> 140
   <option> 140
   <textarea> 142
   <h1></h1><h2></h2><h3></h3><h4></h4> 144
   <head></head> 145
   <hr> 145
   <html></html> 146
   <img> 147
   <marquee></marquee> 149

	`<menu><li></menu>`	150
	`<meta>`	150
	`<object></object>`	150
	`<ol><li></ol>`	151
	`<p></p>`	152
	`<plaintext></plaintext>`	152
	`<pre></pre>`	154
	`<title></title>`	154
	`<ul><li></ul>`	155
	Color chart	156
Chapter 10	**Playing Sounds in i-mode**	**157**
	Sound Formats in i-mode	157
	Creating MFi files	158
	Other ways of creating music	161
Chapter 11	**Programming in cHTML: A Tutorial**	**163**
	Modifying an Existing HTML Document for i-mode	163
	Removing unneeded tags	165
	Trimming graphics to size	167
	Setting text-wrap options	169
	Adding access keys and other touches	173
	Lessons learned	177
	Creating an Online Address Book	177
	Defining the project	178
	Creating the data structure	179
	Directing the flow of user interaction	181
	Building the interface	183
	Authenticating users	187
	Managing account setup	192
	The guts of the application: entering and finding addresses	195
	Troubleshooting problems	199
	Searching and editing the entries	200
	Testing the code for various handsets	209
Chapter 12	**i-Appli: The i-mode Version of Java**	**211**
	Getting to Know the i-Appli Java API	211
	Making Preparations	212
	Creating i-Applis	214
	Yes, You Guessed It: Hello World	214
	Using the low-level API	214
	Using the Panel class or the high-level API	220
Chapter 13	**Programming i-Applis: A Tutorial**	**237**
	What You Want the Application to Do	237
	Building the Back End	239
	Building Your Main Method	241
	Building the Interface	242
	Using the Network in the i-Appli	248
	SoftkeyListeners, Component Listeners, and their Actions	252

	The Complete Source	257
	Testing the Application	263
	A Book Paradigm	264
Chapter 14	**Creating an i-Appli Game**	**265**
	Creating the Project	265
	Drawing on the screen	266
	Processing events	269
	Incorporating sounds	275
	Watching the Game in Action	277
	Planning Possible Improvements	280
	Complete Code Listing	282
Part IV	**Case Studies of i-mode: Implementations and Services**	
Chapter 15	**Case Study 1: Walkerplus.com**	**291**
	Company History	291
	Analyzing the Walkerplus.com Strategy	293
	Applying the Lessons	294
Chapter 16	**Case Study 2: Index Corporation**	**295**
	Company History	295
	Evolving Mobile Content	296
	Moving to the DoCoMo Business Model	298
	God of Love – Creating Marketable i-mode Content	300
	Looking Ahead	304
	Lessons Learned	306
Chapter 17	**Case Study 3: Nikkei**	**307**
	Presence and Profitability	307
	Market-Geared Content	310
	The Lessons	312
	Appendix A: What's On the CD-ROM?	**313**
	Appendix B: A Complete List of Official i-mode Sites	**317**
	Appendix C: i-mode Java API	**385**
	Appendix D: cHTML and X-HTML Basic Tags	**439**
	Appendix E: Emoji Symbol Codes	**449**
	Index	**457**
	Hungry Minds, Inc. End-User License Agreement	**486**

Part I

Definition and History of i-mode

CHAPTER 1
What Is i-mode?

CHAPTER 2
Getting to Know DoCoMo

CHAPTER 3
i-mode Software

CHAPTER 4
i-mode Hardware

Chapter 1

What Is i-mode?

IMAGINE THAT EVERY MOBILE ELECTRONIC DEVICE you could ever want was included in one small handheld package: Walkman, clock, telephone, address book, remote control, camera, e-mail and Web client, Gameboy, and anything else you can think of—all in one small device. Imagine that you are on the subway, listening to music and playing Donkey Kong, when suddenly you remember your appointment with your hairdresser. You use the same device that is feeding you REM and DK to connect to your hairdresser's reservations server to cancel. You get off the subway, walk up the escalator, and wonder what the weather is like outside. Easy—flip open your phone, and in three clicks you are at the weather information. Bummer! Rain again. No problem, call a taxi. You get to the wicket, aim your phone at the sensor, and are billed for your subway fare on your next phone bill. Looking up the stairs at the pouring rain, you decide to have a hot cup of coffee. You aim your phone at the drink button of the coffee you want, and voila! it comes tumbling down. Standing contemplatively at the bottom of the stairs, drinking your coffee and looking at the rain, you think, "Wouldn't it be great if . . ." You aim your phone up at the heavens, press the numbers 4-6-9 (G-O-D), and input the message "Stop this rain." The rain stops, you go up the stairs, and get in your waiting taxi.

All of this is currently possible, and most of it is actually done on the Japanese i-mode system. (Okay, okay, all but the last.) i-mode's architect, NTT DoCoMo (and increasingly, its Japanese competitors as well) see a future in which all mobile electronics and services are wrapped into one device—the mobile phone. It is a new paradigm, though not radical, because most of the current mobile electronic devices such as the Walkman, digital camera, Gameboy, remote control, and so on are already made in Japan. The forerunner of this new paradigm is a service called i-mode.

It has been said that i-mode is not a technology, but instead the name given to a service that happens to use a particular technology for its delivery. This is partially true—i-mode *is* the name given to a service. In this, i-mode seems very close to what AOL is in the United States. Yet i-mode is a set of technologies as well:

- ◆ Handsets (see Figure 1-1)
- ◆ Markup language used to write pages for it
- ◆ Packet-switched network it runs on
- ◆ Implementation of Java

This chapter explores just what i-mode is in terms of the platform the service is delivered on, namely mobile phone handsets, and what exactly the service itself consists of.

Figure 1-1: The first i-mode phones to be released: the 501i series.

Exploring Handsets

In terms of style, size, battery life, sound quality, and features, there is no more competitive market on the planet than Japan for mobile handsets. Nokia failed in its first attempt there because its handsets, the darlings of Europe and the U.S., simply couldn't compete based on features. Motorola never made an NTT DoCoMo-compatible handset. An odd dichotomy exists in what Western media sees as "new" features for European or American handsets that have, in fact, been around for as long as two years in Japan. In a wireless future based around the mobile handset, Japanese manufacturers hold a distinct lead over their European and American counterparts. This section discusses the capabilities of the handsets being used in Japan. Figure 1-2 shows one incarnation of DoCoMo's feature-driven product line — this one is made by Mitsubishi.

The handsets are designed by many manufacturers, based on guidelines set out by NTT DoCoMo for each new model. Certain features, such as the inclusion of i-mode features on all new models, are required; other features, such as color screens, are left up to the manufacturer. The models do not have a manufacturer's label anywhere on them but, rather, are identified by a one- or two-letter abbreviation before the model number. The P503i, for example, is made by Panasonic, whereas the SO503i is made by Sony.

Figure 1-2: The D501i, Mitsubishi's first i-mode handset.

Size and color are the features most visibly different from mobile phones in most other markets in the world, even on phones made by Japanese companies for use in other countries. Most manufacturers are now on their eighth or ninth generation of phone, based on the current digital networks, and each new generation has been distinguished by a lessening in size. Ironically, this shrinking may have come to an end with i-mode, with its increased use of screen space for games, Java applications called i-Applis (see later in this chapter for details, or see Chapter 12 for how to develop i-Applis), and in two generations down the line, streaming media. The Japanese consumer has, however, shown a clear preference for smaller handsets, and this preference will probably force some creative redesigning of i-mode handsets to be small and easy to read. Handsets come in a variety of colors, none of them black.

Shades of silver, gold, pink, baby blue, navy blue, orange, red, purple, and just about every other color besides black are available. Most models of handsets have two or three color variations available. In the case of DoCoMo's phones, because models change so quickly, and because users change phones fairly often, interchangeable faceplates have not taken off to the same extent as they have both in Europe and with DoCoMo's Japanese competitors. For those competitors, model changes are less frequent, allowing custom faceplates to be manufactured and distributed before a new model is out and the faceplate is no longer saleable. Custom straps, antennas, and other accessories are available and used by a wide audience, ranging from housewives to businessmen to teenagers.

> ## Handset Design
>
> The tastes of Japanese consumers, in particular young Japanese consumers, is a subject likely to drive both elderly Japanese and social observers to distraction. Colors, characters, and clear gender definition are absolutely essential considerations in design. Most handsets come in at least two colors (a darker one for men, and a lighter one for women), and many come in more colors. All handsets ship with some sort of character on the opening screen. My wife's Sony model ships with Gumby and Pokey, my own has some little chick (baby chicken, not young woman) that I don't recognize. Hello Kitty is also available. Because color is seen as a sales point, it has been a sort of taboo with handset makers to use black, which is associated with handsets that in the past were purely functional and not very attractive, an image that could be fatal to a handset's sales.

Handsets range in price from 100 yen to 40,000 yen (about US$80 to $320 at the time of this writing). Buying the handsets from a retailer rather than directly from DoCoMo is almost always cheaper, with most popular models costing around 10,000 to 15,000 thousand yen. The shops get 40,000 yen for each customer they sign up, so they have an incentive to sell as many handsets as they can. This incentive has spawned a cycle in which most Japanese, especially the young, own a phone no more than one model behind the current one. And because all current models support i-mode, as do half of the past model-cycle, it is no wonder that i-mode claims more than 27 million users. The fact that handset makers have it good in Japan is a given—they sell about 50 million handsets a year there, thanks to a win-win situation with multiple beneficiaries: retailers, who can maintain revenue streams; handset makers, whose costs are kept down by the sheer numbers of the handsets they can sell; DoCoMo, which can introduce new features quickly; and consumers, who get these new features basically at cost. With a subscriber base of 37 million, it is reasonable to expect that the total number of i-mode subscribers will quickly approach the total number of DoCoMo mobile phone subscribers.

NEC's models have consistently sold better than any of its competitors, and the company pioneered the clamshell-foldable model. The model shown in Figure 1-3 is its last model not based on the clamshell design, and was quickly followed by a clamshell model.

In the safe confines of a large market, handset makers labor away at introducing new handsets every six months or so, confident that the demand for the newest and latest technology will spur demand. This market is a powerhouse of complementary elements: good design, forced on makers by a powerful provider; advanced miniaturization, demanded by Japanese consumers; and long battery life, conveniently provided by the small power requirements of handsets in a crowded island nation. These elements all combine to make Japan the undisputed leader in handset production. When Japan finally adopts the IMT-2000 world standard for 3G (third-generation) digital networks, the world will discover exactly how good the Japanese handset makers are. Those who doubted Japan's continued capability to develop well-designed, technically advanced consumer electronics will most likely be shocked into silence. The handset size, in particular, makes them ubiquitous as no other electronic device has ever been. This ubiquity is at the center of what i-mode is about.

Figure 1-3: The N501i, NEC's first i-mode model.

Are Japanese handsets really that good? In a word, yes. The next section discusses the specific feature sets of i-mode handsets. The photo in Figure 1-4 is of Nokia's entry into the i-mode market. Although it has a lovely exterior design, it does not have a color screen or polyphonic ringing tones as most of its Japanese competitors did by the time it was released.

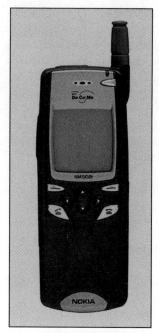

Figure 1-4: The NM502i, Nokia's return to DoCoMo, with its first i-mode handset.

Using the handset for Internet browsing

The i-mode-enabled handsets, which include all currently manufactured handsets, come with the capability of browsing the Internet, though DoCoMo does not actually refer to this capability as "i-mode" but rather as something separate: "Internet." This way of identifying this feature reflects a certain sensibility that I discuss in Chapter 2, but the salient point is that an i-mode handset, in theory, allows one to browse any of the vast stores of information contained on the Internet in HTML format. I say "in theory" because screen size, memory limitations, and the difficulty of viewing sites not designed for i-mode make the promise of access to the Internet more compelling than the actual experience.

The important point of being able to view any Internet content written in a basic form of HTML, however, is not so much that the user can view all of the sites on the Internet, but because most sites, at this point, are anything but simple. They use JavaScript, Flash, and other multimedia features that are not compatible with i-mode. No, the important point is that site developers need to do very little investment in new server hardware or software, educating systems people, and the other investments that serving a site using a new technology would entail: They are already set because cHTML (Compact HTML) is basically just a simple form of HTML, served in exactly the same way, written basically the same, and nearly 100 percent compatible with all of the existing infrastructure – hardware, software, and human.

> ## What is WAP?
>
> WAP comes up again and again in comparisons with i-mode. There is a raging debate about which is better. (I address this debate further in Chapter 4.) WAP is, in a nutshell, a series of protocols used for serving content to mobile phones. Whereas NTT DoCoMo tried to adhere to the same standards being used on the World Wide Web at large, the WAP Forum, a group of mobile phone carriers, handset manufacturers, and content developers, set out to establish new standards that fit the technical limitations of mobile handsets and networks. In its most recent iteration, WAP 2.0, the forum has taken a step back from this approach on the client side, moving to XHTML Basic, a standard arrived at with the W3C Consortium, which defines the HTML standards. This was always a small part of the overall WAP standard, so the official WAP Forum line goes, and the important work of the forum has been to come up with the standards that the networks and back end work on.

This capability to use existing infrastructure to serve Web content is in contrast to WAP (Wireless Application Protocol), which uses a different markup language: WML (Wireless Markup Language). This is a different server protocol that needs a fairly heavy investment on the part of the carrier in a conversion server if users are going to be able to view HTML content on their WML browsers. Systems based on the WAP standard have not been as successful as i-mode, and this may be one very important reason.

Using the handset for e-mailing

The i-mode e-mail system is in many ways an extension of short message service (SMS) and the pager network that preceded SMS, both of which were hugely popular in Japan. (They are still used, but because i-mode Internet mail can both send and receive messages from the Internet, their popularity is rapidly dwindling.) The nature of the packet-based network used for i-mode means that the phone is always ready to receive e-mail messages. Compared with normal e-mail, which a user has to check in order to receive, this is an advantage — and one that most mobile phone owners take as a given because of the legacy of SMS, which was nearly instantaneous.

 I have noticed questions on i-mode newsgroups about sending mass e-mailings to i-mode users. Because doing so actually costs the recipient about three cents, I strongly recommend that you do not even think about it. I hate junk e-mail (known as *spam*), and even people who accept it as a given on their desktop computer do not like having their phone ringing to announce junk mail, having memory taken up by it, or having to pay for it. Moreover, NTT DoCoMo has promised to block any site from ever sending e-mail to i-mode phones if it knowingly sent spam.

A successful use of opt-in e-mail has been a "word-of-the-day" site, which costs 100 yen per month and sends a new English word or phrase every day to recipients. However, I tried an American daily e-mail site that did not work well at all: I couldn't view any of the content because the advertisements at the beginning ate up the entire 500-character limit. DoCoMo cuts off everything after 500 characters.

Using the handset to send mail is not easy, in part because of the lack of intelligent interpretation algorithms incorporated into the sets, an element that most WAP (Wireless Application Protocol) sets do have. The lack of this capability can mean having to hit a number button as many as five times to get the character you want, and perhaps having to use the right arrow if the next character uses the same key. This is a highly inefficient way to write e-mails, but e-mails composed this way usually are only a few words, or perhaps consist of a ringing tone (*chaku melo* in Japanese) forwarded on. An intelligent interpretation system that eliminates some of the key punches, along with a word-wrapping feature, can be expected to be included when AT&T Wireless rolls out its i-mode service next year in Seattle, but the existing algorithm widely used in Europe and the U.S. apparently did not work for the Japanese language, which is why it was not included on the handsets sold in the Japanese market.

People do use the handset to send e-mail, though, and these people with thumbs flying over their handsets' keypads are nicknamed *oyayubi zoku*, which means "thumb tribe." The maximum allowed length of an e-mail is a mere 500 ASCII characters, or 250 double-byte characters. Japanese characters are double-byte, meaning that it takes two bytes of information for one character, rather than the one byte that an ASCII character takes. So the limit is actually 500 bytes, which means only 250 Japanese characters. This is not a lot when compared to most of the e-mails you send from a computer, but plenty when you consider the work of inputting that much with one thumb. There are also attachable keyboards sold for most i-mode phone models if users want to use the handset as their main e-mail client.

In the U.S., carriers such as Sprint PCS and AT&T Wireless do not have limits on how long e-mails, sent or received, can be. In fact, both of those companies have innovative products that allow e-mail messages to be received with word processing documents attached, and allow a user to either listen to the message, or have it forwarded on to a fax machine, at quite reasonable rates. These services are strongly focused at business people, and consequently offer much greater functionality than the basic i-mode service.

On i-mode, subscription services exist that allow users to check their POP e-mail, broadcast e-mail, or extend the functionality of the handset's own e-mail features. This is an aspect of i-mode that will come up again and again – the ability to extend its function by moving the functionality from the handset itself to the network or, in more advanced applications, to a Java midlet (Mobile Information Device Profile Applet) that is often provided by third-party providers.

Using the handset for phoning

Yes, you can call people on these handsets! Amazing though it is, this basic function is still the driving force behind the market. It is the *raison d'être*, the reason for the existence of the handset in the first place. With custom screens, custom ringing tones, i-mode, Java, digital music, and the other features that handset makers are loading onto these little wonders, sometimes we all forget the reason we carry them around in our pockets in the first place.

Because of the close relationship between DoCoMo and the handset makers, a high degree of functionality is included as standard. As of this writing, *all* handsets have the following features: caller ID; phone books with a minimum of a 100-number memory (most have 500); i-mode; an in-handset answering machine; call and time billing information, 10-number history of all calls received and another 10-number history of all calls made; Li-ion batteries generally capable of about a week on standby; and about two hours of talking time per charge.

One notable difference in cHTML and HTML is the use of the `tel.` attribute of the `<a>` tag. Using this attribute, an i-mode site designer can make phoning as easy as clicking a link. This feature reflects the very real fact that the vast majority of DoCoMo's profits are still derived from telephone usage. And it perfectly integrates the two functions of the handset. In the future, if DoCoMo lets them, developers should be able to integrate the various phone-related functions of the handsets with network-enabled Java applets, called i-Applis, to bring even greater functionality.

Another trend that DoCoMo has been experimenting with is its m-stage initiative, which consists of of network-connected multimedia devices that do not have voice functionality. So far, this initiative has found few takers despite a lot of hype and public demonstrations. Though streaming video or music may be cool and the perfect thing for geeks, Japanese consumers have only so much room in their pockets, and something that you can't call home with just won't cut it. These services serve more as testing grounds for future broadband wireless services than as profit-making enterprises.

With broadband 3G service starting in October 2001, consumer products giants such as Sony hope to achieve the promise that m-stage suggests. Sony's first i-mode offering is shown in Figure 1-5.

Figure 1-5: The SO502i, Sony's first i-mode model.

DoCoMo rolled out 3G services in Tokyo in May 2001, albeit on a test basis until October of 2001, when it will be formally available to a general consumer market. So-called third-generation (3G) mobile phones can receive data at 384 Kbps. This makes them about seven times faster than a normal fixed-line connection, and in theory allows them to be used for a broad array of broadband services such as downloading music, video, games, and other multimedia applications. Three models of phones were released, two by Panasonic and one by NEC. The NEC model broke little ground: It is basically a 503i-series phone with faster data connections, which means that it works better than anything previously available, but does not take full advantage of the speed. The two Panasonic models include one videophone model and one data communication model. The videophone model allows communication at 64 Kbps, which is the upload transfer rate. This rate is not particularly fast for a videophone: A popular videophone service that provides English lessons achieves 128 Kbps by using both channels of an ISDN line. The Panasonic videophone represents a beautiful use of the technology, though, and an engineering feat of no small magnitude in its slim clamshell package. The uses of being able to communicate with someone on the move with video are not fully thought out yet, but suffice it to say that telephone sex service operators have already thought of one new application.

Exploring i-mode Services

All current i-mode handsets come with a button or menu called i-menu. This button leads to one aspect of what i-mode is: The contents of this menu constitute the i-mode services provided through DoCoMo by independent information providers (IPs). People can, and do, describe third-party sites that are designed to be viewed on i-mode handsets, written in cHTML, as i-mode sites. There are, then, two i-modes: the one defined by NTT DoCoMo and the one defined only by the limits of the technology and the interests of users.

A brief description of packet networks

i-mode uses a packet-switched network, which is separate from the network used for voice communications. This means that an i-mode handset actually connects to two separate networks. Although at first glance that would seem to be inefficient, costing more for DoCoMo to set up, it is actually one of the keys to why and how i-mode works.

The voice network is billed by connection time. This is how phone calls have always been billed, whether they occur on fixed-line or mobile telephones. The problem with using this model for mobile data connections is its creation of an unnatural rush to hurry up and find the service one wants, get the information, and disconnect—because mobile phone per-minute rates are high just about everywhere. Although this model is fully appropriate for voice communication, which we can easily speed up or slow down, with data services, it creates a pressure at odds with a pleasurable use of the service.

The cost of the packet network, far from being based on the time the user is connected, is billed based on the number of *packets* sent or received. Packets are used on the Internet and other networks to send varying amounts of information from one location to another. The size of packets depends on the networking protocol. On the i-mode system, one packet is 128 bytes. If you were sending a 20-word e-mail, you would be sending one packet. (One byte is equal to one character, if it is in ASCII; whereas Japanese and some other languages use two bytes for each character, and would require two packets.) The cost at the time of this writing is 0.3 yen per packet, though NTT DoCoMo estimates the charge at 1.5 yen, the extra being used for logging on and for the packet transmission protocols, which add extra data to what you are sending to make sure that it gets to the right place. A flat rate of 300 yen per month allows users access to all i-mode content, though the aforementioned packet transfer rates do apply.

At the time of this writing, according to DoCoMo, the average monthly i-mode is 2,000 yen. People generally use the Internet for e-mail and a little bit of Web surfing, and this charge compares favorably to home Internet charges. The experience is admittedly much more limited than desktop Internet use, but the aspect of communication and practical information anywhere on demand seems to be one with some staying power in Japan.

European WAP services have generally used the same (circuit-switched) networks for voice and data, and WAP has been billed based on time connected. That method presents compelling economic disincentives to use the service at all, much less to explore new or unfamiliar services if you do use it. Add this mix to the problem of a limited pool of people writing WML (Wireless Markup Language: the markup language used by WAP for the moment), and i-mode's success does not seem like very much of an accident.

Specific services and an exploration of the revenue model

As of this writing, 1,680 official DoCoMo sites are available through the i-menu. Of these, most charge monthly subscription fees for some parts of their sites. This revenue model is one key to why i-mode has worked so well.

 A listing of these sites is available in Appendix B.

DoCoMo bills subscribers to these sites from 0 to 300 yen per month. It takes a 9 percent service charge and then passes the rest of the revenue on to information providers (IPs). With a market of 27 million i-mode users, an IP with a compelling service that is listed on the i-menu can hope to make a million dollars or more per month. That's peanuts to some of the dot.com crowd, but really good money to many small IPs.

For DoCoMo, this model provides the content that drives the usage of i-mode, which increases the number of packets sent and received, which in turn increases revenues. For the customer, it provides a wide variety of third-party services without requiring users to divulge credit card information over the Internet, and the convenience of paying for these services along with their mobile phone charges. Most motivating for the IPs is the promise of a huge pool of potential customers and a revenue model that will pay IPs to service these customers—something that many, with their roots as Internet startups, have been searching for.

Currently, the following categories of services are available on the main content list of the (Japanese) i-menu:

- News/Weather/Information
- Mobile Banking
- Credit Card/Stocks/Insurance
- Travel/Transportation/Maps
- Shopping/Living
- Gourmet/Recipe
- Ringing Tones/Screens
- Games/Horoscopes

- Entertainment
- Town Information
- Dictionary/Handy Tools
- Mail
- i-Appli Menu
- i-Playstation
- i-Navi Link Site
- i-Townpages (Yellow Pages)
- DoCoMo Billing Guide
- e-billing
- Regional Sites
- Index
- English Menu

These broad categories should give you a good idea about the tone of the content available on i-mode. I have received many e-mails asking me what the attraction for business users is. This is my answer to those writers, and to you readers who are asking the same question: Business people also ride trains, pay bills, trade stocks, get lost, want to know whether they need an umbrella, cook dinner, and wonder whether tomorrow will be better than today. Where are the business applications in i-mode? At the office — not on a publicly accessible service.

 I discuss business use in Chapter 6, but suffice it to say for now that businesses are as interested in using a broadly deployed, reasonably priced, HTML-based system as consumers have been.

i-Appli, the i-mode version of Java

On February 21, 2001, DoCoMo released new Java-enabled models of i-mode-enabled 503i handsets. These handsets allow users to download small Java programs, making the handset, in effect, a very small computer.

A Java midlet is a small application that runs on a "virtual machine." This means that it is the virtual machine, rather than the midlet, that is talking to the processor. Because the virtual machine does the nitpicky work of talking 0s and 1s to the processor, a programmer can write processor-independent programs, leading to the Java mantra "Write once, run anywhere," a capability that is necessary in mobile phones, many of which don't even have true processors but use digital signal processors instead.

If you are in your mid-30s or older and were one of those people who owned a TRS-80, Commodore PET, or maybe a Timex-Sinclair ZX-80, as I did, the fact that the 503i series phones have only 10 KB of memory may seem like deja vu. In fact, my ZX-80 had only 2 KB of memory, and I was still able to program Space Invaders. In those terms, 10 KB is positively luxurious. What have programmers in Japan been able to do with the capability to run little programs on a very small phone? Quite a lot, actually. The following list shows the applications available from the i-menu.

- Games
- My Weather
- Weather Clock
- Foreign Exchange
- Stock Information
- Online Brokerage
- Online Banking
- Map Information
- Ringing Tones/Karaoke
- Screen Images/Clocks
- Entertainment
- Horse Racing/Lotteries
- Variety/Mail

Following are pictures of the Sony handset, which is the first i-mode handset with a TFT screen (see Figure 1-6), and the Fujitsu handset, which was the first handset in the world to support Java (see Figure 1-7).

Figure 1-6: The SO503i, Sony's third-generation i-mode handset.

Figure 1-7: The F503i, Fujitsu's third-generation i-mode handset.

Because at the time of this writing the service has been available for under a year, it is a bit difficult to judge what sorts of applets will be most popular, but I think that games are at the top of the menu for a reason: Most major game companies—Hudson, Sega, Konami, and Namco, to name a few—have an i-mode presence. The expectation on DoCoMo's part is clearly that games will lead the i-Appli charge. Whether this expectation proves to be true is another matter; the success and failure of gaming platforms and games is a hugely fickle matter. It is really too early to tell what sorts of creative ways developers will find to use this smallest of computers, but it is safe to say that much of i-mode's future success rides on i-Applis finding the sort of appeal that i-mode has had.

The revenue model in use by the official i-Appli sites is the same as it is for i-mode sites: A monthly subscription fee of between 100 and 300 yen (about US$.80–$2.40 at current rates) is charged by official i-Appli sites. DoCoMo takes 9 percent, and the remaining 91 percent goes into the site-owner's bank account. The sites also have the option of charging nothing, but for the most part, the sites that charge nothing are selling another product, such as online banking or stock trading, and the Java applications are simply ones to make using their services easier.

Early interest in developing i-Applis was hurt by NTT DoCoMo sitting on the DoJa software developer kit until March 2001. Without the SDK, non-registered developers couldn't reliably test their applications prior to the launch of i-Appli. The behavior of code libraries embedded in handsets was an unknown, so a reliable emulator was essential. Zentek Technology developed the shareware i-Jade emulator, which did a pretty good job, but still had some incompatibilities. In any case, for probably political reasons, DoCoMo chose not to release the SDK until its partner companies had a good head start on everyone else.

The i-Appli standard itself was developed in collaboration with Sun, and its API does not show many incompatibilities with standard Java, though the forms that midlets are required to take, namely jam and jar files, require some reworking of code. Doing so proved difficult without an emulator, and was a source of frustration for many non-registered developers.

One of the things that has become important when talking about i-Appli development, especially games, is the means of moving around on the screen. None offers a cursor; all use some sort of selection method. Figure 1-8 shows a shuttle jog button in the middle.

Figure 1-8: The D503i, Mitsubishi's third-generation of i-mode handset.

A major Japanese i-Appli developers site, http://g-appli.net, in which developers can register the applications they are working on, shows that about 40 percent of applications are games, and 75 percent of downloads are games. DoCoMo seems to have foreseen this situation, and actively recruited major game companies such as Sega, Bandai, Konami, Hudson, and others to provide this fare, and DoCoMo put it at the top of its i-Appli menu.

The first book out in Japanese on how to program midlets – they are not called within the Web browser, so they are not called applets – included Lunarlander, Minesweeper, Snake, Asteroid, and other games popular in the early '80s and revived as i-Applis.

Herein probably lies a strong future for i-Appli development. Presently limited to 10 KB, these programs are, by their minuscule size, fairly simple. In the early '80s, there were books similar to the early books on programming midlets and on BASIC, and a programmer could type in the code and have a game. Figuring out how these games worked, fiddling with the timing (or scoring!), and learning from changes often prompted by bad typing were all ways that would-be hackers figured out what was going on and honed their programming skills. It seems likely that Java, and especially its existence on the most mobile of appliances, the mobile phone, will tickle the imagination of these sorts of people, these hobby programmers and students, and probably encourage a new breed of programmers that in some ways are very much like their forebears.

But they aren't there yet, and the present breed of developers has become dependent on lots of memory, lots of speed, and fairly complicated development tools to do their jobs. i-Appli doesn't really fit into this higher-level paradigm. To do what such developers need to do, most of the power needs to reside on the back end server, and developing Java midlets is more of a way of tweaking and allowing more customization of the user interface.

"You could have graph data sent from a server, and then use Java on the client side to display it," says Kyle Barrow, owner of X-9 (www.x-9.com), a consulting firm in Osaka that develops, among other applications, i-mode interfaces to factory-floor control and information applications. Barrow continues, "But should you be doing this when the server could deliver the complete graph faster and without the need for Java? That's what I have been disappointed with from DoCoMo – they have these i-Applis that draw graphs that just sit there, that don't do anything. Java should be used to manipulate and customise the display of data." Barrow, however, just can't see the benefit in investing in a technology as young as i-Appli. He points out the well-known bug in the Panasonic handset, as well as a lesser-known bug in the Mitsubishi handset, as examples of i-Appli's instability. The present memory limitation has also turned Barrow away from developing in Java for i-mode for the moment. A bug, for example, forced a recall of the Panasonic handset, pictured in Figure 1-9.

J-Phone and au (pronounced *ey-yu*), DoCoMo's two largest competitors, have both announced the development of their own Java specifications, and have promised more memory: J-Phone, 30 KB; au, 50 KB. DoCoMo will follow suit with a 50 KB memory limit in its FOMA 3G i-Appli phones. For developers used to working with the speed and power of a server or even a PC, however, the client-side model of i-Appli is unlikely to be appealing, whatever the memory available.

Figure 1-9: The P503i, Panasonic's third-generation i-mode handset.

"Everything I have seen or heard talked about is either games or something that would be much better done on the server side," says Tim Romero, president of Vanguard (www.vgkk.co.jp), developer of a Java-based, e-commerce package called Mojo, which includes i-mode connectivity. Romero, despite using a Java servlet-based back end, sees no real benefit of using Java on the client side. "I think applets was Java released prematurely. After the real Java got out, people really started using it for server-side programming, for significantly larger systems, and very little attention has been paid to applets. Now, i-Appli, it's applets again." Left unsaid is that in using a client-side model, developers such as Romero and Barrow would lose some degree of control of their revenue-producing product, not something that a developer who wants to keep eating would willingly do.

Even those who profess a love for i-Appli do not see a "killer app" for i-Appli at this point. They see the micropayment system offered through DoCoMo to official sites as an important aspect of developing for i-Appli, and i-mode for that matter. Japanese developers, keen to be the next recipient of instant fame and popularity, not to mention money, would disagree. The promise of a million users subscribing to your site and paying 100 yen per month is extremely attractive, especially to those young hackers learning how to hack games onto their handsets. To business developers, who like to keep control of their core products, and, therefore, would relegate i-Appli to interface tweaking, this promise means nothing.

Although the lack of general business applications may mean not being taken as seriously in Europe and the U.S., Barrow argues that games *are* serious business: "Some of the failure of WAP is that they focused solely on business and ignored the consumer. The game industry is huge, with companies like Sony deriving much of their profits from it, and yet there is this attitude by some Western wireless providers that it isn't serious enough. Billions of dollars in sales seems pretty serious to me."

Indeed, and games are a great way of spending a few minutes on the subway. Figure 1-10 shows the model of choice for doing exactly that.

Figure 1-10: The NEC 503i, the leader in Java-enabled phones for its balance of good technology and very good design.

Chapter 2

Getting to Know DoCoMo

AS THE HIGHEST-VALUED COMPANY on the Tokyo Stock Exchange, the company that *Newsweek* called Japan's only new multinational to emerge in the last decade, and the undisputed leader in the world of the mobile Internet, NTT DoCoMo occupies lofty ground. NTT DoCoMo's success as a company has been built on the popularity of its product offerings, and on its capability to have its standards dominate the Japanese market. With more than half of all mobile phone users in Japan using NTT DoCoMo, and nearly three-quarters of that number being subscribers to i-mode, NTT DoCoMo is seen at home and abroad as the model of a successful mobile telecom.

To get where it has gotten, NTT DoCoMo has at times played rough, using standards to competitive advantage when they worked, and ignoring them when they did not. In this, NTT DoCoMo might be thought of as akin to Microsoft – everyone in the industry, in whispers, has something bad to say about the company. Yet NTT DoCoMo is really the gatekeeper of where most companies would like to be in the mobile Internet space. If a company wants to be in that business, it had better learn to play ball with NTT DoCoMo, or risk failing. So, whispers of discontent never seem to turn to shouts.

The Little Spin-off that Could

The first part of its name gives away its origins; the second part gives away its ambitions. DoCoMo means "everywhere" in Japanese. A spin-off of giant monopoly NTT, NTT DoCoMo is actually valued higher in the stock market than its parent. NTT DoCoMo is actually not its full legal name; it is NTT Mobile Communications, Inc. It spun off from its parent company in 1992, at which time it took over NTT's nearly one million analog customers. This was at a time (and this timing may have played a role) when the U.S. trade representative, Carla Hill, was renewing demands that Japan adopt Motorola's world standard, and open up its market to more foreign equipment makers. An earlier agreement had paved the way for IDO Communications in Eastern Japan and Kansai Cellular in Western Japan to offer lower-cost competition to NTT by using lower-cost American equipment and handsets, based on U.S. standards. This was after it had basically been a country divided by a previous agreement – between the U.S. standard in Western Japan served by Kansai Cellular, and NTT's standard in Eastern Japan.

In many ways, NTT DoCoMo's creation as an independent corporation has its origins in a conflict over standards, and how it had been used up until that point in Japan to stifle competition. Its nine-year history clearly shows how NTT DoCoMo has used standards, and the setting of them, as a tool to stay ahead of the competition. In 1992, NTT's response was typically slow, but also quite practical, in that it recognized that in a competitive market a company couldn't be as slow as NTT was and remain competitive. Rather than making a real effort to reform the whole company, it spun off that part of the company that needed it the most.

A brief chronology

Because i-mode has been developed in one country with growing interest abroad, much of NTT DoCoMo's history and i-mode's background may be unknown to international readers. Getting to know the series of events in the company's history can help put i-mode into a context as well as inform readers before they approach a potential partner. The following chronology briefly outlines important events.

- April 1991: Ultra-small cellular phones, called "Mova," were released. They were based on a proprietary analog technology. With weights of less than 150g, these handsets paved the way for NTT to offer greater services than the car phone services it had been offering until that time.

- July 1992: Sales activities in mobile communications business transferred from Nippon Telegraph and Telephone Corporation (NTT) to NTT Mobile Communications Network, Inc. NTT Mobile Communications Network, Inc. took over NTT's mobile operations. This completed the spinoff of NTT DoCoMo, which began the previous year.

- February 1993: The number of cellular phone subscribers to NTT DoCoMo exceeded one million. The first signs of a consumer uptake of mobile phones began to emerge, although it was still a relatively small market because of the steep US$200 or $300 per month in monthly fees.

- March 1993: Digital service for cellular phones (800MHz) started, based on the Japan-only PDC standard. This was the first widely deployed digital cellular network in the world. Japan, and especially NTT DoCoMo, faced heavy criticism from abroad for not waiting for the European GSM standard and for not adopting the U.S. TDMA standard. Some said, and still say, that the PDC standard was designed from the beginning to be incompatible with any other widely adopted standards: Its use of channels for sending and receiving is opposite that of nearly every other standard.

- April 1994: DoCoMo users were allowed to purchase (instead of lease) their mobile phones for the first time. For the past several years, IDO users could purchase handsets. Digital service for cellular phones (1.5GHz) was also released, and allowed Tuka and the Cellular Group (later to become J-Phone) companies to enter the market, which had hitherto been shared between NTT DoCoMo and IDO.

- April 1995: Digital cellular phone 9.6Kbps high-speed data communications service was released. It actually used a modem plugged into a digital handset, but used the voice network and was charged at still-high voice rates.

- March 1996: Second-generation Pocket Bell pagers were released into an unexpectedly strong market for pagers, based on Motorola's FLEX-TD system. The number of pagers, based on a previous analog system, was already at about six million, and eclipsed that of mobile phone subscribers. The subscriber numbers, however, already started to decline.

- April 1996: The number of cellular phone subscribers to NTT DoCoMo exceeded five million. DoCoMo began to offer retail phone sellers around 40,000 yen per new customer as an incentive to cut phone prices and increase marketing. This amounted to handset subsidies, though a shop could sell a handset for full price and keep the entire 40,000 yen. Most used some part of the money to cut phone prices to attract new customers.

- July 1996: Value-added pager services were added to shore up pager customers. It didn't help – 1996 saw the beginning of a mass migration from pagers to PHS (Personal Handy System) and mobile phones by the youth market segment. It lost some of these upsales to lower-cost competitors J-Phone and Tuka.

- February 1997: The number of cellular phone subscribers to NTT DoCoMo exceeded 10 million. There was incredible growth in the entire market, and NTT DoCoMo cut charges to be competitive with Tuka, J-Phone, and (to a lesser extent) IDO. Average (monthly) revenues per user (ARPU) dropped to 12,570 yen, or about US$100, from nearly 16,000 yen, or about US$130 in 1996. Senior management foresaw that despite the massive growth, something needed to be done about falling ARPU. A team was assembled to sell value-added, text-based services. These services eventually become known as "i-mode."

- March 1997: Packet Data Communications service started. No one was very interested because it had no consumer applications and because of a lack of modems and other compatible equipment. This network, however, was put to use two years later by the i-mode service.

- August 1998: The number of cellular phone subscribers to NTT DoCoMo exceeded 20 million. NTT DoCoMo's incredible growth continued, but it was forced to cut rates again, and ARPU was down to 10,800 yen (about US$85)/month from 12,570 yen (or about US$100) just a year earlier. The i-mode services took on added urgency as subscriber growth rates slowed.

- February 1999: i-mode was released, along with one handset by Fujitsu that supported the service. Competitors J-Phone, Tuka, and IDO were already to market with their services.

- August 1999: The number of i-mode subscribers reached one million. By this time, all four major handset makers had come out with handsets, and early models (such as Fujitsu's) were being given away for free.

- October 1999: The number of i-mode subscribers doubled to two million. Initial worries on the part of content providers and naysayers within NTT DoCoMo began to give way to a feeling that the system could be a hit. Conservative initial estimates helped.

- December 1999: Fujitsu released its second-generation i-mode phone, the F502i, which was the first color i-mode handset. In the next three months, all four major handset makers came out with 502-series handsets — two with color and two without. Color turned out to be a key selling point, and it gave underdogs Fujitsu and Mitsubishi a short-term advantage against top-sellers NEC and Panasonic. Subscribers reached three million by the end of the month.

- April 2000: Nokia released its first i-mode model, the NM502i. It was a grayscale handset using a system different from other phones for ringing tones. The number of i-mode subscribers rose to six million.

- June 2000: NTT DoCoMo, along with top ad firm Dentsu, launched D2 Communications, an ad firm selling advertising on i-mode. Sony released a grayscale i-mode handset, the SO502i; and dual-mode, PHS-PDC phones supporting i-mode were released. New models of *all* DoCoMo phones included i-mode; and models in the basic series, the 209 series, shipped from Fujitsu, NEC, and Panasonic with i-mode built in. At this point, there was basically no difference between phones in the 502i and 209i series. An English site was prepared for the Okinawa G8 summit, and it included CNN, Bloomberg, and others. The number of i-mode subscribers rose to eight million.

- September 2000: The number of subscribers reached 12 million. NTT DoCoMo was told to either improve the i-mode center server operation by the (then) Ministry of Posts and Telecommunications or to not sign up any new i-mode customers. NTT DoCoMo complied, with a new server center operation within a week. It signed an agreement to jointly develop services with AOL. It agreed to help KPN Telecom of Holland develop mobile Internet services.

- February 2001: Java-enabled phones, called i-Appli, were released by Fujitsu, and quickly followed by Panasonic and NEC. This service used the existing i-mode billing and delivery system to deliver miniature pieces of software to a user's phone. Subscribers to i-mode were about 17 million, and subscriber growth reached approximately one million new subscribers per month.

- ♦ May 2001: The general rollout of the IMT-2000 3G network was turned into a field test with 4,500 participants. More than 137,000 people applied to take part in the test. The general rollout in Tokyo was moved to October 2001. Problems with handovers between base stations were cited as the reason for the delay. There were also production delays reported at handset makers Panasonic and NEC.

- ♦ July 2001: The Japanese Ministry of Telecommunications refused to certify the FOMA 3G network as reliable for data transmission due to reliability problems faced in the trial. NTT DoCoMo said that the service would be ready and reliable by launch.

- ♦ October 2001: The world's first IMT 2000 3G network goes online. Handsets initially retail for about US$320 for a standard model, US$500 for a videophone model, and US$200 for an all-in-one PC card model used for data communication. Service is available only in the Tokyo metropolitan area. The number of i-mode subscribers currently stands at 27 million.

Key players

Two books about i-mode, *The i-mode Incident*, by Mari Matsunaga, and *The i-mode Strategy*, by Takeshi Natsuno, give a unique perspective on the inner workings that led to the service eventually called i-mode. The information in this section is mostly taken from these two books.

Keiichi Enoki, formerly a network engineer at NTT DoCoMo and now a vice president there, tells the story of being brought into the office of the president of NTT DoCoMo in the company's Toranomon head office in January of 1997. "Let's use the mobile phone handset to establish a mobile multimedia business," Kenji Ohboshi (who has since become chairman of the board) said. Enoki replied with "That is an interesting business!" but had one important question for Ohboshi: "Where do I get my people from?" "Anywhere you like," Ohboshi responded – a rather surprising answer for the president of a former state-owned company in Japan, in which outsiders are not traditionally welcomed with open arms by lifetime employees cum bureaucrats.

Enoki called a friend (a Mr. Hashimoto) who owned a printing company and who eventually put Enoki in touch with Mari Matsunaga. Hashimoto knew Matsunaga to be an extremely capable leader, with a very good handle on marketing content and an ability to motivate those who worked for her. She had an excellent reputation within the publishing industry, which Mr. Hashimoto was quite familiar with, and in his mind she was the perfect person for the job.

Matsunaga had been an editor-in-chief on various Recruit publications, rotating to a new publication about every three or four years. She was, in 1997, about to end her time at the helm of *Travail*, a job-search magazine for young job seekers. Recruit had asked whether she would be interested in pursuing an editorship at a magazine aimed at active senior citizens. Because most of her career was spent at magazines with a focus on young people, she was understandably surprised. The company, however, was not trying to shuffle her off to an unimportant publication. With a rapidly graying society, the company made a decision to put more emphasis on this market segment, and wanted to assign one of its most experienced and successful editors to the task.

Right at this point, she heard from Hashimoto, an old friend. Matsunaga was inclined to listen to what Hashimoto had to say, though at first she could not believe what she was hearing. Working for a former state-run company was not something that was attractive to her, but her instinctive feel for people told her that she wanted to work with Enoki, and she was finally convinced to take the job. That Hashimoto and Enoki spent as much time convincing her to take the job later proved critical to the form that i-mode took and the success it has enjoyed.

Matsunaga's forte was content (not technical) development, and at a point not very far into the project, she realized that she needed help with the technical aspects of the project. And then she remembered Takeshi Natsuno, whom she met when he interned at one of her magazines as a university student. They had kept in touch, meeting every couple of months or so, and she thought he would be the perfect person to have on board for this job. She gave him a call.

Natsuno started as a regular employee with Tokyo Gas, after graduating from the University of Pennsylvania's Wharton School of Business with an MBA in 1995. During his time at Penn, he bought a Macintosh and signed up with AOL. Compared to the online services of the time in Japan, Niftyserve being the most popular, AOL and its Internet-accessible content were eye-opening. He was asked to write a paper on how the Internet would change business, which forced him to consider the broader implications of handy services such as online airline reservations, credit card management, and others. After returning to Japan and moving up within Tokyo Gas, he quit to become the vice president of a free ISP called Hypernet, which flamed out in late 1997, well after he had joined the i-mode team.

"*Natsuno-kun*," Matsunaga said when she called him. "I am in trouble and need your help."

"Travails at *Travail*?" Natsuno asked ironically, referring to the name of the magazine she had until recently been editor of, but he listened to her ideas with growing enthusiasm.

This is it! he thought, when Matsunaga told him what she was working on, *this has the numbers of users to really be something. This is the thing that will get Japan online. Not the American PC but the mobile phone would carry all the convenient information that an American provider such as AOL carried. This would be the Internet revolution in Japan!*

Enoki's field of expertise is in the engineering of networks, Natsuno's is in fostering content providers, and Matsunaga's is in knowing what users of content are interested in and what sells in the field of content. Enoki's official title is Gateway Business Manager, Natsuno's is Gateway Business Content Manager, and Matsunaga's is Gateway Business Content Development Office Manager. None of the titles is very descriptive, and Matsunaga's title as Office Manager is quite deceiving, considering that she was recognized as Woman of the Year by *Nikkei Shimbun* and Businesswoman of the Year by *Asian Businessweek*.

These three formed the nexus of the team that created i-mode.

The Path Toward i-mode

i-mode came about because of a clear progression of successes and responses to market conditions, combined with the vision and creativity of the people (such as Enoki, Matsunaga, and Natsuno) who were charged with developing the service. This section looks at the causal forces that helped to bring about the conditions for i-mode's success.

Strictly business as usual

In 1992, when NTT DoCoMo was spun off, users did not purchase handsets, but instead leased them at very high rates (for a typical mobile phone bill, well over US$200 per month). This made sense at the time, at least to NTT DoCoMo, because the target market was high-flying executives who rode the stock market and real estate booms.

This business market segment is one who initially spent the money to equip their cars with the new technology, who had adopted the analog handset as a completely necessary status symbol, and who up until that point, was flying high on the back of bloated stock and real estate markets, and was, therefore, the market to go after. That this market had its limits was understood, but the high cost was seen as a strong disincentive in just about any other potential market segment. A limited market seemed preferable to price cuts, which is what all parties recognized were needed for the service to be attractive to general consumers.

Unfortunately, as soon as the company was formed, the boom was over, and it was over in a way that no one at the time imagined—the start of an economic slowdown that has lasted for the entire nine years of NTT DoCoMo's life. This wasn't immediately clear. In 1993, NTT DoCoMo released its first "Digital Mova" handsets, which were fairly lightweight, and which operated on what was really the first widespread digital cellular network in the world. NTT DoCoMo realized that it wasn't going to pay for the PDC network it had just spent a large amount of money building by focusing on a narrow band of high-spending users. By the time the network was fully rolled out in 1994, DoCoMo had begun selling its handsets, albeit at very high prices and with no subsidies.

Catering to a new market

With fairly small handsets, good quality digital network connections, and the prohibitively high cost of purchasing rights to fixed-line phones, NTT DoCoMo was able to grow its subscriber base to five million by 1996. By this time, it was joined in digital networks by IDO and the Cellular Group, both of which had begun to expand nationally out of their regional bases. Both were aggressive in their competition against DoCoMo, financing handset sales to build market share. NTT DoCoMo, typical of its still somewhat plodding approach to business, was rather slow to react, but had the best technology, the best handset makers, and plenty of cash at its disposal.

By 1996, one of its hitherto nascent businesses — pagers — had become all the rage among teens. These teens could be seen lining up in front of public telephones, and, with incredible speed, inputting text messages based on a number-to-text conversion system. The pagers released by DoCoMo were the height of miniature fashion: tiny, one-inch-square models that came in bright colors and carried price tags of 10,000 yen, or about US$100. Users paid only about US$10 per month to use them, but this was not bad for DoCoMo, which never subsidized the cost of pagers the way it has mobile phones. And it gave the company a compelling clue about who the next generation of mobile phone customers would be and about the mode of communication of a generation of users previously only dimly understood.

Two years later, in 1998, this understanding of the youth market paid off with the popularity of the Pocketboard. NTT DoCoMo managed to keep the messaging momentum with the release of the Pocketboard, a very small (about seven inches long by three inches tall by one-half inch thick) e-mail client that could be plugged directly into a mobile phone. This was the first NTT DoCoMo product to connect a user to the Internet; short mail service allowed users to connect only to other short mail users. It was also extremely well-designed: The cord connected the Pocketboard to a mobile phone built into the device so it could never be lost. The Pocketboard was a true consumer product, reasonably priced (about US$100 at the time), and consumer-friendly. It came with a setup program that could be completely done online, and the service was ready within a day. Typical of any product that is a hit with teenagers, it also got a cute nickname, "Pokebo," which is a shortening of Pocket and Board. Until this point, NTT DoCoMo aimed at this market of "mobilers," or mobile Internet users, by trying to attract business users. Its commercials emphasized how busy businesspeople are, and how mobile computing could help them. The Pocketboard, which didn't get the same exposure in NTT DoCoMo's commercials, nonetheless sold hundreds of thousands of units. Specials were given to attract short mail service users, giving a discount if two Pocketboards were bought together. Mail, once again, was the driver of a trend that profited NTT DoCoMo and brought DoCoMo closer to the consumer.

Leveraging the existing user base

Extremely attractive new models of phones, shops that sold them at subsidized and extremely reasonable prices, short mail service, and a groundswell of teenagers who were now able to afford a mobile phone and graduate from their pagers all contributed to this. The role of short mail service was not appreciated much at the time by those outside of NTT DoCoMo, but in hindsight NTT DoCoMo's timing of the release of this service perfectly coincided with the popularity of pagers. It also offered a clear advantage over pagers, namely not needing to line up at a public telephone, and you could see what you were inputting on the screen.

This group of users who were focused on messaging was the seed of Ohboshi's initial idea for a short mail service-based information service. Messaging had clearly shown itself to be the application that attracted the trendsetting young users, and SMS was hugely popular, so why not use this popularity to sell other SMS-based content?

The idea of getting customers to pay extra for content, however, and actually coming up with content that customers would pay extra for were two different things. The only strong area of value-added content then available was short mail service. The cost of this service was not very high, messages could be delivered in seconds, and customers could quickly disconnect from the short mail service delivery access number. Still, this was the seed of the idea: SMS could be used to deliver content to people's mobile phones. People would pay extra to get information sent to them by SMS, and DoCoMo would increase revenues in SMS connections. The short mail service of the time was not connected to the Internet, was not packet-switched, and could handle messages of only 50 characters or fewer. Pocket bells, in fact, were better connected than cell phones, able to receive Internet mail and get news, stocks, and horoscopes. Ohboshi had in mind a similar service for mobile phones.

Matsunaga, especially, knew that to sell something to people, you need to keep several things in mind: who your customers are, what their interests are, and what you can do to improve their lives and in turn make a profit. If you look at most failed WAP implementations, you will see a lack of concern for customers. Looking at NTT DoCoMo's implementation, you will see that very close attention has been paid to customers' wants and needs.

This customer-centered approach, something that is not uncommon in other industries, particularly merchandising, dictated how the other elements of the puzzle fit into place. Because the system had to be convenient, packet-switching was used. Because it shouldn't be too expensive, a price cap of 300 yen, or about US$2.50 per month per content subscription was imposed. Because it had to be easy and profitable for people with good products to offer them to customers, cHTML was used, and the micro-payment system introduced. And for all of these reasons, NTT DoCoMo's own revenue model would be that of simple network provider, charging for the data traffic used.

Going with what works

Since it was spun off from parent NTT, NTT DoCoMo has had many spectacular successes. It has also had some failures, as is the case with any large company. This section takes a look at a few of these failures to see what it is that NTT DoCoMo may have learned.

Following is a list of the things that have not seemed to work for NTT DoCoMo:

- **Business-oriented products.** Since the days of analog handsets, NTT DoCoMo has not had a business-oriented product, besides the basic voice communication handset, which has been a hit with business users. It has tried with mobile Internet-enabled PDAs, executive model phones, and others, but none have reached "hit" status. Business users instead have used consumer-oriented services such as i-mode, Pocketboard, voice mail, and mobile messaging for business uses. One thing NTT DoCoMo has probably learned from its failures is that it needs to understand this market segment more clearly. AT&T Wireless, its U.S. partner, similarly sees its partnership with NTT DoCoMo in terms of its own strength in the enterprise market complementing NTT DoCoMo's in the consumer market. It is safe to assume that NTT DoCoMo hopes to learn from AT&T Wireless about the enterprise market. For now, the focus is on the consumer.

- **Non-communication-centered products.** The eggy, a streaming multimedia product based on the PHS network, was a failure in terms of uptake. It was too big, didn't work as a telephone, didn't have integrated network capabilities built in from the start, and had a screen no bigger than that of a handset's. Something that doesn't work with a telephone, requires a network card, and is too bulky to carry in your pocket wouldn't seem to have a very big market. If, however, there had been more and better content, eggy would have been more successful. When NTT DoCoMo does release streaming media on 3G handsets, which will not be at launch, content will be king. This is a lesson it learned from i-mode, and after the failure with eggy, one even more compelling than before.

- **Car navigation integration.** From the 502 series phones onward, some models of i-mode phone have been able to integrate with car navigation systems to access information from i-mode. A very few users actually make use of this. Although navigation features and television are both popular in the car, too much interaction is apparently needed from i-mode to be very attractive to car users, especially when a user has DVDs full of local area information anyway – in the car navigation system. For e-mail, they may as well just use the handset. Integration seems great, but from an end-user perspective, what does it offer? These are exactly the sorts of questions that Bluetooth-enabled product makers are going to start asking themselves soon. After its likely failure integrating the cell phone with the Sony Playstation, NTT DoCoMo's lesson with the failure of the car navigation system will probably be much clearer – don't create high technical and economic thresholds for products whose actual usefulness don't seem to have been clearly thought out.

For more information on Bluetooth, see Chapter 5.

A list of failures, fairly enough, should be followed by a list of things that *have* worked for NTT DoCoMo:

- **Messaging.** This is the application that got kids hooked first on pagers, then on PHS, and finally on mobile phones and i-mode. Messaging has been a spectacular success primarily because the reason people carry a phone is for communication. This need for communication cannot be underestimated, and addressing this market in a manner that understands and exploits this need for communication will be successful.

- **Consumers.** They are the entire market, and by appealing to a broad section of consumers rather than a narrow group of business users, NTT DoCoMo has built a base of customers nearly a third of the size of the entire market.

- **Games and entertainment.** These two markets, seemingly a surprise to NTT DoCoMo, are two of the most important markets on i-mode today. Introducing Java-enabled phones, 16-tone ringing harmonies, color screens, and other features to make gaming and entertainment more attractive has paid off.

Looking at NTT DoCoMo's successes and failures should provide a good roadmap of where to go next, and perhaps show which areas need strengthening. Learning by success and failure what works and what doesn't can be applied to new business areas. Two of its failures, however, happened despite the fact that NTT DoCoMo should have known better. That it takes risks is apparent from this, and that the key to the success or failure of any product with broad market appeal is in its implementation is paramount.

Strategies and Growth

With a US$3 billion net profit in 2000, a market share of 59% of the entire Japanese market, and a market capitalization higher than any other Japanese company, the logical question is, "Where does NTT DoCoMo go from here?" Some of the answers are clear, stated in annual reports and by NTT DoCoMo's leadership. Other aspects of how NTT DoCoMo is plotting its future are ambiguous; stated plans and actual implementations seem to differ and leaders make contradictory remarks. This section looks at the future of NTT DoCoMo.

Updating the infrastructure

One of the clearest parts of NTT DoCoMo's strategy is its implementation of IMT-2000 W-CDMA-based 3G networks. As this book goes to press, customers in Tokyo have begun using 3G handsets, many of which boast videophone capabilities.

What is 3G? IMT-2000? FOMA? W-CDMA? GRPS? UMTS?

These terms all refer to a different aspect of the third-generation digital mobile phone networks.

- 3G simply means "third generation". Some people also say G3, although this is likely to confuse Macintosh users, who use a third-generation Power PC chip by the same name.

- IMT-2000 is the internationally agreed-upon standard for what actually constitutes 3G digital mobile phone networks, and it stands for International Mobile Telecommunications 2000. It is a standard of the International Telecommunication Union (ITU). It is only a standard in the sense of describing what the network does, however, rather than describing how it does that. You can get more information on this standard at www.itu.int/imt/.

- W-CDMA is a 3G standard proposed by NTT DoCoMo, and supported by Nokia, Ericcson, and other European companies. It is fully compliant with IMT 2000, and was, in fact put forth as the official IMT 2000 standard. It was not accepted because of opposition from the U.S., whose own CDMA standard would have been hurt. It stands for Wide Code Division Multiple Access. It is the clear choice of all European 3G carriers, as well as some U.S. carriers, Asian carriers, and South American carriers. W-CDMA as implemented by DoCoMo has a download limit of 384Kbps and an upload speed limit of 64Kbps. Its theoretical limit is about 2Mbps.

- FOMA is the name NTT DoCoMo has given to its implementation of its W-CDMA network.

- GPRS, which stands for Global Packet Radio Services, is an upgrade of GSM networks to provide packet-switched services. This standard is also referred to as 2.5G, because it is an upgrade of the current second-generation system, but doesn't go anywhere as far as 3G in terms of speed or spectrum efficiency. It is an inexpensive upgrade, which addresses one of the biggest complaints of early WAP users, the cost and speed of using circuit-switched connections for mobile Internet applications.

- UMTS, which stands for Universal Mobile Telecommunications System, is a standard that adopts specific protocols for implementing IMT-2000, and is supported by most of the same companies that support GSM in Europe, in addition to NTT DoCoMo. UMTS is built on W-CDMA..

Much of the focus on NTT DoCoMo's 3G network has been on what sorts of content can be served to mobile handsets. Voice revenues are dropping all over the world, and another revenue model is sought. 3G, with its capability to deliver more and richer content, streaming video, and video conferencing figures heavily in NTT DoCoMo's own promotional material. Of the first two handsets released, one has a built-in video camera for video conferencing.

A May 29, 2001, story in *Business 2.0* asks the question of what we do with wireless video. Several answers are suggested, among them video e-mail, which one person quoted in the story favored. Just as audio e-mail has been less than popular, however, I doubt whether video e-mail will be the killer video app. Because the upload rate on 3G FOMA is currently only 64Kbps (not any better than the ISDN lines that are increasingly common in Japan, nor better than competitor au's CDMA-based Packet One service), videophone applications are unlikely to be quite as cool as everyone expected, with frame rates on the upside of only about 15 fps (NTSC Television uses 30 fps). This will be no more than the eggy was capable of. The eggy was only able to send still pictures, however, whereas the P2010 FOMA phone can communicate using the H.324 videophone standard. That standard is a few years old, most implementations are rather jerky, and the audio often doesn't match the video. Companies such as 8x8 (Eight by Eight), which pioneered set-top videophones for a general consumer audience, have gone on to other things or gone out of business. The problem was twofold: the units had lots of wires and my mom couldn't connect it (actually, I gave her one, an 8x8 unit, which is still sitting in her closet because she disconnected it after every use — needless to say, we didn't use it a lot), and the fact that most people whom you wanted to use it with didn't have one. A cell phone with this capability built in would solve both of these problems. It isn't going to be pretty, and it has a poor track record so far, but video telephony could just be the killer app for 3G. Then again, the killer app on the Internet and again on i-mode has been e-mail, probably the least sexy application.

Another scenario is that we use the devices to do something that we are already quite used to doing on a daily basis — watch TV. Watching television on a mobile phone is about as convenient as one can get, and with the much higher 384Kbps download rate on FOMA, it should look a lot better than the eggy's did. The greatest limitation on prospects for this application in Japan is the rate that DoCoMo plans to charge for FOMA data transfers: 0.03 yen per packet. One packet is 128 bytes. If you were watching your favorite TV show at the maximum rate, you would pay 11.5 yen per second! Okay, okay, actually in the test phase DoCoMo charges on the basis of connection time, between 33 yen and 60 yen per minute, depending on where you are, whom you are calling, and time of day. Just as this model of charging for content nearly killed WAP, however, it is likely to have a negative impact on people actually using FOMA phones in this way. And it would actually be a better option to just include a normal broadcast TV in the handset than to stream at the rate being talked about.

Music is a much better medium for this sort of device, and it clearly fits an existing market — all those people who carry an MD player or CD Walkman or MP3 player. It also presents a much clearer revenue model — 300 yen per song, in Japan, is exactly the scale of charge that the i-mode service was built on. NTT DoCoMo currently offers just that sort of service on its PHS network, as does competitor DDI Pocket. There are still-lingering concerns by copyright holders about digital delivery, but these concerns are outweighed by the potential in revenues. The president of Sony Music Entertainment Japan told a conference in May 2001 that as soon as the bandwidth was there, digital delivery of music was a high priority. Well, the bandwidth is now here.

The untold story is that although 3G may well mean an increased capability to do more with the bandwidth, the fact that more bandwidth is available saves on the cost of infrastructure – more users can be fit on to less infrastructure. A January 11, 2001 story in *Nikkei Business* magazine estimates the current cost per subscriber of PDC network infrastructure at about 130,000 yen, or about US$1,100. The cost per subscriber on the IMT-2000 network, on the other hand, will be only 75,000 yen, or about US$600. Of course, the $1,100 has already been spent, so it only makes sense to take advantage of it as best as they can.

NTT DoCoMo's incredible subscriber growth has, however, strained the capacity of its networks. Fairly soon, the subscriber numbers will outstrip NTT DoCoMo's network capacity. Investment is required, in any case. Though initially more expensive, 3G networks are actually cheaper in the medium to long term, when calculated on a per-subscriber cost, because of their higher capacity. All of this assumes that things will stay the same regarding usage, pricing, and other factors, which is of course unreasonable. In the medium term, however, NTT DoCoMo's investment in W-CDMA network infrastructure is extremely logical, though not for its publicly state reasons of being able to deliver richer multimedia content.

This is typical NTT DoCoMo chutzpah. It needs to upgrade its network; it chooses the most logical course for doing that and then casts that choice as one to offer very cool new services to customers. It gets plenty of publicity from being first, it gets people lining up to be the first to have this cool new technology, and it quickly gets a subscriber base that covers the cost of upgrading its networks. Being able to quickly implement and create a consumer demand for this new network, which it needed to implement anyway, is one key as to why NTT DoCoMo is incredibly strong and very profitable.

 For more information on the nature of NTT DoCoMo partnerships and its i-mode revenue model, see Chapter 7.

Knowing the market

What NTT DoCoMo is offering in its high stock valuation, in its partnerships with foreign carriers, and to its domestic partners, is a clear understanding of both voice and data markets for wireless technology. NTT DoCoMo's success in those markets is seen as proof by those who value its stocks, its help in implementing solutions, and those who seek partnerships with the company, that NTT DoCoMo knows its market, and is able to promote effectively to that market.

There are many, many people, both in Japan and abroad, who believe that this will be NTT DoCoMo's downfall – that the markets are too different in other countries, that its experience in Japan won't help it on the world stage, that its knowledge of its own market is scant, as evidenced by its continued focus on business users.

If you choose to listen to these naysayers, you will dismiss i-mode as a phenomenon that could happen only in Japan and its customers as purely faddish Japanese teens downloading Hello Kitty start screens. You will believe that nobody except NTT DoCoMo is making any money on i-mode, and no business tools exist for i-mode. It is, in other words, a cultural phenomenon that cannot be replicated in Europe or the Americas.

And you would be mistaken. Although there is an element of truth in this view, it ignores the real story, which is that in the same way Japanese consumers purchase televisions, cars, video cameras, and so on from Japanese companies, who then export these same items abroad, i-mode has been well tested in Japan, and is ready for the world stage. To dismiss it as a fad for teens misses the point that most of its users are not teens. To say that there are no business tools is simply untrue, though these are mostly not available in the same way as consumer-oriented i-mode content. To answer the question of whether money is being made on i-mode by companies other than NTT DoCoMo, simply take a look at those companies successfully marketing content. Chapters 15 through 17 do that, and the answer seems to be that some companies make money, and others do not. This is no different from other sectors of a modern capitalist economy.

A refusal to learn anything from NTT DoCoMo's experience with i-mode is a mistake. Many will point out, however, that NTT DoCoMo often does not seem to know its market very well, and relies on its sheer size and power to introduce new products and services.

Expanding into the global market

DoCoMo's move abroad is its least clear move if you take at face value what it is saying and what it said in the past. One reason is that its different reasons are not completely complementary and not always consistent. Its official public reasoning, as stated in the annual shareholder report, is that by sharing its expertise with foreign carriers in which it has taken a stake, DoCoMo can increase the value of its stakes in these companies. Table 2-1 lists DoCoMo's percentage holdings in several international companies. This expansion has been a spectacular failure already in one case, that of KPT, the Dutch telecom operator in which DoCoMo has a 15% stake. The stock has lost most of its value since DoCoMo purchased its stake, and DoCoMo is expected to take a huge write-off. The overall logic of the move is suspect, though, if it is to simply increase the value of its shares in these companies. The telecom sector has performed below that of the S&P 500, and if it were simply for the purpose of making a good investment, an index-tracking fund would be a better investment. And less work for DoCoMo.

Another reason, stated on its Web site, is to enable Japanese users to have access to i-mode service when they are travelling. This is laughable at best. Its tie-up with AT&T will use GPRS, as will that with KPT, neither of which is compatible with the i-mode handsets used by Japanese users. In the long-term view, 3G with i-mode would serve this purpose, and this may be DoCoMo's goal – though with more and more carriers announcing delays in introducing 3G services, it is looking like a long-term prospect.

The truth is probably a little of the stated reasons and a lot of an unstated reason, which is DoCoMo's desire to expand abroad, as well as for reasons of pride. First with Hi-cap, its analog standard, and then with PDC, its digital standard, the rest of the world adopted other standards, and in the case of the U.S., demanded that Japan adopt the same standards. The 3G W-CDMA standard has been adopted as the UMTS standard, and DoCoMo is justly proud of this. But simply being adopted is not enough – it must be implemented. By its stakes in other carriers, DoCoMo's can, and already has been, the voice urging early adoption. i-mode is the carrot, a world 3G standard is the goal. Why this goal is desirable is the missing piece, and I would argue that it probably stems from pride.

TABLE 2-1 NTT DOCOMO'S STAKES IN NON-JAPANESE MOBILE CARRIERS

Company	Country	Stake
Hutchison Telecom	Hong Kong	25.37%
Hutchison 3G	Britain	20%
KPN Mobile	Netherlands	15%
KG Telecom	Taiwan	20%
AT&T Wireless	U.S.	16%
Telefonica Cellular	Brazil	Approx. 3.6%
AOL Japan	Japan	42.3%

Although justly admired at home and abroad as a company with an extremely strong product and an excellent record, its stakes abroad have taken longer than NTT DoCoMo initially thought to produce results.

When the investment was made in AT&T Wireless, the president of NTT DoCoMo said that the service would be ready by fall of 2001. Well, it isn't, and work didn't even really start until fall of 2001 on the technical side of things. Similarly, Hutchison Telecom's service was supposed to be ready early in 2001, and as of publication it is not yet ready. KPN's financial woes have pushed its adoption of 3G further into the future, and it has only begun (in September of 2001) to do the groundwork necessary to roll out an i-mode-like service.

Although all these carriers were undoubtedly happy to have NTT DoCoMo's investment, NTT DoCoMo's approach to them has been to let its partners initiate the rollout of i-mode-like services, and to simply advise these partners on how this needs to come about.

Partnering with NTT DoCoMo could be a great opportunity for these companies to take advantage of DoCoMo's knowledge, especially in developing a successful consumer-oriented mobile Internet service.

DoCoMo is not promoting i-mode as a brand in other countries. Why it is not doing so, considering the publicity generated by i-mode inside and outside of Japan, is a question the company has not clearly answered. That the company has paid for this privilege is obvious by the vast sums it has invested in these companies. That the i-mode service is a solid consumer model is equally obvious by the huge numbers of subscribers using and paying for the service. An experienced and mature consumer products company such as Sony, after launching the Walkman or the Playstation to huge success in Japan, hardly turned around and named them "Soundabout" or "Gamestation" for export, though these names may actually sound better on their own. Why NTT DoCoMo chose to not take advantage of the i-mode brand seems to be out of deference to its foreign partners. Its foreign partners, out of deference to NTT DoCoMo's i-mode success, should be thinking about leveraging a clear world-leading brand.

In the year 2004, when most European countries, Korea, Japan, Taiwan, parts of the U.S., and parts of South America will have UMTS W-CDMA networks, it should be possible for a user to use a handset in any of these places, as is now possible with GSM. Japanese equipment makers, Japanese handset makers, and NTT DoCoMo will play a big part in this. NTT DoCoMo's contribution of the W-CDMA standard, its investments in foreign mobile carriers, and its seemingly selfless advisory roles in those companies it has investments in will have given it a clear and perhaps dominant place at the table where deals are cut, standards are decided, and decisions get made.

Chapter 3

i-mode Software

THIS CHAPTER LOOKS AT THE SOFTWARE OF I-MODE. You look at the underlying operating system, and at the software used in communication and for applications such as Web browsing and e-mail. A clearer understanding of how the layers of software are working should give you a clearer picture of the software framework in which the i-mode handset is operating.

You also look at the kinds of software available on i-mode. You look at cHTML content, as well as Java content, and see the issues that need to be addressed before beginning a development project based on either of these.

The Software Architecture of i-mode

Most people are somewhat familiar with their home PC's software architecture, or at least the vocabulary used to describe it. There is the OS (Windows for most of us), and there are applications such as a browser or word processor. The vocabulary used to describe the software of a mobile phone, however, is somewhat different. In this section, you look at the software architecture of i-mode phones. On a mobile phone, too, you have an operating system. In this case, this is a real-time operating system (RTOS), which is similar to a PC's OS, except that its function is limited to responding in real time to input from a user, the network, or an internal software application. Between the RTOS and application packages are software layers controlling the handset's communication with the network. These layers use protocols such as HTTP and TCP/IP to pass information from user to network, and vice versa.

All that power in the palm of your hand – the system structure

In the case of a desktop PC, the software structure looks something like this:

At the chip level, there is a processor and motherboard.

On top of the processor and motherboard, there is a bios.

On top of the bios, there is an operating system.

On top of the OS, there may be a graphical user interface (GUI).

Finally, there are the applications, which run in the OS or GUI environment.

All of these words are probably familiar, even if your grasp of the specific meaning of each word is not clear.

When speaking of mobile phones, however, very few of us ever hear the technical aspects spoken about in the same way that PC makers throw around the tech specifications of their computers, so we may not be so familiar with the terms.

As with a computer, at the chip level are a processor and a motherboard. On top of them, you won't find a bios; instead, you find a real-time operating system (RTOS) that communicates directly with a processor. Whenever a mobile phone is turned on, this RTOS is also on. It controls what happens when you push the menu button or the power off button. Each handset model has a different RTOS. Rather than do what PC makers do, which is to map the hardware using a bios and then have the bios communicate directly with the processor and chipset, the RTOS performs both the functions of a bios and an OS. For each new cell phone model, a new RTOS, usually based on that of its predecessor with added improvements, is used. Table 3-1 illustrates the software function differences between PCs and mobile phones.

TABLE 3-1 DIFFERENCES BETWEEN SOFTWARE FUNCTIONS

Function	PC	Mobile/i-mode Phone
All calculations, data processing	Processor, such as Pentium	Processor, such as Hitachi
Communicating with the processor and other hardware	Bios	RTOS
Communicating with memory, handling all user input, output, and any interaction between user and machine	OS	RTOS
Controlling the look and feel of user-machine interaction	GUI	Simple Window Manager
Adding software functions to a device	Application	Application

What that means for adding software on top of the RTOS is that it is not so simple as just sticking a disk in a drive. In fact, all software that a handset ships with, including the RTOS, is stored in memory and cannot be changed or replaced without replacing or "flashing" the chips it is stored on. Aside from that, however, the very real problem exists of different handset makers using different RTOSs and different processors. In the case of i-mode, Access has put several software layers between the Netfront browser and a handset's RTOS.

The handset, as it ships, is not modifiable. There are lots of reasons for this, but the key is that allowing any change in basic software introduces instability into the system, in the form of incorrect input, corrupted data, and the possibility of viruses. What, then, is a developer to do? The i-mode architecture contains the capability to add new functions and software without any need to modify any of the basic software. This is what cHTML and Java are used for. Adding ringing tones and screen savers are also examples of this.

The layers between the Access browser and the RTOS

Figure 3-1 illustrates the layers of activity between the user (entering and receiving data via his handset) and the real-time operating system.

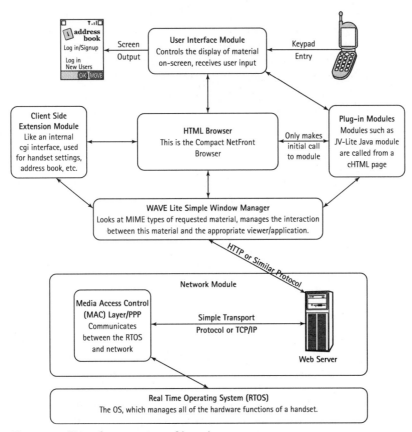

Figure 3-1: The software system of i-mode.

Here's how this all works:

- The first layer is the Media Access Control (MAC) layer or Point to Point Protocol (PPP). This layer defines exactly how the handset *connects* to the network. In the case of i-mode, the MAC is used.

- The second layer is the layer that talks to the network after the connection has been made, using either a Light Transport Protocol (LTP) or TCP/IP. In the case of i-mode, between the handset and i-mode server, LTP is used; and between the i-mode server and the Internet, TCP/IP is used.

- The third layer is the layer that communicates with an HTTP server after the connection has been made and the phone is talking to the network. An HTTP server is the sort used for serving Web pages as well as cHTML content. This layer in i-mode can talk only to an HTTP server.

- For mail, the process must be processed either through a Web-mail interface, or through the i-mode mailserver, not directly through a mailserver on the Web. Nor can one connect directly to an FTP server or through a telnet connection. Although there are people to whom this is an inconvenience, connecting by telnet to a server through a mobile phone is not something most people need to do. Those for whom this ability is important probably have the ability to build their own Web-based telnet or FTP client. It is actually being worked on, and someone somewhere probably is right now checking server logs, rebooting a server, or inputting data in raw SQL via his or her mobile phone.

- The final layer below the application layer is the window manager layer, which is a very basic version of Windows (although comparing it to Apple's hypercard program is probably closer to the mark). Unlike Windows, this layer does not allow a user unlimited flexibility, but controls what is available from each window. This is the layer that most Japanese mobile phones, at this point, use to implement a graphical user interface.

- The next layer is the HTML browser layer. This layer contains the browser that interprets cHTML pages, and displays them. It also contains connections to plug-in modules, such as a Java Kilobyte Virtual Machine (KVM), or (in the future) an MPEG4 playback module or H.323 videophone module. Another sort of module, called a Client Side Extension (CSE), also exists on this layer.

- The CSE is somewhat like CGI in terms of its function – it allows the browser to connect with other internal programs, such as address books and screen savers, and provides a way for the browser to talk to them (for example, to take a picture found on the Internet and make it into a screen saver). This is actually one of the popular uses of i-mode phones, and when people ask me what i-mode's killer app is, I always whip out my phone and show them its startup screen with my son's picture on it.

I am not sure of the reason, but this system rarely crashes. I have actually had i-mode mobile phones from three different makers (Fujitsu, Panasonic, and Sony), and I have not had to restart my phone once. As an earlier adopter, with the very first i-mode phone by Fujitsu, I was still not subject to the sorts of problems early WAP phones had (of crashing and having to reset the phone by removing the battery).

As I mentioned before, all the handset makers except Panasonic use Access Corporation's Compact Netfront software for its i-mode phones. Access products are probably partly to thank, as is the standardization of all aspects of the i-mode architecture (which, rather than being based on a consensus standard like WAP, which is open to interpretation and to some variance in actual implementation, is dictated by NTT DoCoMo).

A strong focus on international standards-based solutions sometimes seems to have the unfortunate side effect of not valuing the customer experience to the degree that is necessary in a service industry, and a competitive service industry like the mobile phone industry has become. This is clear from the WAP standards, in which there are simply so many parts to the standard, and enough room for interpretation in each of these parts in how the standard is actually implemented, that incompatibilities and quirks in implementations nearly guarantee problems. These problems will get worked out when industry standards, meaning standards of implementation, are hashed out between carriers and equipment manufacturers. i-mode has the advantage of already being at this point in its implementation.

 The WAP standard is discussed in more detail in Chapters 4 and 5.

The i-mode Browser and the cHTML Standard

The i-mode system depends on a markup language called cHTML (Compact Hypertext Markup Language) to display information on a variety of different handsets. Like its big cousin, HTML, cHTML uses tags to specify what a certain kind of information is, and the browser software decides how such information is displayed. It is a basic concept that sometimes gets lost in the heat of actually making a page look how you want it to look, but a markup language is not supposed to determine *how* something looks in the final product. That job is left up to the browser. For example, using a tag to make a top-level heading doesn't specify how the browser will present in terms of point size or font family. It just means that this heading is "bigger and bolder" than a second-level heading. The browser has defaults or user-specified options that translate what a top-level heading will look like for that browser.

A markup language is simply meant to tell the browser *what* something is. In the case of HTML, this guiding principle of using the markup language to define text has really taken a backseat to newer technologies that help you force or determine specific appearances in the browser. Technologies such as cascading stylesheets (CSS) and WYSIWYG editors make the tags simply the means to an end. The end is getting a page to look exactly how you want it. This design philosophy is quite different from the purpose for which HTML was developed (providing generic types of text that could translate across any browser). Unfortunately, this trend focused the energies of Web designers in a direction antithetical to the original goal of HTML.

In the case of cHTML, there are many fewer tags. Most don't make a significant difference in how something looks on the page, and the next iteration of cHTML will probably be a subset of XHTML. XHTML is the next generation of HTML. It gets back to HTML's roots of simply using tags to define different kinds of text, and leaves formatting tags entirely out. You should probably use the tags exactly as they were originally meant to be used: as definers of certain types of content. Exactly how the browsers interpret these tags and display them is necessary knowledge, but do keep in mind that although your itch to make your pages look perfect is probably insatiable, try to resist the related urge to use tags improperly to achieve this. Ha! Easier said than done. Indeed.

 See Chapter 9 for details about cHTML tags.

Table 3-2 compares the functions of cHTML to those of HTML.

TABLE 3-2 CHTML AND HTML FUNCTION COMPARISON

Function	HTML	cHTML (as implemented in i-mode)
Tables	Yes	No
Telephone dialing	No	Yes
Forms	Yes	Yes
Sound	Yes	Yes (but as a separate function)

Function	HTML	cHTML (as implemented in i-mode)
HTTPS	Yes	Yes (as implemented, however, the security between the handset and server is open to question, except on the FOMA and 503 series phones)
Access to address book	No	Yes
Access to client serial number	No	Yes (on newer models)

Access Corporation of Japan developed cHTML in cooperation with NTT DoCoMo and its OEM handset manufacturers, and submitted it to the W3C for acceptance as a legitimate compact version of HTML for mobile devices. It is still under review, and, I think, probably will never come out of review. Phone.com (now known as Openwave), the developer of the WAP markup language WML (Wireless Markup Language), made itself a part of the decision. Not surprisingly, it opposes cHTML becoming the accepted standard. Because the W3C works on the basis of consensus, a decision is effectively blocked. The W3C is also working on a platform-independent markup language as the next generation of all the currently used markup languages—HTML, cHTML, WML, DHTML, and many others. This language, called XHTML, will correct many of the problems currently faced by developers who want to develop content separately from how, or on what device, it ends up being displayed. I provide an overview of this language in the section that follows.

XHTML in brief

XHTML is basically the next evolution of HTML, but takes an approach that is drastically different from the HTML you have grown to know (and love?). HTML had its roots in SGML, which has been around since the '70s, and since 1986 has been a standard used for documentation. FrameMaker, a package now owned by Adobe, is an example of a product that uses SGML tags to format long documents in a consistent and cross-platform way. XML, according to the W3C consortium, is meant to be a modern replacement for the venerated standard. The key advantage is the 'X', which stands for "extensible." SGML and XML work by allowing a programmer or designer to define not only what goes in between the tags on a page, as HTML does, but also to define the tags themselves. This is similar to what a word processing program such as Microsoft Word does. It comes with certain built-in styles, such as Normal or Heading 1. When you first open the program to create a new document, those, among others, are what you find for styles. But you can change the way those styles display, or you can add new styles if you like. If you have ever looked at a Word document in a text editor, you probably found it to be basically unintelligible because all those styles and formats have to be included to tell Word how to display the document the next time it is opened. A program such as FrameMaker uses SGML underneath to do exactly the same thing. The big difference is that with SGML, rather than defining exactly *how* to display a certain kind of text, you are giving detailed information about *what* is actually contained. How to display the text is decided within a program such as FrameMaker or another SGML viewer or editor. Figure 3-2 shows what is in a MS Word file. It doesn't make much sense. On the other hand, Listing 3-1 makes lots of sense.

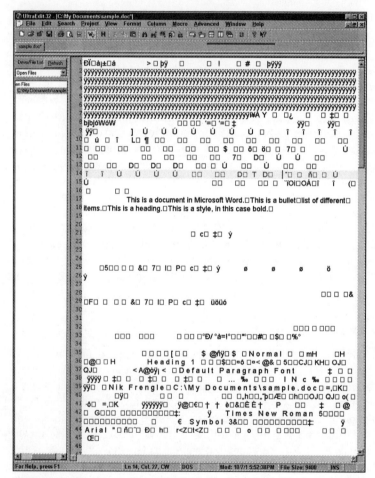

Figure 3-2: What a very short MS Word document looks like in a text editor.

Listing 3-1: A Simple XML Document

```
<?xml version="1.0"?>
<root>
  <intro>This is a document in XML.</intro>
```

```
  <bullet_list>
    <bullet_item>This is a bullet</bullet_item>
    <bullet_item>list of different</bullet_item>
    <bullet_item>items.</bullet_item>
  </bullet_list>
  <heading_1>This is a heading.</heading_1>
  <emphasis>This is a style of communicating something, rather than
a graphical style.</emphasis>
</root>
```

XML is the parent of XHTML. Unlike HTML, which is based on, but not compatible with, SGML, XML's structure allows it to maintain compatibility. XML does not define tags of its own, so it doesn't define what XHTML is or isn't. XHTML instead uses either DTDs or modules to define its elements and attributes. Earlier versions of XHTML used DTDs (Document Type Definitions). The newest version of XHTML, XHTML 1.1, uses modules that include a number of tags of a certain type (for example, forms or tables). XHTML Basic is built the same way, except that the sorts of modules available are more limited, and two of the modules are special versions optimized for mobile devices — the Basic Forms Module and the Basic Tables Module. The other tags are the same as their XHTML 1.1 brethren, though there are far fewer of them.

See Appendix D for a list of all the tags available in both cHTML and XHTML Basic.

Access's Compact NetFront browser Version 3, which is the newest version of the browser used on all the handsets currently available except those manufactured by Panasonic, will support cHTML and XHTML Basic. XHTML Basic will be a standard supported by both WAP (WAP 2.0) and i-mode, and it looks an awful lot like cHTML. This is both good news and bad news. The good news is that you will have to learn only one language when developing for mobile phone platforms; the bad news is that if you currently are using WML, you will be disappointed to learn that the XHTML Basic specification takes a lot more from cHTML than it does from WML. Truly, WML and its card paradigm have been supplanted by cHTML's mini-Web paradigm. XHTML Basic is perhaps not the most preferable option, placing the onus on developers to thoughtfully break up their content into logical pages rather than having that be a basic requirement of the markup language itself (as it is in WML), but it is a lot simpler to develop for.

The bottom line is that if you develop a cHTML application today, it will probably make an easy transition to XHTML Basic. There are already utilities available to clean up HTML for use in XHTML 1.1, and it is almost guaranteed that such utilities will become available for cHTML as well. If, however, you are one of those blighted few who develop for WML, the transition will be a little more difficult. Although the difference between cHTML and XHTML Basic is basically only syntax, the difference between WML and HTML is of a logical and organizational sort — in addition to having a completely different syntax. The rewards for switching to XHTML Basic are great, but it is not exactly a smooth transition.

Later in this book, I cover developing cHTML content using PHP. PHP 4.0, the current version, is an ideal partner for XHTML because it has XHTML-specific functions already included. Perl, too, has libraries available to smooth the transition to XHTML.

Although by the time this book is published shipping models of XHTML Basic-compliant browsers will be near at hand, none are currently available, so this is where we are in the present: learning cHTML.

The differences between cHTML and i-mode HTML

You need to understand the clear distinction between cHTML and i-mode HTML. They started out the same, but have developed some differences as DoCoMo has added certain functions specific to i-mode. Among these are the `utn` attribute, the `istyle` attribute, and the emoji graphic characters. Because i-mode has been by far the most major deployment of cHTML, the two have sometimes been treated as one and the same. There are other implementations, however: L-mode, a wired version of i-mode, also uses cHTML on home telephones and fax machines, and I think you can expect to see a couple of other household devices using it in the future.

Following is a list of additions made in i-mode HTML to the basic cHTML specification, as originally proposed by Access to the W3C.

- The `<blink>` tag was added.
- The `<marquee>` tag was added.
- The `<meta>` tag added support for character encodings, but does not support cHTML's `refresh` attribute.
- The `<ol>` tag added support for the `type` and `start` attributes.
- The `<li>` tag added support for the `type` and `value` attributes.
- The `<body>` tag added support for the `bgcolor`, `text`, and `link` attributes.
- The `<font>` tag was added.
- The `istyle` attribute was added to the `<textarea>` and `<input type='text'>` tags. This allows a developer to specify the default input method; that is, numbers, letters, hiragana, or katakana.
- The `cti` attribute was added to the `<a href>` tag. It adds extended telephony functions.
- The `<object>` tag was added. This is for declaring i-Applis. In future, other objects may also be recognized.

- To call the object declared with the `<object>` tag, the `ijam` attribute was added to `<a href= . . .>` tag.

- The `telbook`, `kana`, and `email` attributes were added to the `<a href="tel:123455566">`. These attributes allow the developer to give a user the choice of storing a telephone number, a person's name, a person's name in kana, and their e-mail address in a user's address book when she clicks on the telephone link.

- The `utn` attribute was added to `<form>` and `<a href=" . . . ">` tag. This is discussed at length in Chapter 9.

Because certain functionality is missing from XHTML Basic (such as precision positioning of text, coloring of text, and other limitations), I think it is reasonable to expect NTT DoCoMo to come up with its own additions to the standard, as it has in the past with other standards. The big difference is that this time there is a clear framework that the company can use to do this legally without breaking the standard: It can simply add a module of tags to perform the desired functions, and add recognition for those tags to the browsers that run on mobile handsets.

That is really what the *X* in XHTML and XML stands for: eXtensible. And extensible in this case means the flexibility to modify the standard to the needs of a particular platform. It is this flexibility, and near-inability to break the standard, if extra tags are declared correctly, that makes many believe that XML will be a standard with a long lifespan, capable of transformation. The downside is the real antipathy that the standard-bearers of XML and XHTML have toward letting page designers have control over how their pages look. This ignores the real artistic and aesthetic needs of designers. To some extent, they get around this problem by using style sheets and declaring new tags. These new tags could do things that the World Wide Web consortium had hoped to get rid of with XML, which is ironic.

In any case, at this point this extended discussion of XML has little to do with what is discussed here, which is the software environment of mobile phones. And at least in the short term, graphics and embedded tables and other precision design techniques are neither available nor useful to one designing for mobile phones.

Running midlets in the browser

As mentioned in the earlier section "The layers between the Access browser and the RTOS," the KVM is a module that can plug into the browser. This is akin to the way Java and other sorts of media are dealt with in the PC world, which is that each sort of content, or MIME type, has a registered application that is used to open that content. In the PC paradigm, however, browsers have come to include Java functionality within the browser window, rather than handing it off to a Java Virtual Machine outside of the browser window, at least for "applets," which are miniature applications running in the browser.

In the case of i-mode phones in the 503 series, which include a KVM and can run small Java programs, the midlets—MID stands for Mobile Interfaced Device, and *midlet* means a MID-based applet—a user can call a Java program in two ways:

- Using the Web page's `<object>` tag to call a midlet. In the case of the `<object>` tag, a module that can handle the specified object is looked for.
- Calling the Application Definition File (ADF, or JAM) directly. The .jam file extension tells the browser to send the file directly to the KVM.

Two files, actually, are necessary to run a midlet — the ADF file and the midlet itself. The ADF file is a very short file that tells the KVM how big the midlet is, when it was created, what it is called, and a couple of other basic things. The midlet is a binary "bit-code" file of the actual program, which the KVM reads and executes. It might include sounds or images within it, or just the program. Its real strength is that it is standalone and quite portable when compared to a Web page or site. After it is on the phone, it should have everything it needs to run. You generally don't have to worry about URLs, missing graphics, missing sounds, or incompatible sound formats — they are all taken care of for you.

There are actually a couple of utilities that convert normal Web pages into midlets. The first one — Hanpake (half-packet) — uses the compression features of midlets to save on the cost of packet charges by converting cHTML pages to midlets. This also allows users to store the page as an i-Appli, and view it offline, something they can't do with a cHTML page.

Speaking in i-mode: Language issues and globalization

What about the nice little things that you can do with your PC (or even your PDA), such as changing fonts? No go on an i-mode phone. In fact, most don't even support different styles or sizes. This is a legacy of its development in Japan, where the kanji fonts are tens or hundreds of kilobytes instead of the few kilobytes it takes to store an ASCII font. It was not long ago that to buy one new font for a PC cost upward of US$100, and that the Apple Laserwriter II NTX-J, which contained five kanji postscript fonts on an external hard drive, cost over US$3,000.

Actually, the KVM uses Unicode internally, whereas the browser and user interface use an older Japanese encoding system called S-JIS. JIS stands for Japan Industrial Standards, which is the standards organization that came up with the encoding. Shift-JIS, or S-JIS, is a newer form of the JIS standard. Thus, it is necessary that all text input and output from the KVM be converted first. You also use Unicode internally when writing Java code. This should be fairly automatic because code is mostly written in ASCII, which doesn't have any real differences with Unicode anyway. Using the cute emoji characters, however, requires that you specify their Unicode encodings rather than their S-JIS encoding. Both forms of encoding Japanese are called double-byte encoding because they take two bytes to encode one character, rather than the one byte it takes to encode ASCII, Russian, Thai, or most other languages with a much smaller character set than either Japanese, Chinese, or Korean. These double-byte encodings mean that if you are using them in an e-mail, for example, the number of characters will be half as many as the number of bytes it took to write them. At the screen level, the characters take up twice the width of ASCII characters, though this isn't really related to them taking two bytes, but to the level of detail needed to render Japanese characters.

Currently, all models of i-mode phones being sold have bilingual Japanese and English menus and browsers. For some reason, this was one of the requirements that NTT DoCoMo demanded of handset makers, and it is one that is quite popular with the non-Japanese residents of Japan, as you can imagine. The actual i-mode menus, however, are generally in Japanese. There is an English menu with a limited amount of content, but to get much out of the i-mode content, one must be able to read Japanese. I use the remote mail service, which allows me to look at five non–i-mode mail accounts, as well as using a train timetable service and my bank's account information lookup service.

These issues of language actually play some role in how i-mode is implemented in the U.S. and Europe. It is completely unlikely that the handsets would support S-JIS, but NTT DoCoMo has stated that one of the reasons for moving abroad is to guarantee Japanese users access to i-mode wherever they are in the world. So, some form of Japanese encoding would be necessary. My guess is Unicode because it is the most fully compatible standard there is today; not just with Japanese, but with just about every other known language in the world. And there already exists the odd situation of phones using S-JIS in one part, and Unicode in another, which is hardly a satisfactory situation. The problem with this scenario of the i-mode standard being switched to Unicode is that NTT DoCoMo has not shown much will to do this domestically; the 3G FOMA handsets are still using S-JIS. It is extremely unlikely, however, that this will be the case in the U.S. or Europe. One other advantage of Unicode, actually, is that it lets U.S. and European users have access to the cute iconic emoji symbol characters. These characters, as used on i-mode in Japan, are useful for saving screen space and conveying meaning in a quick and memory-saving way. Count on the actual symbols used to change from those that are currently used in Japan, though, because many are icons that convey cultural meaning in Japan, but would not necessarily convey any meaning at all to Western users. You can see Appendix E for a complete list, and judge for yourself. Some forum and discussion sites, and AOL in its Instant Messenger service, already allow "smileys," which are different kinds of faces that you can add to messages. Imagine having 168 different ones!

Taking a Look at Content Providers

In talking about the software of i-mode, I would be remiss to not talk about the actual content of i-mode sites. I have made a complete list, albeit one that has likely changed between the time I wrote it and when this book is published, of all the i-mode sites on the official menu. Getting on this list is not an easy process, but for those who make DoCoMo's grade and maintain a useful product, the end results can be very profitable.

Chapter 7 provides details on how these sites became members of the i-mode menu.

The first thing you will notice is the sheer number of banks. In one sense, it is understandable because they all connect to a settlement network, which is a network connecting banks with one another, and used for transferring funds between banks, using a sister company of NTT DoCoMo. The fact is that they have the computer infrastructure in place so that establishing an i-mode site is not a huge challenge. Many of these financial institutions, in fact, especially the Agricultural Credit Cooperatives, have the same computer systems, and seem to have shared with one another their servers and features, so that the services are quite similar. It should be noted that all of these banks, savings and loans, and credit unions have set up i-mode banking more for the future promise it holds, and so that their competitors don't have anything they don't, than because huge numbers of Japanese are flocking to their banks because they have this. Telephone banking has really been around only for a couple of years, and the number of customers using it, while increasing, is still a very small part of the total number of a bank's customers. i-mode banking is actually quite convenient in Japan because most payments are made using bank transfers, which many of the i-mode banking sites can handle. One of the reasons the services probably haven't taken off is the necessity to go into the branch where a customer has his account, and fill out an additional application at most banks and in every case in which bank transfers will be involved. On the other hand, bank transfers not using i-mode can be done through almost any ATM, whether it is your bank or not, your branch or not. Getting people to take the time to do more paperwork may be easier in Japan, but it still is not easy. A couple of new banks that focus on the Internet user (and not having any physical bricks-and-mortar operations) may be able to change this situation. The first step is to take a machete to the normal layers of bureaucracy that see a simple transaction or application, when handled through a bank clerk rather than a machine, requiring between three and seven people to sign off on it.

The next most plentiful category of content is entertainment of one sort or another. This content is conspicuous in my translation in Appendix B because of the rather cute names. Actually, these were quite difficult to translate because many are references to trends that don't exist in the West or are quite specific to Japan. The approach of many of these sites is quite similar to television. In fact, the majority of sites, entertainment or not, call the person responsible a "producer" or a "director." Ringing tones, screen savers, games, horoscopes, and other sorts of entertainment make up the majority of traffic on i-mode, and the approach of information providers to developing this content is extremely trend-driven, and dependent on fads sometimes so quirky as to be completely incomprehensible to anyone over twenty.

One category that has not been particularly well-represented is the tools and utility category, which has only a few listings. One of these services, remote mail, accounts for more traffic than all of the others combined. This is an indicator of the extent to which i-mode is dominated by entertainment. Tools for businesspeople, such as groupware, scheduling utilities, address books, and so on, are completely missing from the i-menu, though people are using them. They connect to these applications, which are hosted at their own companies, rather than on a public network. How much they do this is not something that is very easy to gauge; the information is not public, as it would be if it were NTT DoCoMo providing the services through the i-mode portal. Because users can connect to the Internet, what is offered or not offered on the i-mode menu is not all there is.

Although American companies are arming their workers with PDAs, Japanese companies can save a fair amount of money by using the mobile phones their workers already have. This is quite attractive to companies, for obvious reasons. This promise, however, has probably not been realized to the same extent it probably would be in the U.S. Venture capitalists I have talked to in the U.S. could only talk about B to B. They don't like B to C, as most mobile content companies are, because they can't figure out the revenue model. And they have a point – the most popular content probably still only earns its developer a million or two dollars per month, whereas one big contract providing back office solutions to businesses could pay the same amount. Lots of customers don't always equal lots of money, as more than one Internet startup has found. If you are interested in peanuts (only a couple of million a month, maximum), see Chapter 7 for more information on how to become a registered developer.

Chapter 4

i-mode Hardware

IN THIS CHAPTER, I EXAMINE THE NETWORK that i-mode runs on, along with the specific components of that network. The chapter also covers what distinguishes the proprietary network that i-mode runs on from that used for WAP, and how this distinction will affect adoption of i-mode outside of Japan.

NTT's Packet-Switched Standard

One of the things that distinguished i-mode from early WAP implementations was its "always on" nature. Rather than having to wait to connect to a network, which can take up to a minute or more on WAP, i-mode phones are always connected to the network. E-mails are delivered as they are received, without users needing to log in, and connecting to other services takes no longer than the time it takes to download the menu of those services. This always-on aspect of i-mode is an important feature from a usability standpoint, and it exists thanks to i-mode's use of the packet-switching PDC-P (Personal Digital Communication Packet) network.

This network uses an architecture currently being utilized only by NTT DoCoMo in Japan. Because i-mode runs on this network, it has been closely linked to it. Especially among laypeople, this network is sometimes referred to as the "i-mode network," which is a bit of a misnomer. Although they should know better, some who support competing standards point to this as evidence that i-mode is a proprietary system.

Can a "proprietary" network compete with WAP?

Before getting into a specific discussion of the merits of DoCoMo's packet network and the next-generation packet network likely to use WAP, one extremely important factor should be considered: What is i-mode, and why on earth should there even be a question of compatibility?

i-mode is, for all intents and purposes, a cHTML-based, packet-switched, Java-enabled, mobile-phone-based Internet service. Nothing more; nothing less. None of these characteristics is proprietary: cHTML has been submitted to the W3C for inclusion as a public standard; there is no technical reason why i-mode will not work on any packet-switched network that has IP capability; Java is an open standard, albeit one controlled by Sun; and there are mobile phones in all parts of the world.

To understand how developers approached i-mode, read Chapter 2.

There is a lot of tech double-speak about the superiority of WAP over i-mode, the WAP's open nature versus i-mode's "closed" nature, and the like. Frankly, most of this talk is coming from WAP's side.

"i-mode is a specification," says WAP Forum CEO Scott Goldman. "If i-mode was a superior specification or technology, other companies would have adopted it by now. But 500-and-some companies have gotten behind the WAP standard rather than the i-mode standard. That's got to tell you something."[1]

This statement contains a half-truth: i-mode is not a specification, but the name of a service offered by one company. It contains another half-truth because many of the 500 companies who are members of the WAP Forum are also supporting i-mode (including WAP Forum members and i-mode owner DoCoMo, and giants Nokia and Ericsson). And the statement contains one incredibly inane statement, which is that the companies offering the service are somehow more clever than the 21 million people who use i-mode, which is about two to three times more than WAP users worldwide. The statement smacks of rivalry and bullying, and has been met with either silence or insistence from NTT DoCoMo that there is no conflict at all from its point of view — and that the technical differences between the WAP standard and the i-mode service are ones that can be worked out.

And that is the point of this little diatribe — WAP is a standard that defines exactly *how* content is delivered to mobile users, whereas i-mode is a service that happens to currently use a means of delivery that is proprietary, more out of impatience with waiting for a clear standard to emerge from the WAP forum than a desire to create a proprietary standard. When W-CDMA is standard for 3G networks, as it will be in Japan from this year, i-mode will still be i-mode. When i-mode service (or whatever NTT DoCoMo and AT&T will call the service in America) begins in 2002 in Seattle, it will be delivered on a GPRS (General Packet Radio Services) 2.5G network — the same network used in WAP service — as it probably will be in Europe. But it is still i-mode. Why? Simple — i-mode is a service, not a standard.

Take an Internet example: Ten years ago, the America Online (AOL) software one used to access the service was proprietary and not based on Internet standards. You could not even connect to the World Wide Web. When the Internet standards started to dominate the industry, AOL, still using its own software, also began supporting those standards, and users could access the Internet through AOL. But either way, AOL was still AOL. The content on the AOL portal, the chat rooms, and the services are what made AOL, not the Internet standards. i-mode, like AOL, is a service rather than a standard.

1 As reported by Elisa Batista on Wired's Web site on August 30, 2000, in a story entitled "WAP or I-Mode: Which Is Better?"

The place where i-mode and WAP have butted heads is in the markup language they use: i-mode uses a version of cHTML, and WAP uses WML. There are good points to both, but simply because WML has been approved by consensus by a large number of companies, and cHTML is still waiting for inclusion as a standard, does not make WML the better standard. In point of fact, cHTML is so close to standard HTML as to warrant inclusion in the HTML standard. In the future, i-mode will look very much like the WAP standard: It will use the same networks, the same packet-switching, and perhaps the same server protocols. The markup language, meanwhile, will likely be wrapped into XHTML, and there won't be anything except the i-mode service for the WAP people to sling mud at.

I recently attended a conference on the wireless Internet held in New York. The attendees, many in the mobile phone business, when presented with new technology, often asked questions of the technical sort. For example, how much is its clock rate, what are the packet sizes, what is the network architecture, and is it based on a world standard or a proprietary platform? When asking questions of a business person from a company they were not familiar with, the mantra seemed to be, "So, what's your business model?" Business and technical realities have a premium placed on them, especially in the U.S. All fine and good, but this focus on the technology forgets that a customer experiences these aspects of the product with neither joy nor with any enjoyment.

Japanese consumers do not know the name of the cellular system they use, which is mostly true for American consumers. They do not know what network their mobile Internet service runs on, much the same as American consumers. They do not know what packet-switching is, and I doubt that American consumers do, either.

The needs, wants, and understanding of the technologies are not so different between American and Japanese consumers. NTT DoCoMo's approach to these needs, wants, and understanding of the technology, however, has been significantly more consumer-oriented and service-focused than most other carriers in the world, including most U.S. carriers and its own domestic competitor, au. WAP is a tech standard, whereas i-mode is a consumer product, and i-mode's success is at least partly due to NTT DoCoMo's decision to not discuss, ever, the technology behind the service. It has not printed the word 3G in any of its literature. FOMA is its word for the service, and forget about 3G. cHTML? Pshaw! Call it i-mode HTML. Packet-switching? The company never mentions it, except in the fine print on the mobile phone bill, in which it explains what a packet is, and how your charges are calculated. Because NTT DoCoMo never made the mistake of selling a service that was usually off, it doesn't have to resort to explanations such as "always-on." The proof is in the pudding.

So, WAP's initial slow uptake was probably much more to do with the capability of the company's selling the service to market to consumers, and with the poor implementations on circuit-switched networks and on expensive and buggy handsets, than the WAP standards. The comparisons between i-mode and WAP, in other words, are missing the target when they talk about the technology. It's the consumer, dummy!

One of DoCoMo's phrases from its annual report is, "new services take precedence over technology." i-mode is not rocket science, nor is it especially technologically advanced. What has given it popularity is the way it has met the needs of its users. In this sense, neither DoCoMo nor its customers prioritize the network architecture or the underlying architecture.

That said, an understanding of the current packet-switched network delivering i-mode is helpful for a clear overall picture of what i-mode is at this stage.

A technical definition of the network architecture for i-mode networks

The network used for i-mode in Japan is called the DoPa network, or PDC-P. DoPa is simply the name given to the PDC packet-switched (PDC-P) network architecture used by NTT DoCoMo. DoPa as a service is currently capable of speeds up to 28.8 Kbps. i-mode phones, however, are capable of only 9600 Kbps. Because of other limitations on the phones (such as memory, processing power, and screen space), this was thought to be enough. The slow speeds, however, made TCP/IP somewhat unattractive as a transfer protocol: The small amount of information being transferred hardly justifies the fairly substantial overhead imposed by TCP/IP. And so a network architecture that is somewhat different from straight TCP/IP was developed.

The following lists the behavior of the i-mode network:

1. In this network, between the handset and a Web page sits an intermediary server called a protocol conversion gateway.

2. This server communicates with the handset using a protocol called the light transport protocol (LTP), and then translates the request for a page to TCP/IP.

3. The Internet site then sends the data through the protocol conversion gateway to the handset.

4. If a user pushes a submit button, sending data, this also goes through the gateway.

5. This LTP, then, is used basically to speed things up. It will no longer be neccesary when 3G networks with extremely fast transfer rates are available, and NTT's W-CDMA version of i-mode will use straight TCP/IP.

Figure 4-1 shows the current network's architecture:

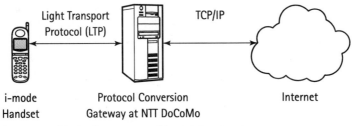

Figure 4-1: The basic structure of i-mode's network.

The astute reader will notice a danger in this architecture: a bottleneck at the protocol conversion gateway. This was indeed a problem at one point for NTT DoCoMo, and the company was told to not sign up any more customers by Japan's Ministry of Posts and Telecommunications until the problem was fixed. Needless to say, this got DoCoMo's attention, and it quickly complied, by setting up a new gateway facility with substantially greater capacity.

 One other thing that using this protocol conversion gateway means is that the length of a URL is limited to 200 bytes, which means that using the `get` method in a form is unlikely to work well if your form asks for very much data.

Though the LTP is not standard TCP/IP, it is a typical NTT DoCoMo tweak of a system to get more out of it, whether it is based on a specific standard or not. This works for DoCoMo because it is the market leader in both Japan and the world in terms of mobile Internet subscribers.

The i-mode server

Though the previous section describes what is happening in the most basic transaction between a user and the Internet, the majority of users do not use this method, accessing one of the i-mode sites instead.

When i-mode was conceived, it was thought that using the term "Internet" would be a turn-off. Instead, it was a phone that had all sorts of available services, features, and functions in a place called "i-menu." If you want to use the Internet, this is in a completely different area on the phone. In actuality, the Internet and i-mode are one and the same, with the i-menu simply serving as a portal, in effect. The requests for information from the i-menu, however, do not travel over TCP/IP as requests to the Internet do. Instead, they go to an "i-mode server." This server is really just an HTTP server, serving pages to mobile phones. When this server points a user to another site on the i-menu, they are sent directly to that site. This happens by either using the Internet, or (in the case of i-mode banking) by dedicated lines. Figure 4-2 shows two different ways of accessing content.

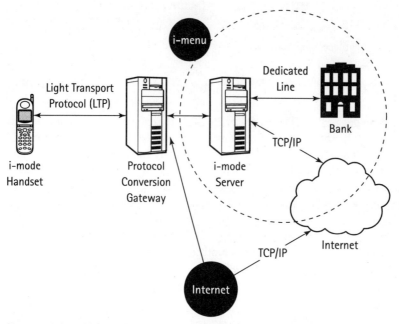

Figure 4-2: The flow of information on an i-mode network depends on the presence of an i-mode server.

On the handset itself, these two ways to access content are differentiated, even if content provided by official information providers actually resides on an Internet-connected server.

For security reasons, i-mode banking is not connected through the Internet. Neither NTT DoCoMo nor the banks are describing what sort of verification process is used in i-mode banking (as you would expect), but chances are it would be vulnerable to hacking if it were placed directly on the Internet. It does not hurt that NTT Data owns the dedicated-line banking network structure, called ANSER, but it is probably safe to say that even not being connected directly to the Internet does not protect a bank if security on its end is lax. There have been no reported incidents yet, but Japanese companies are often loath to report such incidents, so this is not very surprising.

An i-mode server is actually eight different kinds of servers. For simplicity, in Figure 4-2 I have bunched them all together, but for a clear understanding it is necessary to look a little more carefully.

The i-mode server functions as a relay server between NTT DoCoMo's i-mode network (the PDC mobile packet communications system: PDC-P system) and the

Internet under i-mode service. This server includes three service functions: relaying site information on the i-menu (i-mode menu) from the information provider (IP) server; handling Internet mail or i-mode mail; and relaying access to the Internet.

The i-mode server actually consists of multiple server systems, each of which fulfills the assigned functions to realize various services. NTT DoCoMo describes the system using the following terms and definitions:

- I-MAX (Interface-Mobile Access Exchange). This server connects M-PGW (Message-Packet Gateway Module) and other MAXs in the i-mode NSP (Network Service Provider) center. Load-sharing management for other MAXs is performed. When it connects to the M-PGW, it uses user information transfer protocol and network management. This server is also referred to as the Protocol Conversion Gateway.

- C-MAX (Contents-Mobile Access Exchange). This server receives information from the IP (information provider) and transmits it to a subscriber. This server is also called an i-mode server, or an i-mode portal server. To reduce IP message process load, the server can send a distribution request to information providers for multiple subscribers at one time. When the C-MAX receives the response, it can distribute it to multiple subscribers in a short period of time.

- M-MAX (Mail-Mobile Access Exchange). This server receives e-mail transmitted by a user using the mail function of an i-mode mobile phone or SMTP protocol of an outside domain. It also stores received e-mails in a "to e-mail" box. When it receives a message, it sends a notification to the destination i-mode mobile phone.

- B-MAX (Business-Mobile Access Exchange). The SAT (Sales Analytic Terminal), MNT (Maintenance Terminal), and BRT (Business Remote Terminal) connect through this server.

- D-MAX (Database-Mobile Access Exchange). This server collects and analyzes marketing data of i-mode services.

- N-MAX (Name-Mobile Access Exchange). This manages i-mode mail account names used by i-mode mobile phones.

- U-MAX (User-Mobile Access Exchange). This is the database of i-mode service subscribers. It stores basic subscriber information and user application information.

- W-MAX (Web-Mobile Access Exchange). This is the content server built up in GRIMM. It can provide common menu or regional menu contents, and has display functions such as My Menu, which is customized by users.

Packet-Switching Using DoCoMo's PDC-P Standard

DoCoMo had several specific requirements for i-mode: It had to be available all the time, it had to be reasonably priced, it had to deliver e-mail immediately, and ideally it would make use of DoCoMo's current infrastructure and PDC networks. PDC-P, a technology DoCoMo had sitting on a shelf for some time, was just the ticket. It perfectly fit the needs of i-mode, and didn't add a huge amount of infrastructure. DoPa was the name of this network, and it has had other — though much more minor — benefits for DoCoMo's business. They have wired drink machines with a self-contained interface, which allows a company to monitor when a machine needs attention. When a machine needs drinks, it can send off a quick message to a control center, saying which drinks to bring. For short text messages such as i-mode and these drink machines, PDC-P is perfect.

There are two methods of data transmission: Packet-switching and circuit-switching. In the case of circuit-switching, which is used by telephone companies for voice calls, when a user picks up her phone and dials, a circuit on a switch is dedicated to handling that call. This means that there must be as many circuits as there are calls. There is not a circuit for every user because most of the time the phones are not used. One of the tricks of the phone business is figuring out the minimum number of circuits needed to serve a certain number of customers. When the phone companies fail at this, you get the "all circuits are busy" message. The advantage of this method of switching is its reliability and simplicity — a user's call is transmitted as an unbroken series of (usually analog) sound waves. It is really the only option for transmitting analog sound, since sound is transmitted in waves, and has to have a continuous connection or else it is unrecognizable.

Packet-switching, on the other hand, which can only be used with digitized data, works in the following manner:

- It takes the data to be transmitted, and breaks it into chunks called packets.
- Each packet is a certain size; in the case of i-mode and many other packet networks, the size is 128 bytes.
- Each packet is given a "header," which tells the network where the packet should go and which part of the entire transmission the packet contains.
- The first bit of information is used to determine how a packet is transmitted. A packet network, using a router, determines which route is the most efficient one, and sends the packet that way. Compared to the "dumb" switches used in circuit-switching, the routers in a packet network need to be "smart" to be able to determine this.
- The second part of the information in the header is which part of the particular packet is in the whole. This is used on the receiving end of the network to put all the packets together in the correct order before sending the data to its final destination.

Generally, the advantage of packet-switching is that it utilizes network resources in the most efficient way.

In the case of the PDC-P network, routing is not a huge concern because the network is private, and the number of routes the data can take is limited. Even so, there is one definite advantage to this ability in the context of the mobile phone: When moving between cells (a cell is the area covered by one antenna), the handoff is quite simple from one cell to another because each packet has the information of who it belongs to and where it is going, independent of which cell it is being sent through.

Figure 4-3 depicts the network structure used in the PDC-P network.

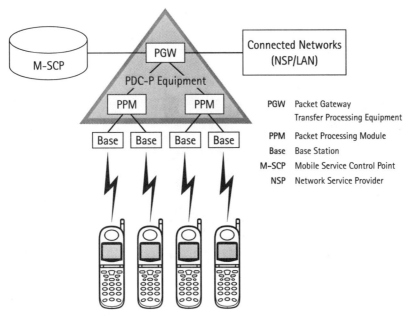

Figure 4-3: A specific view of what happens at the PDC-P network level of the i-mode implementation.

In the case of i-mode, Figure 4-3 basically shows how the packets get from the handset to the protocol conversion server or i-mode server that is on the network.

The PDC-P system consists of the "PGW," which provides functions to connect to other networks such as LANs and the Internet, access the M-SCP, and interface with the connected network and the "PPM," which carries out packet transmission and reception with the mobile phone or other device via the base station.

The PDC-P packet system can be used to connect to any network. In the case of the DoPa service, various Internet service providers (ISPs) provide connections for their customers through the PGW and onward to the Internet. In the case of i-mode, NTT DoCoMo fulfills this role of ISP, using the eight kinds of servers previously mentioned to manage users and contents. In this sense, the i-mode service is not only portable, and easy to use in terms of features and content, but also an end-to-end solution, requiring a user to pay only one bill. That is one of its advantages over its sister,

DoPa, in which a user is responsible for both packet charges and any charges incurred to the ISP.

When the i-mode servers and the PDC-P network are put together, you see the relationships shown in Figure 4-4.

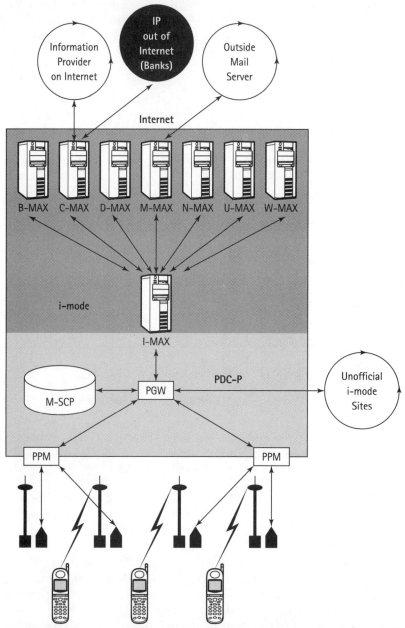

Figure 4-4: Bringing the whole picture together, this figure illustrates the relationship between the PDC-P network, the i-mode infrastructure, and the Internet.

Because connecting directly to unofficial sites on the Internet does not use the i-mode system, billing for those sites is up to site owners, and is not accomplished through NTT DoCoMo's billing system, as are charges for information providers.

The environment in which DoCoMo currently provides i-mode services is currently proprietary. That does not mean that i-mode itself is proprietary; in fact, it is not. There are two different and distinct parts of this system – the network element and the central office element. The network is the physical system of antennas, base stations, local exchanges, and central office equipment designed to take a signal from one point to another. In the 3G systems of the future, these elements will be standard, and a broad range of wireless providers will be using similar sorts of equipment.

The other part of the i-mode infrastructure is the central office system of servers providing content, mail services, access to the Internet, user-management features, and so on. In many ways, these servers are no different than those used in a big Internet service provider. Because DoCoMo is by far the biggest service provider in Japan at this point, this is no surprise.

One of the beauties of the i-mode system as it has been developed in Japan is that a developer doesn't really have to worry about any of the logistics of the network. The protocols are seamlessly switched, error correction is seamlessly taken care of, and exactly *how* data moves from server to user or vice versa is not something developers need to be concerned about. This fact doesn't sit well with people who want to understand exactly what is happening in detail before they embark on something, but it suits DoCoMo and its customers very well indeed.

Part II

The i-mode Environment

CHAPTER 5
The Direction of Mobile Network Development

CHAPTER 6
The Audience

CHAPTER 7
The Developers: Getting a Piece of the i-mode Pie

Chapter 5

The Direction of Mobile Network Development

BY MANY IN THE TELECOMMUNICATIONS INDUSTRY, Japan is viewed as somehow different from the U.S. and Europe in regard to mobile phone development. When I speak to people about i-mode in the U.S., the general consensus is that it is a 'Japan thing,' and won't take off in the same way with American mobile phone users. They may be right, although I think that most consumer behavior is universal, and that given the same degree of entertainment and usefulness in their services, American consumers would respond.

This chapter discusses some of the factors that have shaped the current state of the mobile phone development in Japan, the U.S., and Europe, and where that development is headed. Because this book is about i-mode, which is Japanese, I base this discussion on the Japanese perspective.

The Revolt Against the Fixed Line: If a Home Phone Costs This Much, Let's Go Mobile!

A fixed-line telephone in Japan costs a customer 79,800 yen (approximately $700) just to have the right to install. Incredible as it seems for an advanced country such as Japan, a telephone customer still must purchase the "right" to a telephone line from NTT. This system is a relic of a time when Japan was impoverished and a telephone at home was considered a luxury, one that the fat cat who had it could darned well pay for! The fact that this system exists today is a testament to NTT's ability to thwart change, and to the Ministry of Posts and Telecommunications' (MPT) complicity in a system that has no benefit whatsoever for consumers.

A mobile phone, on the other hand, costs between nothing and 30,000 yen (less than $300), and the monthly basic charge is nearly equivalent to that of a fixed-line phone. Unlike in the U.S., a mobile phone user does not pay anything for incoming calls. The concept of "airtime" as in the U.S. is unknown. A user only incurs charges of any kind, whether the charge is against their airtime or in actual cash, for outward phone calls. Mobile phones use their own area code — 090 — and callers to that area code know that they are calling a mobile phone and will be charged a higher rate for the call, whether the person is next door or across the country. Users who use their phone only for receiving calls can join a plan that actually costs less per month than a fixed line does, even if they talk on the phone 24 hours a day — as long as all of those calls are incoming. The person calling them, however, could end up paying a lot.

In the U.S. and Europe, the situation is even better for mobile phone users who actually do make lots of calls. In the U.S., you can find plans for $30 to $50 per month that provide more minutes than most people need. And all over Europe, you can find pre-paid phones with no fixed monthly fees. Voice has become a commodity, and the price of that commodity has fallen drastically in every major mobile phone market. This is a boon to consumers, who get a go-anywhere phone for a price that approaches that of a go-nowhere phone. In highly mobile modern societies it is not difficult to see why mobile has become so common.

In Japan, in addition to paying a huge fee for the privilege of having a phone line, customers pay 3 yen per minute for local calls. This really adds up in the case of a heavy Internet user like myself, with a 20,000 yen ($170) per month phone bill being fairly common, with most of the charges for local calls made to my ISP.

Things are now changing. NTT has introduced a fixed-price plan for Internet use, whereby users pay about $35 per month for unlimited access to their ISP. This is a lot better than paying $170 per month, but it is too late for those who were turned off by fixed-line Internet and have turned to their mobile phone for their basic Internet needs: They can get mail, check the basic information they need, and do it without having to buy a computer or pay tons more money to NTT. That is the story, anyway.

The truth is probably a combination of factors. One is definitely that the growth in mobile phone use is connected to the high cost of fixed line phone cost. The growth rate in mobile phones has been higher than that of fixed-line phones for several years, in real terms. Sixty million people have mobile phones, which means that nearly one of every two Japanese has a mobile phone. Many who don't have mobile phones will either die (the rate of mobile phone usage is, unsurprisingly, much lower among the elderly), or get them in the next five years. Already, the real cost for long distance calls on mobile phones is cheaper now than it was on a fixed line just three years ago. People are willing to pay a little more in connection charges for the convenience of being connected wherever they go, especially when they had to pay very little for the mobile phone in the first place.

There is also much less of a negative attitude towards carrying mobile phones than there seems to be in the U.S. The phones are lighter, generally private phones rather than work phones, uniformly equipped with silent mode so that they vibrate rather than bugging everyone around you with ringing, and have long-lasting batteries. Having a mobile phone in place of a fixed-line phone is a rational decision for a large segment of the population, and having one in addition to a fixed-line phone an equally rational decision for another segment. In terms of cost, convenience, and features, the mobile phone in Japan offers better value than fixed-line phones do. A user may pay minimally more per month, but the added convenience makes a mobile a better value.

One behavior remarked upon by a European mobile phone application developer in Japan is that while Europeans tend to turn their mobile phones off at night, Japanese leave them on pretty much all the time. The batteries will last literally three weeks on standby mode, so there isn't a perceived need to turn the phone off to save battery power. All new models include a built-in answering machine, so just because the thing is on doesn't mean it needs to be answered. Unlike the fixed-line caller ID service released only two years ago in Japan, the caller ID on a mobile phone works with a user's address book to display the name of any caller in a user's address book, which makes call screening easier.

Search For the Killer App: A Build-Up Before the 3G Storm

In some ways, DoCoMo's approach to its business and the future of that business is much more like that of an Internet company than that of a telephone company: DoCoMo is looking for the killer app that will bring in new revenue and give it marketplace advantage, rather than simply managing a commodity product. This is one of the things that make DoCoMo so attractive to investors, and it was definitely this kind of thinking that resulted in i-mode. In the development of 3G products, NTT DoCoMo is far ahead of the competition: It already has a mobile Internet service (i-mode) which reaches nearly 28 million subscribers; and it has a fully operational 3G network on which to develop.

Even before the 3G network came on line in October, 2001, however, NTT DoCoMo had been busy developing 3G applications using its PHS network, which boasts speeds of 64Kbps, much faster than the 28.8Kbps the i-mode/DoPa network maxes out at. Take a look at some of the efforts that are being made at developing 3G content:

- **M-stage:** This service consists of two sides: video on demand, and music. The video service has not been hugely successful, but the music service has shown promise, with new handsets released that can download and store music directly as well as make phone calls.

- **Sony 502wm:** This handset is not a PHS handset. It operates on the normal network, and stores music on Sony's memory sticks in the ATRAC format used on MDs. It is a little bulkier than a normal handset, but boasts i-mode, a color screen, and the ability to function as a digital Walkman (the "wm" on the end stands for Walkman). It can't download music from the network, however, requiring a user to have a PC to link up to.

- **Network Gaming:** This market is not one NTT DoCoMo is doing anything directly with, but one that has come about because of the dynamism of content developers working on i-Appli games. The present breed don't involve a lot of interaction with other players because of speed issues, but developers are already developing games which use video and other rich media.

- **PlayStation:** Related to Internet gaming, NTT DoCoMo and Sony have forged an alliance to connect Sony's PlayStations to the Internet using NTT DoCoMo's phones. Until Sony comes up with its own handheld game device, however, how much market there will be for this usage is unclear.

- **Location Based Content:** Though DoCoMo said some time ago that an important aspect of FOMA would be its location tracking functions, nothing more has been heard of this in terms of content or technical specifications. NTT DoCoMo however, has two affiliates developing the infrastructure for serving location based content, and marketing that system to businesses.

This combination of a huge player in a capital-intensive industry, with huge financial resources, massive market capitalization, and a willingness to take risks has meant some interesting products, some more successful than others. One massive example is 3G itself, which is a hugely expensive undertaking, and where being first-to-market will give DoCoMo very little aside from bragging rights, while forcing it to take a disproportionate share of the risk by being first in an unknown, untested service.

DoCoMo is betting a lot that being first in the area of 3G will help the company extend the overall level of its networks and services from the current three- to four-year lead over the U.S. to one so insurmountable that turning to DoCoMo and its handset and equipment suppliers will be the only viable option. Actually, DoCoMo is nearly fully focused on its own market, and it wants a new revenue-producing product on the way before that project is necessarily needed, in order to meet projected decreases in revenues expected when growth in the number of subscribers slows and competition for current subscribers forces carriers to slash prices.

So, for DoCoMo, perhaps more important than the question of what the killer app will be for 3G phones is the question of what the company can do to raise its revenues. With a download speed of 384Kbps, which is much faster than the connections of the vast majority of fixed-line users, simply using the 3G network as a high-speed alternative to a fixed line could be attractive. Sony has said that it was waiting for 3G networks to be widely available before introducing Internet capabilities to the PlayStation 2. Ironically, Access, the same company which developed i-mode's cHTML markup language, has recently started offering a browser and Internet connection package for the PS2, sold entirely through fixed-line Internet providers.

DoCoMo's FOMA site lists the following applications as being compatible with FOMA:

- i-mode
- Videophone service
- High-speed data communication
- Streaming media (initially only video)
- Multitasking (being able to access Web sites while speaking on the phone)

As I mentioned in Chapter 2, NTT DoCoMo has capacity problems, and this is as much of a driving force behind its need to invest in 3G as any of the above applications. It is trying very hard, however, to develop new services which will pay for the massive expenditures that building its 3G network will require. The case in the U.S. is substantially different, in regard to the need for more capacity. Though the subscriber base has grown phenomenally in the past five years, these users are spread out over a much bigger piece of the available radio spectrum, and over a much wider geographical area. Capacity alone might not drive U.S. carriers to 3G for a decade.

Unlike Japan and Europe, there are still vast parts of the U.S. not covered by digital networks. For network providers in these areas, it has simply not been economically feasible to build the infrastructure. When it will become economical is anyone's guess, but probably later rather than sooner.

In both the U.S. and Europe, new entrants have bid on and won 3G licenses; they will likely be first, or at least second, to market. No 3G infrastructure currently exists, but building 3G is the most sensible option for a new entrant — even if their bid conditions don't require it. Base stations may cost two or three times as much as conventional ones to build, but they have five or six times the capacity. So the per subscriber cost is actually less.

In the case of Europe, a spectrum has already clearly been allocated to the new 3G networks. In the U.S., although companies have bid on the assigned spectrum and won the bids, the FCC has not yet been able to relocate the present users to other spectrums, and so has not been able to allocate the spectrum it has already sold.

In terms of spectrum auctions and allocation, Japanese carriers have a much better deal than their European and U.S. counterparts: The allocation was granted for free. In March 2000, the Ministry of Posts and Telecommunications (MPT) announced that it would accept applications for three 3G licenses. If more than three carriers applied, the licenses would be decided on the merits of the carriers. Since there are only four carriers in Japan, and one, Tuka, was expected to fail, it is no surprise that only three applications were received. Everyone was happy.

The MPT clearly recognized the massive amount of investment that would need to be made in 3G, and considered that this investment was good for equipment makers, good for consumers, good for the carriers, and good for the country. Sure, valuable spectrum was being granted for free, but due to their massive investments in using that spectrum, most at the MPT believed, and still believe, the carriers deserve to be granted its use.

What all of this comes down to is that there are very real geographical, political, and economic differences among the U.S., Europe, and Japan that affect how 3G will come about, what will motivate carriers to roll it out, and when this will happen. The key point in all of this is to keep in mind that while killer apps are, well, killer, they aren't the whole or even the most important part of the story in any country. This is frustrating from the standpoint of a developer or consumer, but unavoidable as long as governments, geography, and economic reality are involved. None seem likely to disappear as important factors in the future of mobile communication.

Because of its relatively compact geography, business-friendly government, and strong electronics industries, Japan looks to have more factors in favor of a rapid expansion of 3G networks than the U.S. or Europe. It is likely that all of Japan will be covered by one or even two 3G networks before the U.S. has coverage even in the top ten markets. Europe, where carriers are renegotiating high spectrum fees, is less predictable in terms of time, but looks clearly aimed to have major countries covered in the next three years or so, though recession-hit telecom companies may slow this down.

Wireless: A David versus Goliath Struggle

Because of the cost of obtaining a fixed-line phone and the decades-long practice by NTT of charging extremely high rates for long distance and international calls, the consumer use of the phone has been somewhat retarded in Japan. One reason, undoubtedly, that family fax machines have been such a common item in Japan is that it is much cheaper to send a letter by fax than to talk to someone by phone. Extensions of that practice have been e-mail, and its mobile predecessors the pocket bell-pager and short mail.

Fiscal 1997 actually saw the number of NTT's fixed-line subscribers decline. In that year, the number of subscribers was around 60 million. This number has declined every year since then, and sometime in late 2000 or early 2001, the total number of wireless subscribers exceeded that of fixed-line subscribers. In this growth phase, an important point of wireless communications is coming to light — whereas one family might share one home telephone, the mobile phone is a much more personal mode of communication, with a potential much greater than that of the fixed-line phone.

Japan has lagged far behind the U.S. in terms of Internet usage, in large part because of the prohibitively expensive local telephone charges. While 3.33 yen per minute does not sound like much, heavy Internet use is punished severely, with a 50-hour per month user racking up 10,000 yen, or about $80 in charges, in one month. Schools, which are often overcrowded and underfunded in Japan, could not afford to train students in how to use the Internet because of the cost of phone charges, as well as the cost of upgrading older computer equipment, much of which is based on a proprietary system from NEC called PC98, which is incompatible with modern software. This problem, and the overall slow rate of adoption of the Internet in Japan, together have created a situation in which the MPT has been increasingly critical of NTT and pushed hard to get NTT to reduce rates to make the Internet more attractive. Pressure from the U.S. Trade Representative did not help. NTT's response to this situation has been typically one part obfuscation, one part pleas of innocence, and all parts change-averse.

DoCoMo started, to some extent, without NTT's baggage and has shown a will to resist letting its connection with NTT slow it down. This is walking a tightrope, since NTT still directly owns part of DoCoMo itself, as well as having large stakes in and direct control over regional DoCoMo companies.

In any case, with competitors such as cable radio company Usen promising to offer telephone services over its broadband Internet networks in future, DoCoMo is less a threat to NTT than others. A complete divestiture of NTT DoCoMo would bring NTT a huge cash infusion, some of which could be used to pay for some of the restructuring and infrastructure improvements needed to compete strongly in the future. Having NTT DoCoMo stock at its current lofty level is of benefit to NTT as the largest shareholder, and as the largest shareholder, NTT will do nothing to harm DoCoMo in any way.

L-mode, and the i-moding of the Fixed-Line Internet

In an ironic twist, NTT has developed a new service for fixed-line phones that basically copies what DoCoMo has done with i-mode on wireless phones. This service, called L-mode, is slated to begin in June 2001.

Using a markup language nearly identical to cHTML but with the addition of three tags, L-mode browsers will be included in fax machines, cordless phones, and public terminals. The service initially planned to use 1-900 numbers (in Japan, 0990), but there was a huge outcry at this blatant attempt at profiteering by NTT, and those plans have been dropped in favor of using local access numbers in the same way as normal ISPs are used.

The content provided on the service is very much like that of i-mode, with banking, shopping, recipes, and other such information available. NTT has about 1,000 content providers. If you take a look at Appendix B, you can see that banks, credit unions, and agricultural cooperative credit unions make up a huge number of the i-mode sites (502 by my count). You can expect that L-mode will use the same connections to these institutions as i-mode, so it's likely that half the available sites will be banking sites.

The stated aim of the service is to narrow the "digital divide." That NTT has had a lot to do with creating this divide by its usurious local phone charges and phone line charges has not escaped many critics, including the MPT. The service was delayed on the orders of the MPT because of concerns that NTT was using its monopoly power to force customers into using the company as an ISP. L-mode service is solely available through NTT, so MPT and others feared that NTT would take customers from ISPs that did not have NTT's monopoly power. There were also fears, and indeed continue to be fears, over NTT's limiting of portal information providers.

In many ways, these objections mirror ones that have been leveled against i-mode. The big difference, however, is that in the case of i-mode and NTT DoCoMo, there was no question of DoCoMo exercising monopoly power – the company has around 50 percent of the mobile phone market, compared to NTT's

99 percent share of the local market, and NTT DoCoMo has strong competitors who do not depend on DoCoMo in any way for their infrastructure, whereas all fixed-line competitors must pay NTT interconnection fees. NTT has successfully put these fears of the MPT to rest, but many see this as more of a reflection of their still-cozy relationship than evidence that NTT will really change.

The target market for L-mode is housewives, elderly people, school children, and others who find using either a PC or i-mode handset confusing or inconvenient. One interesting aspect of this is the equal parts cooperation and rivalry with DoCoMo in this project: NTT clearly aims to stop bleeding customers to DoCoMo's wireless handsets, and yet has approached this aim by copying everything, from the name to the service line-up to the revenue model, directly from DoCoMo. There is not much DoCoMo can do about it, either, since no matter how much the company wishes otherwise, it is still controlled by NTT. In reality, DoCoMo probably isn't very worried about NTT taking customers from it; more than half of i-mode users also have a PC at home that they use to connect to the Internet, providing a very different audience than that at which NTT is aiming.

The ironic element in all of this is that NTT's idea and implementation are truly compelling and probably will do much to narrow the digital divide. The hard part to swallow is that it is NTT doing it.

As far as developing content for L-mode, it is nearly identical to developing content in the early days of i-mode. There are no color screens on L-mode fax machines and telephones, which *do* come with the ability to download ringing tones, which is quite an interesting addition to the normal home phone. The process for becoming a service provider is much the same as i-mode's, requiring a detailed plan and dedicated support staff. Although not totally up-to-date, an information package about L-mode is available in English from NTT East (www.ntt-east.co.jp/Lmode/english_dl.pdf), which discusses the concept in quite a bit of detail.

One extremely attractive point to L-mode is that developers have an even broader market for their content. Because of the minimal differences between L-mode and i-mode, it's quite easy to leverage work already done for i-mode. The obvious cautions are that content with a strong mobile element will be less attractive to home users. The opposite is not necessarily true, however. Recipes, which would seem to be very home-centered content, are nonetheless quite popular on i-mode, probably because working people start thinking about what to have for dinner when they are in the supermarket, and having a list of ingredients is handy. This sort of thought is required to be successful at either i-mode or L-mode.

Other networked appliance plans

There is already an Internet-enabled microwave oven available in Japan, which can download recipes from the Internet and knows exactly how long to cook and on what settings for that particular recipe. Japanese consumers are probably more likely to buy such a novelty product than U.S. consumers. This microwave runs a version of the NetFront browser, which also runs on i-mode and is fully HTML capable. That it currently makes use only of the ability to download recipes is not so important to either the manufacturer or the consumer that buys it: The manufacturer is using the oven to test the feasibility of connecting home appliances to the Internet, and consumers are probably tickled by the idea of their oven having its own Internet address and being able to use their i-mode phones to access it, tell it what to do with that stuff they left in it when they left this morning, and when to do it. That this novelty wears off after about five times because they can't be bothered to leave stuff in their microwave every time they leave the house in no way diminishes their investment – it still works just as a microwave oven should, and the fact that it could connect to the Internet if you wanted it to is still true.

That is the novelty side of things, which often is the precursor in Japan to actual practical implementations of a new technology. That they get people to pay for the novelty is a tribute to Japanese consumer-electronics makers' marketing prowess, and often helps to pay for basic research into new technologies.

A standard has been proposed, and already been adopted by many companies, called Bluetooth. A consortium of over 2000 companies has agreed on this standard, which is used for devices to connect with each other wirelessly over very short distances (up to 30 feet).

Bluetooth-enabled computers and mobile phones are already out in Japan, and it has been predicted that the cost of adding Bluetooth to a product will fall to about one dollar. It has also been predicted that at that cost, almost every manufacturer will add Bluetooth as a matter of course. But simply adding Bluetooth to a product is not all the makers have to consider: What functionality will be available to Bluetooth clients, and how to implement this functionality seamlessly into the product, will add costs that are in the short-term more than simply adding a Bluetooth chip to every appliance made.

The do-all mobile phone, however, plays a significant role in some of these products. In the first stage, already upon Japan right now, the mobile phone can act as a kind of universal remote control for all of the Bluetooth appliances in the house. In actual proximity to the appliance, the mobile phone can send the remote control commands directly. Away from home, a Bluetooth home server can receive commands from a Web-connected mobile phone and pass them on to the appliance. There is no reason this has to be done via mobile phone, but with half of Japanese and Americans, and more Europeans possessing one at this point, nothing with a number pad and display is more likely to be found in your pocket than a mobile phone.

The problem with all of this high-fallutin' talk about networked appliances and controlling them by remote control using a mobile phone, is that for pretty much all home appliances a human operator is really necessary. Sure, you can start your coffee using an Internet-connected or Bluetooth-enabled coffee maker, but who put the coffee in the thing in the first place? Or your microwave can download a recipe from the Internet, but who mixes the ingredients and puts them in the oven in the first place?

Though manufacturers may very well start adding Bluetooth to all new appliances, it is likely to be about ten years, when robot maids are within the realm of commercial feasibility, before any of these features are actually used in a significant way. As a nifty universal remote that automatically knows what features it controls on what devices, a Bluetooth-enabled mobile phone is quite handy. As the means of directing all of the actions of appliances in a household, though it will soon be fairly simple to do, it won't be useful until other breakthroughs are made.

In terms of the U.S. and Europe, where appliances are expected to last decades rather than years, there will definitely not be a day in the near future when your entire home will be controlled by anything as fancy as Bluetooth. Home automation and the concept of total digital control over all appliances is more compelling in Japan, which has one of the lowest birth rates, and highest longevity rates in the world. The elderly will be excellent customers for products that make their lives easier.

The mobile phone in all of this is really the network client, and remote control, if that is needed. Anytime a connection is needed to the network by any of these items using Bluetooth, and your phone is nearby, the other appliances will request a connection, which your phone will deliver, perhaps completely unbeknownst to you. It will be the killer all-in-one remote, probably with some security features to prevent something like when someone switched our channels from outside our window during the Barcelona Olympics. Very annoying.

Any application that takes advantage of these features of a mobile phone will likely have some success. I had an idea a few years ago to use the pager network, and to stick pager functionality into an all-in-one remote control. A user would use the Internet to send a message to the remote to tape a certain program at a certain time when they were away from home. It was a pretty good idea, and I nearly bought Motorola's pager application development kit, but got busy with other things. With Bluetooth this would have been extremely simple. I could simply have done an Applet telling my phone to connect to my home's Bluetooth server, which could tell my VCR to get to work. Of course there is the question of who puts the correct videotape into the VCR.

Thinking of the word "appliance" in a smaller sense will help you understand what Bluetooth quite likely *will* be quite successful at – connecting things that now connect by wires. A computer to its monitor, modem, printer, keyboard, and mouse. A VCR or DVD player to the sound system, television, and cable box. A stereo to its speakers. A lamp timer to its lamp. And so on. These wire-replacement uses, especially in products that get changed within a ten-year time frame, will probably have taken strong hold in several years, mostly sold as packages of things that communicate with one another, rather than things that we buy separately. At first this will be because of varying implementations of Bluetooth, and later because it will simply be easier to get speakers and stereos from the same maker to communicate properly.

Because mobile electronics that use wires, computers, and other small appliances are really not specific to any country, the spread of Bluetooth is really not specific to Japan, the U.S., or Europe. Japanese consumers tend to pay a premium for higher-end items, whereas Americans and Europeans purchase based more on price, which will very slightly slow Bluetooth's spread in the U.S. and European Union compared to Japan. That will be momentary, though, and the days of matching yellow terminal with yellow wire, the correct remote with its appliance, are probably numbered. Hallelujah!

Chapter 6

The Audience

IN THE NEW PARADIGM in which the mobile phone carrier is no longer a utility, but is instead a service business, consumers of their services can no longer be thought of as simply "users" of the service. A service that expects in future to derive much of its profits from data transmission and selling mobile content to its users needs to think of those paying for its products as an audience, much as is done in the entertainment industry.

i-mode's audience is what has made it a success, and what will make or break any similar services in the future for the U.S., Japan, or elsewhere. In the telecom paradigm, introducing new features was a matter of upgrading the network, requesting new features from handset manufacturers, and other "hardware" changes. In the new paradigm, in which new features take the form of software delivered to an audience, introducing new features necessarily includes marketing. For marketing, and, indeed, for understanding who the audience is for your product, a clear understanding of this audience is important, and that is what I attempt to help you do in this chapter.

Who Uses i-mode and for What?

The fact that almost one in four Japanese now possesses a mobile phone with i-mode capability, according to NTT DoCoMo's current stated i-mode user base of around 28 million, would seem to invalidate the question — everybody seems to be using i-mode. The simple fact that 28 million Japanese have i-mode phones, though, doesn't mean that they are using all the i-mode services available.

Young people – the standard-bearers of i-mode

Though economically the least powerful group, the young in Japan have a very real power to determine trends and often affect the direction of consumer product development. As mentioned before, the killer app, starting with the pager, short mail service, the pocketboard, and finally with i-mode, has been text messaging. A commodity in the personal computer world, this application has nonetheless had the most value for consumers, and can be expected to continue to have value for them. The young were the first to use messaging on mobile devices as a major lifestyle choice, but they were definitely not the only ones to use it.

One of the most significant statistics I found was that in a survey of the general population of Japan (which necessarily excluded children and others not in the mainstream buying public), 62.3 percent of respondents had a mobile phone. A home fax machine was second, with 58.7 percent, followed by a PC, printer, notebook computer, MD player, digital camera, cable or satellite TV, video game machine, and digital video camera.

For those who doubt that the mobile phone is the right device to deliver certain kinds of content, such as music, Japan's population offers an emphatic yes! With this level of penetration, delivering content to this device would be completely logical if doing so were possible. And unlike most of the other devices listed, the method of delivery is clearly built in to the device.

Speaking of statistics, another one is quite interesting, and that is how many people are actually using mobile phones that have text-messaging capabilities. This seems clearly dependent on a user's age. Of mobile phone users surveyed by Hakuhodo, those in the youngest group, from ages 19 to 29, used this capability about 68 percent of the time. Those in the oldest group, from ages 50 to 74, used it around 10 percent of the time, with male users accounting for *all* users among the respondents of the survey by Hakuhodo Institute of Life and Living.

The surveys show quite clearly that a lot of what i-mode is being used for is e-mail. This usage is true across all age groups. A survey of mobile Internet-enabled phone owners showed that only 26.1 percent of them actually used the phones to surf the Internet, meaning that most people who use the Internet capabilities of their phones at all use them for e-mail. This should hardly come as a big surprise, considering that the same is probably true of users of personal computers, for whom e-mail is still probably the most important application. But it clearly puts into perspective what i-mode is actually being used for, and that is mail in a majority of cases. The fact that DoCoMo is adding to its ARPU (Average Revenue Per User) as much as it is from i-mode is a good indicator of the power that mail has in driving traffic.

This popularity of mail also explains the popularity of sites that use it quite well, such as Tsutaya Online. Tsutaya is one of the largest video rental franchises in Japan. It also is one of the most popular sites on i-mode. A large part of Tsutaya's popularity is its heavy use of e-mail. The site e-mails users when their favorite actor is on a talk show or when their favorite group releases a new album. If you request it, the site will e-mail information about the day's television lineup. You can request nearly any kind of reminder or entertainment information. This personalization, combined with the ease of use of e-mail, certainly goes a long way toward explaining the popularity of the site. Coupons that can be used in Tsutaya's stores are e-mailed to users, giving them a good reason to visit the stores and spend money. When the Playstation 2 was first released, and obtaining it was possible only through a waiting list, Tsutaya's site got lots of traffic because it accepted users' reservations at its online site. To actually get the Playstation 2, if you were selected, you had to go into its store. It was a very well-done clicks-and-mortar business model, and i-mode members played a big part in this.

Slightly older statistics showed that about 40 percent of i-mode users were over the age of 40. Those statistics are no longer applicable, however, because every new DoCoMo phone has i-mode capability, and most users pay the 300 yen per month to enable the feature, if for no other reason than they get a discount on the phone if they sign up for i-mode when they buy it.

How businesses have implemented i-mode-based solutions

Chapters 16 through 18 present some case studies to demonstrate how content providers have gone about providing their content to i-mode users. i-mode and other Internet phones are able, though, to access much more than information and entertainment. Because users can access the Internet, they can interface with the systems used to run their businesses. To what extent users are doing so is not yet known; surveys of business use have just started to ask these kinds of questions, and the data is not yet in. Two uses that are emerging in Japan are ones that are ripe for tapping in the U.S. I discuss these without using the companies' names because I was asked to maintain confidentiality.

The first example is using handheld units to communicate on the factory floor. Having a mini-HTML browser has allowed engineers at several large chip fabricators to receive information critical to operations. They can monitor equipment, be apprised of any problems on the largely automated production line, or access information from back end databases as reference or for presenting data in graphs or charts. The one problem with this use has been radio frequency interference because mobile phones use a fairly high-powered transmission to connect to its network. Newer versions of PHS phones can view cHTML data, although DoCoMo has chosen to not call them i-mode for fear of down-selling the i-mode service to the cheaper PHS platform. For factory use, in which radio interference is a concern, though, PHS has a strong presence. Using these handsets not only to communicate by voice from all around the factory, but also to view data from anywhere in the factory, is a boon, enhancing productivity.

Another use is for a restaurant chain, which uses a point of sale (POS) system to collect and send data to its headquarters about sales, customers' orders, stock levels at various locations of the restaurants, and dates and details concerning a shipment. This chain's head office management employees spend a lot of time in actual restaurants, and do not always have the time or inclination to check the full range of data. Using i-mode phones, they can view just the necessary information they need to know while they are on the road. The i-mode phone is unobtrusive, the battery doesn't run out the way a notebook's does, and checking the company's data is a quick, simple task that takes less than a minute.

Several companies, including NTT Data, NEC, Just Systems, and Lotus either have i-mode versions of their products or have included the capability to synchronize an i-mode handset with their existing products. Access to schedule, address, intranet, and other groupware applications is at the heart of these products, and anecdotal evidence suggests that they are being deployed. Who is using the i-mode features, and for what reasons, is nearly impossible to discern at this point because in many products, no difference is made between other sorts of remote access and i-mode access. Small applications such as address books and schedules seem to have an advantage over larger applications such as spreadsheets — simply because of their suitability for use on the road. Most salespeople at this point have a notebook PC for use in booking orders, entering contact information, checking available stock, and accessing other basic information necessary to perform their jobs. A good i-mode application would enable a user to do the things that take just a little time and would be better performed using i-mode than getting a notebook computer out, set up, connected to a mobile phone, and finally connected to the groupware server. Insurance agents would not use their i-mode handsets to enter orders, which might have more than 50 fields to fill out, but very well might use it to check a current customer's information, which does not necessitate any data input. There are real limits to what i-mode can be used for, and those limits must be considered when implementing an application. The usefulness of having quick, convenient, inconspicuous, and ubiquitous access to business systems, however, ensures that the HTML or Java-enabled mobile phones will have a clear place in business.

Applications that enable a business or industry user to achieve something on the fly are excellent prospects for successful i-mode applications. Alerts from a database, checking information on company systems, and even visual factory-floor applications that pinpoint problems are all good candidates. Those applications that include very much data entry are probably not very compelling on the i-mode platform.

How Does i-mode Fit the Needs of Its Audience?

As a service, i-mode must address the needs of its audience. Some of the services that make up i-mode, such as e-mail, are ones provided by the carrier NTT DoCoMo. Many more are ones that DoCoMo never thought of, provided by third parties. Identifying a need or desire for a service, and then marketing that service, are things that both DoCoMo and the information providers are busy working on. In this section, you see the main uses i-mode is put to.

Messaging, the little app that made i-mode

As mentioned previously, messaging is really the driver of i-mode and other mobile Internet implementations. That e-mail is a commodity, and a cheap or free one in the wired Internet world, means nothing in terms of how useful it is. The Japanese Prime Minister as of this writing, Koizumi, sends a weekly message to millions of people who sign up for his "mail magazine," or "mailmaga" for short. Housewives send group e-mails to family members, telling them to all be home early tonight for a family party. Girls and boys exchange mail messages before meeting in person. E-mail is changing the way that people communicate in Japan as much as it has done so in the U.S. (perhaps even more because e-mail has attained ubiquity in like measure to the mobile phone).

E-mail itself has been around since I was a kid, and even before. I remember using it for the first time in the late 1980s as a university student. I was sitting in the computer lab, which consisted of DEC terminals connected to a DEC mainframe running VMS, a quite popular mainframe system at the time. I had to spend a couple of all-nighters finishing a project for my beginning programming class, writing Pascal code. Many of my fellow inhabitants of the basement lab in Laird Hall were what we call "geeks." I hesitated to use the word until I observed one particularly amazing behavior: Though they were sitting within a few feet of one another, many of these people would send e-mail to one another! I couldn't believe it, and at the time I put it down to being geeks. In retrospect, I still think that, but this sort of behavior is increasing in Japan. Here is the dilemma: Do you spend five minutes, or even ten, composing an exceedingly simple e-mail message, or do you just call and work it out in one minute? More and more Japanese teens prefer e-mail. And not just teens: Those with personal computers, including myself, I must admit, often prefer the medium of e-mail to a phone call.

Only three years ago, scores of people gathering at meeting spots and talking on their mobile phones were common sights in Japan. Now, it's common to see some talking on the phone, others with thumbs flying over their keypads, and still others reading messages on their mobile phones. When my editor sent me an urgent yes or no question last week, instead of taking out the notebook computer that I have been carrying around, I just typed into my phone, "No, Danny. –Nik." That typed message took a mere minute to do; my computer would have required that long to boot up. Sending the message cost only about a penny, which is probably 1/30th of what connecting my computer to my cell phone using my cellular modem would have cost because that charge is based on time. But, at least for me, economic reasons are less compelling than those of simple preference – talking to someone on the phone is sometimes preferable, but it often takes a longer time, and if I am busy a short e-mail makes more sense. Or perhaps I just don't feel like talking right now.

To be fair to the teenagers in Japan, most spend an excess of time both e-mailing *and* talking. I expect that in the next year or two, a gang or two of university entrance exam cheaters will be caught using mobile e-mail to exchange answers during the exam. The discovery will probably come about because one of them forgot to turn off the ringing tones, and will receive a call or an e-mail that will cause the phone to ring and the proctor to investigate. Where was the mobile phone when I took the SAT?

Customization and the need for uniqueness

The first application to really spark interest in i-mode was e-mail. The ability to send mail not only to friends with the same kind of phone, but also to anyone on the Internet, was a big hit. For younger people who had been using other kinds of text messaging, e-mail capability prompted a natural move to i-mode. For people who had been hesitating to access the Internet, a mobile phone provided a reasonably friendly interface to the Internet, which many people previously regarded as too complicated to set up and use. E-mail brought about the first wave.

The second wave occurred with three developments: Downloading ringing tones became possible; companies and people at large became aware that they could download new startup screens; and color screens were affordable. Suddenly everyone, or at least everyone under the age of 20, had their own ringing tones, and often a specific ringing tone for each of ten or so friends. Screen savers with cute animated characters, the latest teen heartthrob, or sports stars also became increasingly visible.

These ringing tones and startup screens are saved in memory, and are limited to no more than 10 KB each. Different models of phones have different limits, but current models support at least ten harmonic ringing tones (up to 16 voices), as well as many more single-voice tones. There are memo functions that can be used to store an average of ten startup screens, assuming that the memo memory isn't being used for voice memos.

All of these uses fit into the category of customization. For the last four or five years, young users have been fitting their phones with antennas that glow when the phone is in use, often with cute characters adorning them. Mobile phone straps have become a fashion accessory, and bags with special mobile phone pouches are increasingly common. Makers often sell new phones on the basis of how many colors of flashing lights a user has to choose from when the phone rings. In the days before color screens, users could also select the color of the backlight.

Where this need to personalize their "*keitai*" (Japanese for mobile phone) comes from is anyone's guess, but most users have customized their phone in one of the ways I've mentioned. As for myself, I have always said that the killer app of i-mode is being able to have a picture of my son, Christopher, on the start screen. I can't personally think of anyone I would rather see when I pop open my clamshell P209is. I could take or leave music. I have Japanese friends, though, who love the music, and buy books on how to program their own ringing tones. Many video stores have a machine that you can plug your mobile phone into, pay 300 yen, and download new ringing tones.

If you take a look at the sites on the official i-mode menu in Appendix B, you will notice that many are ringing tones sites.

Japan is a society in which school children all wear the same uniform, ride the same kind of bicycle to school, and follow a myriad of school rules, even during summer vacation. Being strange in any way marks one as "estranged from his fellows" (*nakama hazure,* in Japanese). The personalization that mobile phone accessories allow is a necessary outlet of personal preference, something that is too often not valued in Japanese society as a whole. Ironically, as is often the case in the U.S. and elsewhere, this drive for distinction is taking on an aspect of conformity.

Sony's new model of au phone, even if it is a model or two behind those that company has out for DoCoMo (with no TFT screen or Java browser), has something that is proving to be even more attractive: changeable "skins." This feature is similar to that of a model of the Apple Powerbook that was available several years ago.

Feeling feline today? Choose a leopard skin motif. Is cute your style? Your skin can have hearts. These skins can easily be changed, and a couple of companies already are selling original motifs cut to the correct sizes.

No one should underestimate the power of this drive for customization, personalization, and uniqueness. It is one of the facts of the market for i-mode, and one pushing it in new directions — directions that DoCoMo, you, or I can't even imagine. Truly.

While teens in many countries decorate the outsides of their phones with skins and other decorations, i-mode has given these same teens the ability to make their mobile phones unique in the way the inner functions display and sound as well. All of this customization seems to be telling us a couple of things:

- Teens in a regimented culture feel even more compelled to show their uniqueness, sometimes in small ways (such as what their ringing tones sound like, or sometimes in the length of their skirts).

- Sound is important. This phenomenon is new in Europe and the U.S., but unique ringing tones have proven popular all over the world. As a delivery mechanism for these tones, i-mode is probably the single largest source.

- Technology is sometimes cool, but good design is often even more important. Though most reviewers rate the Panasonic phones technically inferior to Sony's, their design is considerably more attractive, and they sell much better. NEC, with good technology and good design, dominates the Japanese market.

- The little things count — what comic character is on the default startup screen often determines (along with the color) a buyer's choice when they purchase a mobile phone.

Chapter 7

The Developers — Getting a Piece of the i-mode Pie

THOUGH WE TOUCHED ON IT BRIEFLY IN CHAPTER 1, a further exploration of the revenue model used for revenue-sharing between NTT DoCoMo and its official information providers (IPs) might be helpful.

At a conference I recently attended in New York, there was general disbelief at this revenue model and its applicability outside of Japan. For the wireless provider, taking only 9% of revenues, as DoCoMo does, seemed too low. And for content providers, not having any guaranteed revenues, which would be paid to them directly by the wireless provider (regardless of usership), was considered unacceptable.

This can be (and probably will be) the subject of an entire book on the difference in the way NTT DoCoMo has approached the provision of wireless content versus the approach of its competitors. It is really stark, and the best analogy is this: Most European and U.S. carriers see their wireless content offerings as a package of content, like a magazine. A magazine pays content providers regardless of readership, and is then left to try in as many clever ways to sell that content to as many readers as it can. But any one piece of content is just a small bit of the pie, so the demands on it are fairly minimal: It has to be true, available, and something that the wireless provider thinks its users will like. Conversely, the wireless content providers (the writers, so to speak) are under no real motivation, after initially selling their content to the wireless company, to improve it, to get user feedback to know what those improvements should be, to offer users more value the more they use it, and so on. Sure, there is the longer-term self-preservation consideration, but a content provider in this model does not "own" its own customers, so finding what they want and what it takes to please them is neither easy to do nor particularly profitable.

A Further Exploration of the Revenue Model

Using the same sort of analogy, NTT DoCoMo would be a newsstand that sells many different magazines. Each magazine costs a maximum of $3.00, and even if the only way to purchase it is a one-month subscription, this subscription can be cancelled at any time. If a user becomes bored with a certain publication, she simply removes it from her reading list (the My Menu list in i-mode), and is never charged again. This model places the onus on content and customer relations directly with the content providers, and offers a compelling encouragement to them to come up with new and interesting content. Providers who can do that will be profitable, and those that cannot do that will fall flat. Rather than being tied to one particular basket of information services, a customer is given the responsibility and freedom to choose what he wants, and, of course, to pay for what it is that he has chosen.

People pay for their local newspaper, which is delivered to their home every morning or evening. They pay for cable TV, and sometimes for video rentals. People pay for magazines and books. They pay for access to the Internet (usually) and for goods that they purchase at www.amazon.com. And yet, there are those that say an i-mode service would not work in the U.S. because people don't want to pay. Sure, no one *wants* to pay for anything, including wireless services that may entertain them or make their lives easier. But to extend that to mean they *won't* pay for them if they are offered is quite a different proposition, I think.

Quickly, look at how the revenue is taken in:

1. A user sees the need for, or the interest in, a certain service. She can find this service by going to the Menu List menu on the i-mode menu and going through a couple of category menus.
2. The user reads the blurb that the service uses to describe its offerings, or tries out the part of its site that is free.
3. The customer decides to sign up for the service.
4. She presses the sign-up button.
5. The customer is asked for her password, which she sets herself.
6. The customer is asked if she is sure she wants to sign up.
7. The service is registered on the My Menu area of the customer's startup screen, and the customer uses the service.
8. The customer is charged 300 yen per month, or about US$2.50, on her mobile phone bill, which is paid by automatic debit from her bank account every month.
9. If she doesn't like the service, she is given an opt-out option on the first menu every time she accesses that service; if she chooses this option, she will never be charged again.

Charges for subscriptions to i-mode services appear on a mobile phone billing statement as an aggregate of all charges from official services. Because of the process outlined above to sign up for an official service, there seem to be few mistakes that end up on users' statements. A free phone call to DoCoMo will answer the question, if it comes up, as to what the charges were for. Customer service is always polite, prompt, and aimed at rectifying whatever the problem is. There seem to be few with the system, but when they do come up, they are dealt with well, in my experience.

There is one other revenue model that everyone in New York was talking about, and that I hear few people in Tokyo talking about – advertising. This sponsorship or clickthrough model has been somewhat less than successful of late in the wired Web. Perhaps there are site owners out there getting much better clickthrough than my anemic four out of ten thousand (no joke), but the rate is quite low in general.

Much has been made of location-based advertising. For example, suppose that you are walking past a McDonalds, and beep! you get a message telling you that there is a special on Big Macs waiting for you inside. A barcode comes up on your mobile phone screen, saying that a cashier at McDonalds can scan in for your special offer. For McDonalds, rather than tossing out a bunch of ten-minute-old hamburgers, this offers a reasonable and efficient use of sales strategies. If you were actually hungry, this might make sense.

The problem with this kind of advertising is that even if it might be appreciated, there is the very real risk of backlash because no one wants to be inundated with ads as he walks down the street. In the Japanese i-mode model, too, I would actually pay the transmission charges for receiving the ads, which is also likely when GPRS comes online in the U.S. and Europe.

The current i-mode revenue model shares responsibilities and benefits equally: DoCoMo is responsible for the i-mode network and menu, and benefits by charging for actual traffic on the network; IPs are responsible for listening to customers and developing compelling content to attract customers, while benefiting by keeping 91% of revenues generated by their sites; and customers are responsible for choosing and paying for content, and benefit by having this competitive content readily available on their mobile phones. Neither of the other two models are as balanced, and neither have a proven track record similar to that of i-mode.

New entrepreneurs – a little bit of revenue multiplied many times equals a fortune

In Chapter 16, you learn about Index, a company that has grown to be a million-dollar company solely on the basis of its wireless offerings. This is the dream for many an entrepreneur. With more than 350,000 subscribers to its Love God service, each charged 300 yen per month (about US$2.50), Index takes in about 95 million yen per month, or about US$830,000. And that is just on one service! It has other services as well. There are about 20 people working on the Love God service, so even after their salaries, which are any Japanese company's highest costs of doing business, well...you do the math, and find out why there are more than 40,000 sites vying to be included on the official i-mode menu.

These are not companies that are in the red, either. In Japan, there is no money growing on trees if you say you are involved in the wireless Web, nor is an IPO likely to bring in some of the valuations that were seen on the NASDAQ at the height of the Internet boom in the U.S. A content provider's revenue model is quite clear: create contents that are salable, sell them, and get paid for them. There is no ambiguity: The way to profitability lies in creating good content. Oh, and one other important (but by no means trivial) thing: Become an official content provider with a billing relationship with NTT DoCoMo.

The requirements were spelled out (in Japanese) in a document released by DoCoMo in March 2001. Suffice it to say that the requirements were released after considerable pressure was put on DoCoMo to justify its choices and somehow open up the process. Many of the things that were important in choosing content are discussed by Mari Matsunaga and Takeshi Natsuno in their books, though these are in Japanese. The basic requirements are that it is content that customers find entertaining or useful – content that is deep in the sense that if a user visits every day, or even more often, she will find something different. They should have some experience as a content provider, and have technical support staff available in Japan to answer questions in Japanese.

Looking at the list of official content providers, you see a vast number are banks. I don't think any of them charge for their basic services, and I doubt that they are getting paid by DoCoMo. Their real incentive is to provide better customer support.

Developers' relationship with DoCoMo

As I mentioned previously, the process by which DoCoMo chooses content providers is subjective, which leads critics of DoCoMo to scream that this is just another example of a monopoly playing unfairly by a set of rules that is not transparent. In response to these charges, DoCoMo came up with an inclusion standard in March of 2001. I am including this for the benefit of those who are thinking of applying to be information providers, as well as those interested in how DoCoMo deals with these sorts of issues.

NTT DoCoMo Standard for Inclusion on i-mode Menu

In this document, DoCoMo states the standard it has enacted for the inclusion of a site on the i-mode menu (henceforth called the "i-mode menu inclusion standard"). In addition, the i-mode menu inclusion standard consists of a statement of principles about content inclusion, i-mode menu site standards of morality, and a content inclusion standard.

Statement of principles about contents

In this section of its inclusion standard, DoCoMo reserves the right to do the following things:

- ◆ To examine in advance the contents of which inclusion on the i-mode menu is requested, to determine the propriety of inclusion.
- ◆ To determine the propriety of inclusion on the i-mode menu.

- To judge whether contents carried on the i-mode menu completely fulfill i-mode menu site standards of morality and the content inclusion standard.
- To continually evaluate whether sites fulfill i-mode menu site standards of morality and the content inclusion standard.
- To revise the principles of morality or content standard at any time. These may change, and sites may be refused inclusion based on the subjective business judgment of DoCoMo. Changes to the principles of morality and the content inclusion standard may be based on changes in the environment that surrounds the i-mode user's needs, the situation of society and i-mode, any change in i-mode menu management objectives of DoCoMo, and so on.

i-mode menu site principles of morality

DoCoMo sets certain moral standards for accepting new content, as well as for providers when they are renewing content. These standards are listed below:

- Contents must be sensible, and must not act to spawn the distrust of i-mode users. DoCoMo give the following examples of contents that cannot be carried:

    Contents whose purpose or meaning is not clear, or is ambiguous

    Contents that contain falsehoods or inaccurate representations

    Contents that are contrary to publicly accepted norms and customs

    Unscientific or superstitious content that causes i-mode users to be puzzled and creates unease

    Sites that solicit membership into political, religious, or other organizations, or that solicit money for these organizations

    Contents that work on an i-mode user's subconscious in a way that cannot be perceived (subliminal material)

    A site that represents a guarantee from or connection to DoCoMo where no guarantees or connections exist

    A site with a possibility of having a remarkably bad influence on society

    A site with a possibility of giving many i-mode users displeasure

    A site containing nude pictures or sexual material that offends many i-mode users' sensibilities

- Contents must not lack in grace, and others must not be slandered or their honor damaged. DoCoMo gives the following examples of contents that cannot be carried:

Contents with a possibility of slandering others or infringing their privacy, or that could cause distrust, defamation, and disturbance in operation

Contents that discriminate unfairly according to race, nationality, occupation, sex, circumstances, thought, belief in a principle, mental capacity, physical disability, and so on; or promote discrimination

A site that denies or slanders unfairly the service that DoCoMo offers

- Contents must meet social ethics and not be in violation of related laws. DoCoMo gives the following examples of contents that cannot be carried:

    A site with a possibility of recommending, affirming, or promoting statutes or criminal misconduct of others

    A site that promotes or deals in obscene material and/or juvenile pornography, prostitution, and/or juvenile prostitution

    A site with a possibility of gambling, or affirming or promoting the selling of a lottery, and so on

    A site that performs pyramid schemes or multilevel marketing plans

    A site that deals in goods that came to hand by the crimes of theft, burglary, fraud, extortion, embezzlement, or misappropriation of others possessions, and so on

    A site that deals in goods that infringe on the following rights of others: patent rights, a utility model patent, a design right, a trademark right, copyright, and portrait rights of others

    A site with the possibility of affirming or promoting use of a psycho-stimulant, an illegal drug, a psychotropic drug, marijuana, opium, a toxic substance, or other powerful drugs

    A site with the possibility of injuring international goodwill

    A site that breaks the public elections law

    A site that states personal names of others

    A site that uses the name, the portrait, the trademark, the work, and so on, without notice, or without obtaining a rightful claimant's consent

- Contents must not infringe a healthy education of youth. DoCoMo gives the following examples of contents that cannot be carried:

    A site with the possibility of inflaming faddish behavior and excessive customer interest

A site that acts in an antisocial way, or that has the possibility of injuring the life and the physical safety of individuals, such as sites promoting violence, without considering the healthy education of youth, using affirmations and expression that glorify violence or injurious activities

A site that does not take adequate measures, such as appealing for caution beforehand when youth may be likely to copy something where safety of the life or body may be injured

A site expressing that which is contrary to healthy, socially accepted ideas, and that spoils character

Content requirement 1: It realizes a value for the i-mode user

The content carried by DoCoMo on the i-mode menu should be indispensable in what it offers to an i-mode user. The following criteria are used by DoCoMo to judge contents' value to i-mode users:

- It is a service that is intelligible and is easy to use for an i-mode user. Suitable explanation is made about the contents of service, charges, the usage, use agreement, and so on. The structure is suitable for the information for which an i-mode user subscribes.

- A site offers content of continuing value to an i-mode user. There is a sufficient amount of information and sufficient depth to be of more than passing interest. DoCoMo gives the following examples of contents that cannot be carried:

    Sites without a sufficient amount of information

    Sites with very little types and quantity of information, causing users to wait for screens to load

    Fortune-telling with very few items

    Dictionaries with very few words

- An i-mode user expects new information, even when accessed frequently. DoCoMo gives the following examples of content that cannot be carried:

    Collections of ringing tone melodies to which new songs are not added suitably often

    News in which the renewal of information is not suitably carried out and that does not have up-to-date news

A balance between the information charge and the contents should be present.

Content requirement 2: Miscellaneous

Besides the above requirements pertaining to content freshness, depth, and usefulness, the following requirements also apply to content:

- ◆ A broad range of individual users should be attracted to using the service. It is not a service for a specific user group. DoCoMo gives the following as examples of content that cannot be carried:

    A service only for users belonging to a specific company, organization, or club

    A service for a small number of users

    A site by an individual managed as a hobby

    A service that shows neighborhood, small-scale events

- ◆ It is a service that i-mode users can use in comfort.

- ◆ Guidance to a site or contents outside of the i-mode menu is not the main purpose of a site.

- ◆ It fulfills the following regulations when content contains the element of a commercial transaction:

    It does not offer a place to conduct deals or transactions between individuals.

    A content provider has actual experience in the same kind or a similar business. For example, the provider has a current wired Internet site, which offers the structure of the commercial transaction and settlement of accounts on the home page.

    Sufficient explanation is made so that an i-mode user can understand the site and the method of transaction.

    A content provider corresponds pertinently and responsibly to inquiries from interested third parties and i-mode users. For example, a call center for responding to inquiries from i-mode users is installed.

- ◆ It has satisfied the following regulations, when contents contain the element of a community:

    It is managed pertinently so that slander of other members may be avoided and trouble does not occur in relation to the protection of individual privacy, especially concerning services aiming at formation of a community between the users of a service's contents, such as chat or a bulletin board. For example, there is software on the system that prevents input of a telephone number, address, and so on.

    A service provider must have sufficient capability to carry out, plan, create, edit, and offer the contents.

The delineation of responsibility is clear.

- A site uses information gathered from i-mode users to offer higher value-added products and improve its service

- A site with the capability to continue and offer contents will have the following characteristics:

    A site that plans its service based on permanency

    A site with features that an i-mode user uses continuously

These standards, it should be noted, are mostly self-regulating. Any content provider caught violating them would risk being kicked off of the i-mode menu. Considering the considerable challenge of getting on the menu in the first place, this is not something that many would like to have happen.

There is no specific mechanism for user complaints of violations of these standards, and most users are not even familiar with the standards: They simply assume that anything on the i-menu will be appropriate. Because information providers provide telephone support, most complaints are solved on that level, between user and information provider. Complaints to NTT DoCoMo's customer service number will certainly make their way to those in the Gateway Services section, those responsible for i-mode content.

Clear violations are rare, if not non-existent. However, there is considerable interpretation involved in following these standards, and proposals from new content providers are expected to follow the rules to the letter, while current content providers have some wiggle room in the interpretation. This is standard business practice in Japan – before you know you can trust someone, he must jump through a number of hoops to prove he is worthy of your trust. After you have decided to trust him, the trust is nearly, but not quite, unconditional.

Unofficial Sites

Unofficial sites are sites that do not meet the standards set out by NTT DoCoMo, and/or have either not been submitted, or submitted but not accepted on to the official i-menu for whatever reasons. These outnumber official sites by 30:1. These sites are accessed using the function available on all i-mode phones that allows users to connect directly to Internet sites by URL.

Many of these sites are *deai,* or dating sites. These have been very popular with younger users, and are forbidden by the standards for official sites. Other sites are not in violation of any standards; they have just not been accepted on to the menu for whatever reason. NTT DoCoMo is notoriously picky about selecting which sites to accept on to the menu list. Below are some of the disadvantages to not being listed on the official menu.

Continued

> **Unofficial Sites** *(Continued)*
>
> > A site cannot charge through NTT DoCoMo's payment system, which charges users on their mobile phone bills.
> >
> > A site has a nearly impossible time attracting people browsing for a certain kind of information, and who usually use the official i-menu to do this.
> >
> > By requiring a user to input a URL, which can be rather time-consuming using a phone's number pad, these sites are not as easy to get to as official sites.
>
> One of the questions that comes up again and again is how these sites can make money. Many, if not most, do not make money. Since many are personal hobby sites, their owners don't really care. Many others are little more than advertising for companies.
>
> There are a couple of different ways that serious content sites can charge for content:
>
> > Use an independent portal, such as Giga, which uses its own numbering system for sites, which is much easier to enter than an entire URL. Giga also rates content, and offers a way for independent content providers to charge for content.
> >
> > Use convenience store payment. Major convenience store chains have opened up their systems to online retailers. A user can show a number to a convenience store clerk, who inputs it into a cash register, which is connected to the payment system. The user pays, and the site that charged him is notified.
> >
> > Use a credit card. This is not very popular, because the most likely users of pay services on non-official sites are young people, who often do not have credit cards, and because of security questions, which make credit card companies loathe to approve merchant accounts for mobile commerce sites.
> >
> > Use 0990 numbers. These are like 1-900 numbers in the U.S., and are used extensively by dating sites.

Benign Protector or Malignant Monopolist?

The standards, as stated by NTT DoCoMo, are hardly objective. But do they need to be? The point of the standards is to state the way that NTT DoCoMo has been making subjective decisions until now, and how it plans to make subjective decisions on content from now on. As a company competing in a competitive market with three other strong competitors, why should NTT DoCoMo be expected to act like a modern monopoly in terms of opening its system to any and all comers? There does not appear to be a pattern of favoring any particular companies or individuals, nor of using its power to force other companies out of business.

Where there may be some problems taking this model to a Western market is in the explicit and implicit censorship of material. What may be considered matters of taste in Japan, such as avoiding talk of religion or politics, constitute restrictions on American freedom of speech and freedom of religion when stated explicitly. Actually, because Japan's constitution was written by Americans to closely resemble the best parts of the U.S. Constitution, it violates the Japanese constitution as well. The Japanese Supreme Court, however, is far from literal as to its interpretation of those freedoms, and with about one-twentieth the ratio of lawyers practicing in Japan as are practicing in the U.S., it is fairly expensive to make a principled challenge based on constitutional grounds. And constitutional cases are usually not decided for decades.

That said, however, it is not clear that NTT DoCoMo's intent is to restrict anyone's freedom, nor to control expression. The inclusion standards should probably be taken for what they are, which is DoCoMo trying to create a non-hostile, user-friendly environment for conducting business, for participating in a community, and for perusing the contents offered. These aims are benign in the sense that they are for the benefit of the user, and are therefore beneficial to NTT DoCoMo.

In the U.S., the route to becoming an official provider lies through AT&T, and it looks as if information providers for its PocketNet service have an inside track because AT&T has said that these providers will be providing content for its i-mode-like service. It is doubtful that AT&T would explicitly state standards such as NTT DoCoMo's, because this could cause legal troubles in the U.S. That content providers provide value for their customers, however, is obviously in the best interest of AT&T. How it ensures that a certain standard is met is not yet clear, but suffice it to say that in this area, Japanese and American cultures are different enough that it is guaranteed to be different from the way NTT DoCoMo ensures standards.

AT&T Wireless is 16% owned by NTT DoCoMo. This partnership's first fruit is expected to be an i-mode-like service to be released sometime in the first half of 2002 in the Seattle area, on a GPRS network. Beyond that, neither AT&T Wireless nor NTT DoCoMo is saying too much. Neither have information on their Web site about the future service. Before July 2001, both sides were waiting for the break-up, which saw AT&T Wireless spun off as an independent company. Since then, NTT DoCoMo has reorganized its U.S. operations, transferring headquarters to New York to be closer to AT&T Wireless's new headquarters, and consolidating the two companies it had been doing business under into one company.

A senior DoCoMo executive told me that it was extremely unlikely that AT&T Wireless's i-mode-like service would actually be called i-mode. AT&T Wireless has previously linked its tie-up with DoCoMo to its PocketNet service. A spokesman says that AT&T Wireless and DoCoMo's cooperation will likely result in three different portals — one for business users, one for consumers, and one specifically for young users. This same spokesman says that AT&T is very interested in learning what it can from DoCoMo for the consumer and youth-oriented portals.

Preparing a Partnership Proposal

Besides the standards stated previously, there are the concrete steps that NTT DoCoMo requires content providers to go through in order to provide content accessible from the i-menu.

The first point is to create a content proposal. Following are the elements that need to be included in the proposal:

- Outline of the plan – describe content, refresh and update information, and discuss what type of content will be provided.

- Detailed content description – from the top menu, what can you do? Include a detailed site map and menu categories.

- Screen images, flow charts – explain and show how the content will look on an actual mobile phone display from the top screen down through the various levels.

- Depth of content – explain your database of content in terms of its source and how deep it goes. Be specific.

- Corporate outline – provide an outline of your company.

NTT DoCoMo also offers some hints to prospective content providers, summarized as follows:

- Be sure that the content is concrete and detailed. This is not a conceptual plan, but one that will actually be implemented. Clearly demonstrate what a user can do using the service on i-mode with details. If you are proposing a service that already exists on the Internet or other media, clearly define that service, and explain the situation in those other media.

- Demonstrate the special characteristics of your content. If the proposed content does not already exist on i-mode (be sure to do your homework!), explain the originality of the concept. If the proposed content does already exist on i-mode, clearly show the advantages and differences over existing content. What is the draw for i-mode users to visit your site?

- Be sure to include all the elements described previously when preparing your plan: outline, detailed description, flow chart, depth of content, and corporate information.

- Keep in mind the following essential points for i-mode content:

    Continuity. Does your site have content that will keep users coming back time and again?

    Freshness. Does your content update often enough so that if a customer visits the site, even a couple of times a day, she can find something new?

Depth. Is there sufficient content available to satisfy more than a cursory use of the content?

Clear benefit. What benefit does your content offer a user? Have you made that clear?

In addition, a content provider must offer several customer and technical support levels, depending on whether their sites are free or paid, and whether they are offering the content in Japanese or English or both. Table 7-1 outlines these levels.

TABLE 7-1 RESOURCES REQUIRED TO ATTAIN OFFICIAL INFORMATION PROVIDER STATUS

	Japanese/ English Premium	English Premium	Japanese/ English Free	English Free
Content Planning Documentation	Provide both English and Japanese documentation	Provide English documentation	Provide both English and Japanese documentation	Provide English documentation
Customer Support	Customer support number in Japanese and English	Customer support number in Japanese and English	Customer support number if possible; otherwise, e-mail, fax, or overseas phone support in Japanese and English	Customer support number if possible; otherwise, e-mail, fax, or overseas phone support
Technical Support	Technical support staff must be able to operate in Japanese, and there must be at least one contact in Japan	Technical support staff must be able to operate in Japanese, and there must be at least one contact in Japan	Technical support staff must be able to operate in Japanese, and there must be at least one contact, but it can be overseas	Technical support staff can be overseas, and English is okay

After you have met all the previous requirements, it can still take as long as six months or more to find out whether your site has been approved or not. As you can see by the requirements, it is not easy to become an official site, nor is it a task for the impatient. The advantage is, of course, that the content revenue model is very clear and easy for both you as a developer and for the customer. In some sense, this is probably NTT DoCoMo testing the resolve of proposed content providers. If it has to wait six months, it really means that a company has to want to be in the business for the long haul to bother with the application process. Fly-by-night operations are not likely to wait that long, and guys sitting at home after work programming computers in their dens will most likely have moved on to their next project in that time.

These hobbyists and fly-by-night operations are antithetical to what NTT DoCoMo wants content providers to be – customer-centered, able to judge what will fly in the marketplace, able to listen to customers and hone their products based on feedback from customers, and having a very strong focus on providing a quality product. Though the contents are not being offered by NTT DoCoMo, there is a strong feeling that because they are being offered on NTT DoCoMo's i-menu, its reputation is also at stake, and it is a very real possibility that it will be called if there is a problem.

As much as free access and less-stringent rules are hoped for by non-official sites, what they really want is a place on the i-mode menu and a slice of the profits. But the situation would probably be quite different if there were 40,000 sites on the i-mode menu: navigation would become fairly difficult; 20 sites with the same kind of material would lead to a situation in which users wouldn't know which one was best; DoCoMo would be hard-put to deal with complaints over billing or service; and the popularity of i-mode would, in all likelihood, decline.

Is NTT DoCoMo discriminating? Yes! In the good sense of that word, it is very discriminating – so its customers won't have to be quite so discriminating themselves.

Part III

Developing i-mode Applications

CHAPTER 8
Discovering the Lost Joy of Coding Small

CHAPTER 9
cHTML: The Language Used for Creating i-mode Pages

CHAPTER 10
Playing Sounds in i-mode

CHAPTER 11
Programming in cHTML: A Tutorial

CHAPTER 12
i-Appli: The i-mode Version of Java

CHAPTER 13
Programming i-Applis: A Tutorial

CHAPTER 14
Creating an i-Appli Game

Chapter 8

Discovering the Lost Joy of Coding Small

THOSE OF US WHO ARE OLD ENOUGH, or who started programming young enough, to remember the Altair, Kaypro, Commodore PET, TRS-80, or any of the other early personal computers, will in many ways be reminded of the limitations a programmer had to work under in the past. But just as the feeling of fitting an entire Space Invaders game into a measly 2 KB was sweet, so is that of fitting the same Space Invaders game, this time with 256 colors and sound, into 10 KB. Even sweeter is the usefulness that comes with having such programs available on a device that we can always carry with us.

In a sense, the current i-mode handsets are to mobile phone functionality what the early PCs were to PC functionality: Basic, and still accessible to hobbyists and people interested in the technology. One of the beauties of i-mode's present limitations is the accessibility it gives a casual programmer, who can write applications without especially complicated debugging or rigorous coding practices. These limitations actually give an advantage to an amateur working at home in the evenings, because developing applications so small requires shortcuts and hacks that would be antithetical to someone else understanding your code. These limitations also impose certain requirements on us, such as having to use smaller color palettes, reduce the number of sounds or graphics, throw out certain functionality that programmers would otherwise like to include, and so on.

This chapter shows you several pointers for creating code and graphics that can display well on a typical handset. I also discuss editors and emulators that can help you create and test the code as near to handset size as possible.

Working with i-mode's Memory, Storage, and Screen Limitations

In addition to the issues of speed and screen size, various memory and storage limitations apply to i-mode content, as follows:

- Graphic: 10 KB maximum
- Page: 10 KB maximum

- i-Appli Java application: 10 KB maximum
- Number of Java applications that can be stored at one time: five

In addition, there are the limits of cHTML:

- No tables (except on a couple of handsets)
- No cookies
- No JavaScript
- No support for JPEG (except on a couple of handsets) or PNG
- No actual support of font sizing (the tag is available, but no handsets support it)

And finally, there are limits imposed by the phone hardware:

- 9600 bps connection speed
- Colors are limited, either to 256 or 4096, except on one model
- There is no mouse to use to navigate, so you use the arrow keys in a way similar to the Tab key or Backspace key
- The built-in Java virtual machine does not allow any connection between midlets and the phone's functions, such as address book or telephone functions

Some of these limitations are more important than others: 10 KB for a page to be displayed on a tiny screen, using no frames, JavaScript, or other size-increasing features, provides more than enough space; whereas 10 KB for a Java midlet is quite restrictive.

Coding compactly

As mentioned previously, the absence of JavaScript, as well as many common tags available in HTML, should automatically reduce the size of cHTML pages drastically. Really, even 10 KB is probably too much to have on a page, unless a high proportion of this content consists of links or invisible text such as tags or comments. Many developers use What You See Is What You Get (WYSIWYG) tools such as Adobe GoLive or Macromedia Dreamweaver to design Web pages. Even though at one time developers may have been able to hack away in a text editor to get reasonable pages, GoLive and Dreamweaver offer greater functionality and speed than developers could hope to achieve in a text editor, pausing to view their creations every so often in Netscape (and believe me, *nobody* used IE!). Nobody is asking developers to give up all this usefulness, but this same usefulness has exacted a very real cost in terms of page sizes. By reducing the number of tags available, cHTML forces developers to give up certain features and functionality that they have become accustomed to using. The WYSIWYG tools are simply not as useful in this paradigm. One area in which these tools would be useful, if they were reliable, is in previewing pages. Unfortunately, they are not reliable, which forces developers to preview outside of these tools anyway. May as well use a text editor.

So, coding compactly in terms of cHTML pages is not a huge factor. i-Appli is another matter altogether. Calling classes not included in the DoCoMo or Java C2ME packages, meaning classes that you have written yourself or have obtained from free code libraries, causes the JAR file, which is the file used to store the bytecode of a midlet, to include the size of the class in its 10 KB limit. Thus, even the small amount of overhead incurred by coding in good Java form and clearly separating functions into classes, may leave you still just short a kilobyte in some situations.

Trimming graphics by color

Keeping graphic sizes down is a proficiency many Web site designers have acquired. The tools for doing this have become quite good, too. I use Adobe ImageReady, but Macromedia Fireworks or any number of lesser-known, image-editing software packages should be able to help you keep your graphics small.

The first area for you to consider is how big a palette you are using. Anything more than 256 colors is definitely a waste because this is as much as most handsets can display. I would actually aim at 16 colors for nonphotographic graphics. Doing so will save you lots of space. One of the biggest uses of colors is *dithering*, making a blue edge meet a white edge, for example. Most image-editing software automatically smoothes this edge by dithering from blue to white, using progressively lighter shades of blue. On an i-mode handset, this is really not noticeable anyway. I removed these shade gradations from my palette, and saved about 50 percent on the size of my file. Knowing exactly which shades can be safely cut out is sometimes difficult, but most image editors allow you to view your image in real time as you cut something, and you can always undo your action if the color was being used somewhere very noticeable. Figure 8-1 shows a window from Adobe ImageReady, showing the palette that was created by default.

Figure 8-1: You can reduce graphic size by eliminating gradient colors in your palette.

All the colors between blue and white can be safely cut, and were cut, bringing the file size down by about half. On a 256-color phone, no discernible difference showed up between an image that had these colors in its palette and one that did not, except that the one without the unnecessary colors loaded faster!

Bringing graphics down to screen size

Another area in which you might skimp is the actual size of the image. You have, for example, about 100 pixels horizontally. But I do mean *about*; pixel quantity varies by model. Say that you have a graphic that contains 50 pixels across. You could use the `width` attribute of the `img` tag to set the horizontal sizing of the graphic at 100 percent. If a handset were 120 pixels across, that would be the size of the graphic. If it were 80 pixels across, that would be the size. You do pay a price in terms of quality for blowing small graphics up. But if the graphic is well defined to begin with, the image quality shouldn't suffer too much. The quality of most of the screens in different series and on different models varies, but none (with one possible exception, which has in any case been recalled because of a software problem) have what you could call great picture quality. Attempting to having great pictures on your site will be an act of futility and will certainly go unappreciated by users downloading pictures at a pokey 9600 bps.

Figure 8-2 shows us the result of the code in Listing 8-1.

Listing 8-1: A Simple cHTML Page to Display the Graphic

```
<html>
<head>
<title>Saving Space</title>
</head>
<body>
<img src="small.gif" width="100%" height="100%">
<img src="big.gif ">
</body>
</html>
```

Figure 8-2: The figure on the left shows the smaller gif image blown up; the figure on the right shows the gif image already sized at the larger size.

The biggest difference between these two graphics is the "jaggies" in Figure 8-2. The left-hand figure was only 50 pixels wide and 50 pixels tall. The right-hand figure, whose edges are noticeably smoother, has a size of 100 pixels by 100 pixels. Both of these graphics, which are a blue color, use a palette of only two colors, white and blue, lessening their size considerably. The left image is only 255 bytes, nearly exactly one packet, whereas the right-side image is only 577 bytes. I personally prefer the quality of the one on the right, but it takes three packets of data, so I would need to think about that. Probably a resolution of about 75 pixels, which comes out to 408 bytes, is a good compromise. I do this using the same cHTML code above to display (see Figure 8-3).

Figure 8-3: A medium quality of 75 pixels wide.

This technique — making graphics the smallest possible size and then using the height and width attributes of the <img> tag — has the additional benefit of allowing you to fit the graphics correctly to the size of the screen that the content is being viewed on. You will notice the trade-off in this technique, which is having slightly jagged edges and the distortion of the proportions. You can get around this latter drawback by using only the height or width attribute, in which case the graphic keeps its proportions, if this is important to you. In the previous example, I thought that keeping the whole thing on one screen was better. Figure 8-4 presents the same graphic, this time using the width attribute to size the graphic to 100 percent of the width of the screen.

Figure 8-4: Using only the width attribute results in the height re-sizing proportionally, taking it off the screen.

The example that fits on one screen, despite its distortion from circle to oval, has the advantage of not requiring a user to scroll. This is another consideration, which really goes without saying: even though only the width will actually be forcefully resized to fit the screen, scrolling to see a graphic is annoying and counter-productive to the reasons we normally use graphics – to convey a message in an instant, in a visual way.

Connection Speed, Limitations, and Considerations

As of this writing, i-mode is being delivered on the PDC-P packet network. This network has an actual capability of 28.8 Kbps, but i-mode runs at 9600 bps. This rate is about one-quarter of the minimum of the 33.6 Kbps that most people get on their home computers. Because of the limits on the sizes of files, however, this speed is usually sufficient. Actually, the i-mode servers, not the connection speeds, often cause the bottleneck, holding up speedy delivery of contents.

This highlights an important difference in the way people use services on i-mode. People often use entertainment and mail services while waiting for someone, riding the train, or stepping outside for a smoke, and speed is not an overriding issue. But when users need information immediately (for example, train schedules), speed is really important. Speed considerations divide into three areas: how fast a page loads; how many pages users need to go through before getting what they need; and how long the back end server takes to process users' requests.

I did one service that was a Japanese-English dictionary. I wanted it to be as easy as it could be, so I decided to not require users to choose which language they were going to input, nor, in the case of Japanese, which alphabet they would be using. The server-side script, I decided, should do all of that, and the user should be freed from inconvenience. Good thought, and someone who knows SQL better than I may have been able to get it to work as planned, but I wasn't able to. The problem was in the sheer amount of data, and the existence of three different systems of writing in Japanese. While a search through the entries of one alphabet took about 6 seconds, three alphabets took a whopping 18 seconds! A user logs onto this site to quickly find a word, and is then kept waiting 18 seconds. Not a good thing. But by simply asking the user what sort of data he was entering, I could have cut the time by at least two-thirds. This illustrates the important trade-offs that end up getting made when balancing two extremely important aspects of developing applications for i-mode: Making the application as easy as possible while realizing that the user is using this application on the run, and speed is crucial.

> **Speed Is In the Eye of the User**
>
> Still another consideration relating to speed is the perception of speed. Juergen Specht, the CTO of Nooper.com, made a good point about this. European phones, he said, often did not tell users to wait or give them any indication that the phone was at work on their request. Users responded by pushing buttons, which often led to a phone actually crashing, necessitating that the phone be completely reset. NTT DoCoMo has taken care of this perception of time for you — all models have a moving symbol to indicate that content is loading. If you know that a page is slow to load, you can also use a "loading" page.

A server-side model: Getting around i-mode limitations to provide rich content

The online dictionary project used an SQL server called MySQL to store and serve the dictionary data. Getting the data from the server to the i-mode telephone is another bit of sophistry many of you may be familiar with, called server-side scripting. Even in the case of normal HTML, served to computers with gigahertz speeds, sometimes over network connections with megabyte speeds, the use of server-side scripting is prevalent. In a pure i-mode environment, in which you have no other real options to add functionality, it is essential. I say "pure i-mode" to distinguish i-mode from i-Appli, which does allow you to add functionality, though with some limitations on memory that still makes server-side scripting attractive.

Often called CGI programming, after the Common Gateway Interface of the original Web standards, server-side scripting has evolved from being done almost entirely in Perl to JavaServer Pages (JSP), Active Server Pages (ASP), Miva, PHP, and others. My own experience relates to Perl and PHP, with PHP being my preference at the moment. The combination of PHP and the MySQL database is powerful, flexible, and free. Together with the free Apache server, this combination is unbeatable. This combination generally runs on some flavor of Unix, but recent Windows versions are quite stable. They should also work on Apple Macintosh OS X, which runs the BSD Unix kernel below its smiley surface. Because OS X ships with Apache built in, simply installing PHP and MySQL shouldn't be a problem.

Using server-side Perl, PHP, ASP, JSP, and other scripting languages to bring richness to i-mode

Where PHP and other scripting languages excel is in offering a rich set of scripting options unavailable on whatever platform is being used, and offering an easy HTML-based interface to those options. Nowhere are fewer scripting options available from the browser, and so nowhere are server-side scripts handier than in i-mode. And code libraries with thousands of free scripts are available. These scripts are for the most part written for platforms with a richer set of HTML options, but can be modified to fit your i-mode development needs.

To a large extent, these scripting languages serve as intermediaries between the browser and an SQL database: They take the user requests for data and pass them on to the SQL server, and they take the data from the SQL server and format it into HTML (or any other markup language, for that matter) for the user. In the process, scripting languages may do some processing of the data, manipulate the form that a user's input takes on its way to the server, or format the results from the database on their way to the user. Because this book is about i-mode, I focus on the user-script-database paradigm, which provides a rich experience for the user and a flexible, cross-platform environment for the developer. This paradigm also works within the constraints imposed by i-mode.

Because my own experience is with PHP and MySQL, that is the platform I use for examples. Both of these are included on the CD-ROM included with this book and on many rental server services. Because this book has a different purpose than to teach PHP and MySQL, I limit the scope of the coverage of those topics, especially of PHP. MySQL is a nearly ANSI-compliant version of SQL, so most of the issues discussed regarding SQL should be applicable to other flavors of SQL. Equally, PHP is a scripting language like many others, including ASP, JSP, Perl, Miva, ColdFusion, and others. These languages have the capability to communicate with an SQL server, and can be used in the same way as PHP has been used in our examples.

I would be remiss if I didn't at least acknowledge another very important server-side language (although it is not a scripting language): Java. Java servlets — small applications that run on a server — seem to be a natural fit with i-Applis, and indeed do very well with them. The problem with Java, though not a necessarily fatal one, is that if you want to use the functionality of the servlets in an HTML setting, you will probably end up using JavaServer Pages (JSP), which introduces another thing to learn. PHP combines the two functions. As noted in Chapter 13, PHP is used quite capably with i-Appli. Another advantage, if you don't plan on using a dedicated on-premise server to host your site, is that many rental server hosts include PHP in their services. Java on the server is still pretty rare, so if you plan to host, you will have to look hard for a provider.

Using SQL databases to host dynamic content

If you develop a serious site, using a database is not just a good idea, it is nearly a requirement. In the case of i-mode, cookies are not available. Many standard HTML developers have come to depend on cookies for many different purposes, not the least of which is identifying a returning user. In i-mode, you can get a distinct serial number from a user's phone passed in the user agent using the utn attribute. A similar way of doing this for official sites exists, but those sites have signed nondisclosure agreements, and aren't saying what it is.

And as long as users remain with DoCoMo, whether they change their phone or not, their serial number stays the same. You heard that first here – this is definitely *not* something that DoCoMo wanted to make public. Further, by looking at the URLs of official sites, I can say certainly that NTT DoCoMo uses a different serial number than the UTN, though it is also 11 digits. In this case, the company's customer service trumped its need for secrecy. My reasoning is purely deductive and anecdotal: When I bought a new phone, the company transferred all my phone numbers and other information from my old phone to my new one. I was surprised, though, when my i-mode menu and services recognized me as the same user. Only two circumstances could have made that possible: My old serial number was transferred to my new phone, or another sort of identifying number was transferred. For security reasons, DoCoMo has put this information under an NDA, which I thankfully haven't signed.

The entire issue of the utn attribute is, in fact, highly political. As a simple technical matter, it is very easy to achieve: Simply tell the server to stop truncating the USER_AGENT data sent. But NTT DoCoMo is concerned about sites that it has no control over having access to sensitive information about users. This concern is probably commendable, but becoming an official information provider is quite a long process and not for the faint of heart. The concern is also tied in with NTT DoCoMo allowing users to access other portals and allowing developers access to the company's micro-billing system. A good guess is that the present system is not very secure, and NTT DoCoMo is afraid that making the information public would put users in jeopardy. The new FOMA phones use a sim card with strong encryption, which is probably what DoCoMo is waiting for before it opens the system.

Using this serial number as an identifier, you can store whatever user information you need in a database. The script can identify a user's preferences and dynamically generate a page based on that preference. Following is an example of a site serving horoscopes. The SQL code in Listing 8-2 creates a user database.

Listing 8-2: Creating the User Database in SQL

```
CREATE TABLE users (
  uid int(11) NOT NULL,
  name varchar(30) DEFAULT '' NOT NULL,
  email varchar(30) DEFAULT '' NOT NULL,
  birthday date NOT NULL,
  sign varchar(10) NOT NULL,
  PRIMARY KEY (uid)
);
```

A good PHP script, called PHP MyAdmin, allows you to simply cut and paste this kind of SQL command into a window. If you don't use something like PHP MyAdmin, you need to get to the MySQL command line and type the command there. It is powerful, but using it is not especially easy if you are used to using the Windows or Mac GUI. After you have set up an SQL database, created a database and tables, and populated the tables with any information you consider to be a necessary prerequisite, you need to build your interface.

The cHTML code to initially register a user's information would read like Listing 8-3:

Listing 8-3: User.html

```
<html>
<head>
<title>User input</title>
<body>
Please enter<br>
your info:<br>
<form action="users.php" method="post" utn>
Name
<input type=text size="10" maxlength="30" name="name" istyle="3">
<br>
E-mail
<input type=text size="8" maxlength="30" name="mail" istyle="3">
<br>
B-day
<input type=text size="10" maxlength="10" name="birthday"
istyle="4">
<br>
<pre>Ex. 1978-03-18</pre>
Sign<select name="sign" size="1">
<option value="Aries">Aries
<option value="Taurus">Taurus
<option value="Gemini">Gemini
<option value="Cancer">Cancer
<option value="Leo">Leo
<option value="Virgo">Virgo
<option value="Libra">Libra
<option value="Scorpio">Scorpio
<option value="Sagitarius">Sagitarius
<option value="Capricorn">Capricorn
<option value="Aquarius">Aquarius
<option value="Pisces">Pisces
<pre>    (YYYYMMDD)</pre><br>
<input type=submit name="submit" value="Send">
<input type=reset name="reset" value="Reset">
```

Chapter 8: Discovering the Lost Joy of Coding Small

```
</form>
</body>
</html>
```

The code in Listing 8-3 would display the windows shown in Figure 8-5.

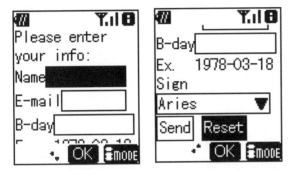

Figure 8-5: The user sign-up screen

The PHP script, which takes this information and inserts it into the database called users.php, is listed in Listing 8-4. Although PHP developers often throw everything into one script, including the HTML, good coding technique that clearly separates what you are doing is a better practice. It also allows those of you who are performing your scripting in another language to ignore or separate the PHP from the rest of what I am talking about.

Listing 8-4: The Code for users.php

```
<html>
<head>
<title>Processing User Info</title>
<body>
<?php
$utn=getenv(http_user_agent);
if (strlen($utn)>21)
{
$user=substr($utn,-11,11);
$connection = mysql_connect("localhost", "username", "password" );
mysql_select_db ("dbname");
if (mysql_errno())   {
print "I am sorry.<br>The following<br>occurred while";
print "<br>connecting to<br>server:";
echo mysql_errno().": ".mysql_errno()."<br>";
}
else {
```

Continued

Listing 8-4 *Continued*

```
$result=mysql_query(" INSERT INTO users VALUES
('$user','$name','$mail','$birthday','$sign')");
if (!$result) {
print "I am sorry.<br>There was a<br>problem getting";
print "<br>your information<br>into our database.";
echo mysql_errno().": ".mysql_errno()."<br>";
}
else {
print "You are now<br>in our database!<br>Please go to";
print "<br>our page to see<br>your horoscope.";
}
}
mysql_close ($connection);
}
else {
print  "You can't register<br>unless you ok";
print "<br>sending your<br>info. Try again!";
}
?>
</body>
</html>
```

The PHP script in Listing 8-4 may not make much sense if you are unfamiliar with PHP, but it does two things pretty simply: It gets a user's handset serial number by separating it from the rest of the identifying user agent string in the HTTP header, and it puts this information into a user database. The fact that the script opens and closes with raw, normal HTML is one of the main advantages of PHP over Perl. Though Listing 8-4 makes use of a fair amount of PHP's scripting features, you also can make a PHP page almost entirely HTML, interspersing PHP mini-tags with data in them with the HTML.

Figure 8-6 shows my input viewed in phpMyAdmin.

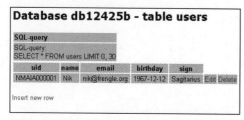

Figure 8-6: The database, viewed in phpMyAdmin.

This small project may seem to be a lot of work doing something that many programmers are used to doing with cookies, often using nothing more than a WYSIWYG editor and filling in a few options. This project, however, is useful and necessary with i-mode.

The `utn` attribute is available only on 503-series and 210-series phones or later. For earlier models, you are simply out of luck unless you become an official DoCoMo information provider and receive the information automatically with the `user agent` variable. For a non-official site, this information is cut off in the tail of the `user agent` variable, so you can't see the information. Because the attribute is part of the i-mode browser and this browser is not updateable, the absence of the `utn` attribute means that the handset cannot indicate to the i-mode server that it wishes to have this information sent on rather than cut off. By the time you read this, at least half of i-mode users should have either 503- or 210-series phones. That's probably not a bad market to aim at. You will have to fix the previous code, of course, if you want to ask other users for their 11-digit mobile phone number to use in place of their serial number.

Having a user's information in a database rather than a cookie provides the advantage of not limiting your use of the data to simple responses to user page requests, as cookies do. You could do "happy birthday" mailings, age-specific promotions, Chinese fortunes, and so on.

Chapter 12 covers the use of PHP, MySQL, and cHTML together.

Using Emulators and Editors

In days of yesteryear, programmers would input data on punch cards. That was not fun. We have come all the way to the WYSIWYG editors of today, such as GoLive or Dreamweaver. But most WYSIWYG editors have not caught up with the mobile phone, and you would be lucky if what you see is what you get on more than one handset. Because most readers of this book probably do not live in Japan, emulators are a neccesary addition to your arsenal of tools. In this section, I discuss those tools.

Finding an emulator

Especially for the many readers and others reading this book outside Japan, developing applications or sites for i-mode requires that you can test those applications. Because the only way to make Java applications or cHTML pages accessible to your handset at this point is through a connection to the Internet, and because this connection currently doesn't exist outside of Japan and Hong Kong, your only real choice is to use an emulator to display a page. None of the emulators available currently support all of the functions of a 503-series handset. Basically, two kinds of emulators exist: Java and cHTML. The Java emulators from DoCoMo do a pretty good job of emulating a general i-Appli experience, as do those of i-Jade. Both are available in the Java folder on the included CD-ROM.

The CD-ROM also includes several emulators for cHTML, but they have several drawbacks:

- They don't display emoji characters reliably.
- They don't have color screens, whereas more than 80 percent of i-mode phones do.
- They have problems displaying pages in exactly the same way they are displayed on a real DoCoMo handset.

In view of these problems, you may wonder when you read the next chapter how I obtained the screenshots presented in it. I found one very good emulator, i-mimic, that is available online only. It attempts to emulate the popular P209is-series phone, which happens to be the same one I possess, and so I can say that it fully succeeds in its attempt to emulate this model. You can view it at the following site: `www.x-9.com/mimic/`.

It even emulates the user agent correctly! This capability constitutes a very big plus, but the emulator has one minus as well: On that particular series, several of the tags and attributes available on the 503 series are missing, including `utn`. For most applications, though, these features are not used, and the i-mimic emulator is an excellent resource.

I have included a symbol font on the CD-ROM with this book as well, which contains all the emoji symbols used on i-mode handsets. You can use this font to preview your fonts in a normal Web browser, though you will need to make sure to replace the characters with the correct decimal or hexadecimal codes before you upload your page for viewing on i-mode.

Part of the problem with any emulator is that it does what i-mimic does — emulate one particular handset. Because the manner in which pages are displayed between handsets varies somewhat, even a totally accurate emulator such as i-mimic is limited to displaying how one particular handset displays something. For the casual user, this limited display should be enough, but if you are designing fee-based, universally accessible content, it isn't good enough. Your easiest option is to move to Japan and buy 20 or 30 handsets. Whoa . . . hold that thought. Moving to Japan is anything but easy and is far from cheap, and the newest 503 handsets run 29,000 yen (about US$225) each.

> **Surrogate i-mode Testers**
>
> Thomas O'Dowd, along with Juergen Specht, moved to Japan and bought a *lot* of handsets. Specht and O'Dowd also have a business called Nooper Labs, which basically tests an application on every different handset available and then e-mails you the screen shots. The company also performs other sorts of testing, such as "man-on-the-street" tests, in which it asks a sampling of random users to try the application and give feedback on usability, presentations, level of interest, and so on.
>
> I don't know how much Specht and O'Dowd charge for their services, and I am not promoting their company, but at least you know that such a service is available in case you're interested. Probably a service such as this would be useful for a company that wanted to make a full push into the Japanese market but didn't have the resources to send someone to Japan just to test out its application. Using this service would present an imperfect solution, but may be the only way to ensure that your page looks as it should on every different handset. The Nooper link is `http://nooper.co.jp/labs/?l=en`.

I don't know how carriers in other markets view developers or what resources those carriers generally provide developers, but DoCoMo has not treated independent developers in a way that has been helpful. Although DoCoMo possesses emulators that it uses in-house, these emulators have not been released to the public. DoCoMo does not give developers a discount, or even free phones, and the company has been quite reluctant to release even such basic information as how much memory the FOMA phones will have (30 KB, which I have learned from sources outside NTT DoCoMo).

Part of the problem with emulators, it must be said, is Unisys' patent on the GIF format, which limits the ability of sites to give emulators away and imposes enough of a financial burden to discourage creation of any sort of tools that could invite any notice by Unisys' pack of lawyers.

Coding with editors

As mentioned previously, one of the brilliant things about cHTML is that it is actually just HTML. You can use the same editors that you use for normal HTML. The problem arises with how attractive or useful these editors are if you can't use their WYSIWYG grid layout schemes, which you can't do with cHTML because it does not support tables. Adobe GoLive is nice in the sense that you can use it to create both standard HTML pages and cHTML, but you lose most of its functionality when you use it with cHTML. This limitation is not an inherent limitation in the tools, but rather relates to cHTML itself.

I have included the free WAP Profit editor and emulator on the CD-ROM. This does a pretty good job of speeding up the process of writing pages and presents an easy way to include emoji characters in your pages. WAP Profit has a few problems with its preview mode, however, so you should make sure to look at your pages on a real handset or through i-mimic. One of the nice things about the handset in the WAP Profit preview is that it automatically performs text wrap, saving you the hassle of putting <br> tags in all over the place. I wish that i-mode handsets did this! Because automatic text wrap is a common feature on WAP phones, from which WAP Profit obviously derived, it is not a surprising addition. This feature is not yet available on actual i-mode handsets.

Working with the language of i-mode

I am actually ignoring an entire body of software when I say that no good emulators exist. Considering that i-mode is a Japanese system, it would be reasonable to expect that Japanese developing tools are needed to develop content for this system. They are, albeit not to the extent one would expect. One of the better tools I found is called simply i-mode Tool, but it has a problem: Although it will install on non-Japanese versions of Windows, it won't display Japanese text, even when using NJ Star. NJ Star is an add-in Chinese-Japanese-Korean (CJK) front end display processor that allows people using non-Japanese versions of Windows to view Japanese. As a workaround, I guess I could recommend that you do what I did, which is to set up a dual boot system with Japanese Windows on one drive partition and English Windows on the other. Doing this involved quite a lot of work, though, and ignores the fact that most people can't read Japanese, so even if they had the dual-boot system, could properly run the software, and could view the Japanese, they wouldn't know what they were seeing anyway.

Following is a simpler solution to the problem of viewing pages with emoji: On the CD-ROM with this book, I have included a font: eimode.ttf. This is a roman symbol font that has emoji characters that will work on any system, guaranteed.

This will, of course, not allow you to see your page in exactly the same dimensions as you would see it on a mobile phone screen, nor to use any of the i-mode-specific tags. No emulator allows you such options, though, and in the area of image handling, only the i-mimic emulator handles them correctly, and a standard HTML Web browser does fine with them. So, maybe build a page with a frame that is the size you think a cell phone would be or with a dynamic HTML or XML window that does the same thing.

 If you like the eimode font but need Japanese, you can use a good font called Keitai-Font from a company called Enfour. You can check this out at www.enfour.co.jp/media/. Be aware, however, that the software that the company advertises as being able to get you on a proxy server to test your pages doesn't seem to install on English versions of Windows. If you want to develop Japanese content, though, one would expect that you would be using a Japanese system. If you aren't, you probably should be.

As mentioned previously, DoCoMo puts out new phone models about every six months, and with each model series, between five and ten companies release handsets. There are simply too many models to emulate well, and so little time passes before a new model comes out. A company such as Access can more fruitfully direct its efforts toward developing new products for new platforms than toward maintaining emulators for each model. Probably, in the future, a descriptor file will be available that describes the specific behaviors of a handset model, and can be used by an emulator to emulate that model. The framework for this sort of emulator hasn't been created yet, though. A note to any aspiring application programmers among you: Your efforts, if you were to do something like this, would be very much appreciated and probably well paid.

See Appendix A for a listing of editors and emulators that appear on the CD-ROM and information on how to access them.

Coding small and coding with an emphasis on usability on a small screen does not have to mean a lack of functionality. By using the elements such as graphics and forms that we do have available, you can create useful rich content that has an impact. Using server-side scripting and a database, you can move much of a site's functionality to the back end, leaving a well-designed and menu-driven front end.

Viewing pages, although difficult to do exactly for every handset, should become easier. Dealing with S-JIS Japanese encoding will not be necessary when i-mode gets to America, and although it will probably lose the cute emoji symbol characters, the service will be much easier to preview on a standard Web browser.

Chapter 9

cHTML, the Language Used for Creating i-mode Pages

THIS CHAPTER DISCUSSES CHTML-SPECIFIC TAGS and how they differ from standard HTML in use and output. The examples are all on the CD-ROM, and can be viewed using one of the included emulators. They can also be viewed using any browser, but the cHTML-specific tags will not be displayed properly.

File Formats Used in cHTML

Unlike HTML, there are some important sorts of files that cannot be used in cHTML. Chief among these are JPEG files. For graphics, only GIF files are supported on all phones. This is because JPEG requires a fair amount of processing power to decompress images — power you are unlikely to find on a mobile phone. Some 503 models, however, can support JPEG, and as processors become more powerful, it could become widespread. As of this writing, there is no Western text specification for cHTML, but it is safe to assume that it will be either ASCII or Unicode. Smart money would say Unicode, simply because this would give NTT DoCoMo increased flexibility, and its handset makers wouldn't have to change the text format for each country. At this time, however, only Shift-JIS (S-JIS), a Japanese text format, is supported. Don't worry! When using English, it is exactly the same as ASCII, so you probably won't have to make any adjustments to how you format your text.

The GIF file properties supported by different series of phones are listed in Table 9-1:

TABLE 9-1 GIF FILE PROPERTIES SUPPORTED BY VARIOUS HANDSETS

	503i-Series	502i, 209i, 821i	501i
Non-interlaced GIFs	Yes	Yes	Yes
Interlaced GIFs	Yes	Yes	No
Transparent GIFs	Yes	Only color models	No
Animated GIFs	Yes	Yes	No

Using cHTML Tags in i-mode

Aside from a few additions, cHTML looks *exactly* like HTML. That means it has the same beginning and ending tags — `<html></html>` — and all the other structural tags are exactly the same. In the explanations of each tag in the examples, there is nothing structurally different from standard HTML, and only a few tags and attributes are different. How these tags and attributes behave, however, is somewhat different. This is not an aspect of the markup language, but of the browser. (See the notes at the end of the chapter for an explanation of the use of [2] and [3].)

&XXX;

This is actually not a tag, but a way of displaying special characters. It is basically the same as standard HTML, but it is used a lot more and requires being done by hand, whereas most standard HTML editors do this for you. The & designates either punctuation characters, such as an ampersand — `&amp;` — or so-called *emojis*, which are i-mode-specific picture characters, such as `&#63889` (which is the character for a heart ♥) or `&#63647` (which is the symbol for the sun ☀).

See Appendix A for the contents of the CD-ROM, including emulators, and see Appendix E for a listing of the *emoji* symbol characters.

I have also included a symbol font with all of the emoji characters, which is included on the included CD-ROM. Listing 9-1 shows a typical use of these *emoji*.

Listing 9-1: A List Of Horoscope Signs, Using the emoji Picture Symbols

```
<html>
<head>
<title>
Horoscope list
</title>
</head>
<body>
Today's horo-<BR>
scope<BR>
Select your <BR>
sign from below:<BR>
<a href="aquarius.html">&#63665;</a>Aquarius<br>
<a href="pisces.html">&#63666;</a>Pisces<br>
<a href="aries.html">&#63655;</a>Aries<br>
<a href="taurus.html">&#63656;</a>Taurus<br>
<a href="gemini.html">&#63657;</a>Gemini<br>
```

Chapter 9: cHTML, the Language Used for Creating i-mode Pages

```
<a href="cancer.html">&#63658;</a>Cancer<br>
<a href="leo.html">&#63659;</a>Leo<br>
<a href="virgo.html">&#63660;</a>Virgo<br>
<a href="libra.html">&#63661;</a>Libra<br>
<a href="scorpio.html">&#63662;</a>Scorpio<br>
<a href="sagitarius.html">&#63663;</a>Sagitarius<br>
<a href="capricorn.html">&#63664;</a>Capricorn<br>
</body>
</html>
```

The screens on the mobile phone should look like those in Figure 9-1.

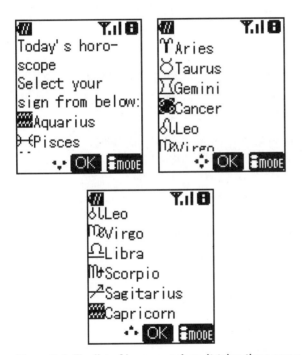

Figure 9-1: The list of horoscope signs. It takes three pages of scrolling to get through.

The use of these emoji is an important way of keeping content short and easy to read. In a weather report, for example, the symbol for sun (☀) instantly conveys the meaning, and takes up little screen space. In the previous example, the emoji were used as buttons to click on, allowing users who used the service every day to quickly scroll down to the symbol they associated with their sign and then jump to the page with a horoscope. I have to admit, I don't totally see the relationship between all of the symbols and their signs, but if I used this service every day that wouldn't matter — I would have the horizontal bar with the diagonal arrow in my head as meaning Sagittarius. The fact that it was written beside the button doesn't hurt either.

The other use of emoji is as free decoration. Unlike GIF images, they take up no memory; therefore, they are "free." Creative i-mode site designers take advantage of this to enhance the look of their sites without adding to download time or risking using up too much memory, which was one more reason for using the previous example.

`<a></a>`

The anchor tag is very much the same in cHTML as HTML, though cHTML adds some values not normally used with this tag's attributes. Those attributes are shown as follows. This tag takes the following attributes:

- `name` The value of this attribute should be what you want to name your anchor point.

  `name="anchorpoint"`

- `href` The value of this attribute can be a URL or file path, a `mailto:` value, or a `tel:` value.

  `href= "tel:09022223333"`

  Using this example, along with placing the call to the number, puts the information in their phone's memory. Quite useful.

  `href= "tel:09022223333" telbook="Nik Frengle" kana="gobbledeegook" E-mail="nik@eimode.com"`[3]

- Another option for the `href` attribute is a mail address, plus, in the case of 503-series phones, the subject and body modifiers.

  `href="mailto:service@eimode.com"?subject`[3]`="DoCoMo Story"&body`[3]`="Send more info"`

- `cti` The value of this attribute can be telephone numbers or tones (# and *); commas, which introduce a two-second pause; or forward slashes (/), which wait for a user's input.

  `cti= "0367891234/,,#234"`[2]

- `accesskey` The value of this attribute can be numbers 0-9, or # or *. Note that on some phones # and * are not supported.

  `accesskey="1"`

- `utn`[3] This attribute, which works only on 503-series or above phones, asks for a header from the phone, giving information on what version of cHTML it is running, what model of phone it is, and the serial number of the phone in the following format: *DoCoMo/1.0/X503i/c10/ serNMAIA000001.* Because cookies are not included in cHTML, use of the serial number of the phone to identify a user can be done using this attribute.

Chapter 9: cHTML, the Language Used for Creating i-mode Pages

The next sections explain some of the more unique uses for these attributes in i-mode.

DIALING VIA HYPERLINKS

The `href` attribute should be familiar, but the link probably won't be — it is a telephone number.

```
<a href="tel:09022223333"  cti="09022223333/,,#234">
```

It is used in the same way as `mailto:` is used in standard HTML. Phone numbers have a maximum length of 24. By clicking on whatever is between the link tags, a user's phone dials the number. This is quite handy for customer support, address books, and any application in which having the user call is useful.

There were a lot of cases of pranksters not telling a user that what he was clicking on was a phone number, and i-mode users inadvertently dialing emergency services and other prank numbers. DO NOT DO THIS! Always identify a phone number link as such.

The `cti` attribute allows you to introduce pauses; for example, with voicemail systems, or automated switchboards. Listing 9-2 shows how to produce a dialing link as it appears in Figure 9-2.

Listing 9-2: Using the Telephone and cti Tags in the Anchor

```
<html>
<head>
<title>
Customer Support
</title>
</head>
<body>
To phone our<br>
Support Staff<br>
Click <a href="tel:0355553333" cti="0355553333/,,#234">here!</a>
</body>
</html>
```

Clicking "here" in Figure 9-2 causes the phone to dial the number in the reference, switching momentarily out of the i-mode browser to the phone. When the phone call is finished, a user is returned to the browser.

Figure 9-2: A simple page using the telephone value and cti attribute.

CREATING LINKS FOR THE KEYPAD

As mentioned in the bulleted list for this section, the <a> tag can also make use of the keypad using a line like this:

```
<a href= "http://www.anysite.com" accesskey="1">
```

Again, the first part of this tag is one that anyone who is familiar with HTML is intimately familiar with. The accesskey attribute, though, is specific to cHTML. It gives you use of all number keys (on some models, the # and * keys as well). Using the previous example, a user could simply hit the 1 key on the phone's number pad and then be taken to the link. This is quite useful, saving a user from having to use the arrow keys to move up or down between links and then click. It is quite common for this to be used in conjunction with the emoji character for the keys. Listing 9-3 produces the results shown in Figure 9-3.

Listing 9-3: A Simple Page Using the accesskey Attribute

```
<html>
<head>
<title>Area Selector</title>
</head>
<body>
Please select<br>
the region you<br>
are in, below:<br>
<a href="northwest.html" accesskey="1">&#63879;</a>The Northwest<br>
<a href="california.html" accesskey="2">&#63880;</a>California<br>
<a href="southwest.html" accesskey="3">&#63881;</a>The Southwest<br>
<a href="midwest.html" accesskey="4">&#63882;</a>The Midwest<br>
<a href="other.html" accesskey="5">&#63883;</a>Other Places<br>
</body>
</html>
```

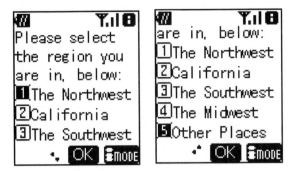

Figure 9-3: The list, using hard links on each option, along with access keys.

For example, if Figure 9-3 were the first screen of a weather information site that you made regular use of (even if you lived in the Midwest, which takes scrolling down to get to), you would probably remember that it was number 4, and you could simply click the 4 key on the keypad. In Japan, the numbers with squares around them have come to indicate that a user can just punch that key on the keypad. It isn't required, but is a convenient symbol to a user that there is an access key. The `accesskey` attribute is used with other tags as well.

<base>

The `<base>` tag specifies an absolute URL that acts as the base URL for resolving relative URLs. The `<base>` tag has only one attribute:

`href="http://eimode.com/i/"`

The value of this attribute should be a URL such as

`<base href="http://eimode.com/i/">`

The previous code sets all links relative to this base, the `eimode.com` Web site directory structure, so you don't need the full path to pages or graphics every time you reference them, as long as they are in the base directory.

It also allows code such as `<a href="">`, which takes you to the default page in the directory.

<blink></blink>[2]

This tag is a kind of blast from the past for those of you who remember writing pages for Netscape Navigator 2.0 (it wasn't that long ago). It was so cool to have things blink, so everyone did it. Welcome to i-mode 2.0 and deja vu all over again. This tag is handled somewhat differently by different phone models. The blinking stops on some, and continues ad infinitum on others. Don't worry too much about annoying viewers because they can probably get it off the screen by scrolling.

<blockquote></blockquote>

This tag is used for quoting within text.

```
He said<blockquote>this is an outrage, I can't believe it</blockquote> after
they kicked him out of the bar at midnight.
```

On i-mode phones, this tag offsets the text within by two characters on both left and right sides. For example, Listing 9-4 shows the results shown in Figure 9-4:

Listing 9-4: Use of the blockquote Tag

```
<html>
<head>
<title>
Outrage
</title>
</head>
<body>
He said
<blockquote>this is an outrage, I can't believe it</blockquote>
after they kicked him out of the bar at midnight.
</body>
</html>
```

Figure 9-4: This is how the blockquote code in Listing 9-4 displays.

You will notice that this brings your text width between the quotes down to only 12 characters. I just couldn't be bothered with inserting line breaks anymore. This is not an uncommon affliction, and a lot of English sites have succumbed to this form of laziness. Don't you fall into it! I actually did this to show exactly how hard it is to read text this way. Imagine reading a whole story with words not hyphenated or pushed to the next line, as I have on CNN's site. Awful!

<body></body>

This tag is exactly the same in HTML and cHTML, defining the body of the text to be output to a page. This tag takes the following attributes:

- bgcolor   For background color

  bgcolor="red"2

- text   To define text colors

  text="green"2

- link   To define link colors

  link="blue"2

These values can be one of the color names or a RGB color in hexadecimal form of one of the colors in the 256-color palette. See the end of the chapter for color names, and how to format hex color names.

 Using any of these modifiers on a black-and-white handset (Yes! They are still out there!) are ignored; they are displayed in black. You can designate colors either by their 16 names or by hex.

This tag inserts a hard line break in text. The <br> tag has only one attribute in cHTML: clear. It determines word wrap as follows:

<br clear="all">

The possible values for this attribute are all, left, or right.

MAKING CLEAN BREAKS IN I-MODE

Although it states in the cHTML specification that the <br> tag is used to determine how text wraps around an inline image, there are a few tricks to actually getting it to work. The first is that the image needs to use the align attribute. Without the attribute, text will align with the bottom, and there will be no need for this tag. For example, Listing 9-5 produces the results shown in Figure 9-5:

Listing 9-5: Using Line Breaks with the clear Attribute, but without Using Alignment on the Image

```
<html>
<head>
<title>Happy</title>
</head>
<body>
<IMG src="9-5.gif">Happy People<BR clear="all">All happy people<br>
Jack<br>
Jill<br>
Benny Hill<br>
</body>
</html>
```

Figure 9-5: The text goes to the bottom in this example from Listing 9-5.

ALIGNING BREAKS

By simply using the `align` attribute, though, as shown in Listing 9-6, you can produce a better result (see Figure 9-6):

Listing 9-6: The Same Code as Listing 9-5, with the Addition of the align Attribute to the Image

```
<html>
<head>
<title>Happy</title>
</head>
<body>
<img src="9-5.gif" align="left">Happy<br> People<br>
   clear="all">All happy people<BR>
Jack<br>
Jill<br>
Benny Hill<br>
</body>
</html>
```

I get the result shown in Figure 9-6, which looks much better.

Figure 9-6: A much better result, with the text aligning.

You will notice that I made one minor change beside the `align` attribute, which was to add a `<br>` tag between `Happy` and `People`. This is the one tricky part of using these tags — you have to figure out how much room you have based on how big the graphic is and how wide your screen is, which varies depending on the phone model. Give yourself extra room or risk being wrapped ugly!

This tag is used exactly the same as its HTML cousin; it centers the text placed between the tags. Listing 9-7 shows you how to center text on an i-mode screen (see Figure 9-7).

Listing 9-7: Using the <center> Tag to Place Text Horizontally in the Middle

```
<HTML>
<HEAD>
<TITLE>Center</TITLE
></HEAD>
<BODY>
<CENTER>MAJOR</CENTER>
Damage has been<br>
done by forget-<br>
ing to brush<br>
after every<br>
meal.
</BODY>
</HTML>
```

Figure 9-7: The topic heading is centered at the top of the page.

<dir><li></dir>

This is the directory tag, which uses line tags in between to define a directory list. This tag is an old one, and not commonly used any more. It has been replaced by <ul>, and in i-mode it creates exactly the same output. (See the section on <ul> at the end of this chapter.)

<dl><dt><dd></dl>

These three tags create a definition list, <dl>, the title of the definition <dt>, and the definition itself <dd>. These tags exist in the HTML 4.1 specification as well, so you may be familiar with them already. In i-mode, they display somewhat differently than in an HTML browser. The code in Listing 9-8 creates the screens shown in Figure 9-8.

Listing 9-8: Creating a Definition List

```
<html>
<head>
<title>
Geek Dictionary
</title>
</head>
<body>
<dl>
<dt>Dweeb
<dd>young hyper<br> person who<br>may mature<br> into a Nerd
<dt>Nerd
<dd>technically <br>bright but <br>socially inept<br> person
</dl>
</body>
</html>
```

Chapter 9: cHTML, the Language Used for Creating i-mode Pages   135

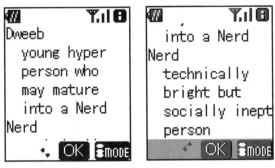

Figure 9-8: A definition list, with two definitions.

 On this model of i-mode phone (the P209is), the definitions are indented by two characters, narrowing the width to only 14 characters. If you were using a server-side scripting language, such as Perl or PHP, and were using a text-wrap routine, you would want to take this into account.

`<div></div>`

This is a block-level generic style container. The `<div>` tag has only one attribute, `align`, as shown here:

`align= "right"`

Possible values for this attribute are `left`, `center`, or `right`.

This is a tag used in normal HTML, and can be used in cHTML as well. Although only the `align` attribute is available, it is useful because it saves having to use the `align` attribute in the paragraph`<p>` tag or in the header tag`<h1>`; allows changes in alignment without unnecessary waste of space like the `<p>` tag with its one line of empty space; and allows new paragraphs to be started without losing the alignment. For example, Listing 9-9 would be displayed like Figure 9-9.

Listing 9-9: Using the `<div>` Tag with the align Attribute

```
<html>
<head>
<title>
Div Test
</title>
</head>
<body>
<div align=right>
```

Continued

Listing 9-9 *(Continued)*

```
To break para-<p>
graphs while<br>
keeping your</div>
<div align="left">text aligned<br>
use this!
</div>
</body>
</html>
```

Figure 9-9: Using a <p> tag within a <div align=right> tag is allowed, as shown.

This screen tag actually highlights the real beauty of the <div> tag on i-mode. You don't need to use the <p> tag, which wastes a whole line of space, to align text on the right.

<font></font>[2]

This tag allows you to define how text looks. In cHTML, this ability is limited. The <font> tag has only one attribute, color, as shown here:

```
color= "fuchsia"
```

The color can be one of 16 names or a hexadecimal RGB value. See the chart at the end of this chapter for a full list of names.

This tag is not available on 501-series phones, but this is to be expected because none of them were color, anyway. Black and white handsets ignore this tag, and display the text as black. I wish this book had a massive budget and had color pictures on every page, and I could show you what this tag looks like, but well, such is life.

<form></form>

Forms allow users to send data to a script on a server. In cHTML, they are the basis of all interactivity between user and application.

This tag takes the following attributes:

- `action` This must be a URL. `mailto` and `tel` values do not work.

    `action="/cgi-bin/submit.cgi"`

- `method` Values can be either `post` or `get`.

    `method="post"`

    This tag and its attributes function exactly as they do in normal HTML. See code Listings 9-10 to 9-12 for more on forms and form-related tags.

- `utn`[3] This attribute, which works only on 503-series or above phones, asks for a header from the phone, giving information on what version of cHTML it is running, what model of phone it is, and the serial number of the phone in the following format: *DoCoMo/1.0/X503i/c10/serNMAIA000001*. Because cookies are not included in cHTML, use of the serial number of the phone to identify a user can be done using this attribute.

<input>

This tag falls within a form and defines what sort of input a form field takes, depending on its attributes. This tag takes the following attributes:

- `type` The name should be one recognizable to the script it is being sent to; that is, the name of a variable.

    ```
    type="text | password | checkbox | radio | hidden | submit
        | reset"
    name="recipient"
    ```

- `value` This is used with the check box type. Possible values are `yes` or `no`.

    `value="yes"`

- `size` This is used with the text and password type. This value probably should not be more than 14 because this is as much as will fit on one line of most handsets.

    `size="number of characters"`

- ◆ `maxlength`   The maximum length a user's input can be. Up to 512 can be entered, but the `textarea` tag should be used in that case.

    `maxlength="4"`

- ◆ `checked`   For use with the checkbox or radio type. If you leave off this attribute, the check box or radio button is unchecked.

- ◆ `accesskey`   A value for 0–9, or *, or #.

    `accesskey="1"`

- ◆ `istyle`   Possible values for this attribute are 1, 2, 3, or 4.

    1 sets the input method to Japanese "hiragana" double-byte characters.

    2 sets the input method to Japanese "katakana" characters, which on i-mode are a one-byte subset.

    3 sets the input method to Roman characters; the ones that you are reading at this very moment.

    4 sets the input method to numbers.

    `istyle= "4"`[2]

These form-related tags are basically the same as their HTML equivalents, but there are some important differences, as well as some things to keep in mind when using them on a mobile phone handset.

The `<input>` tag has a wide range of choices available. Divide them up using the `type` attribute:

`<input type=text>`

This creates a text box, as in normal HTML. The extra attributes used that are unique are the `accesskey` attribute and the `istyle` attribute (see the `textarea` function for a description of this).

`<input type=password istyle="4">`

This creates a password field. I should say that although this would seem like a good choice from a security standpoint, and a given in HTML coding, not many people are using this input type with cHTML. The simple reason is that inputting characters using a phone's number pad is cumbersome enough without having what you are inputting concealed. This is totally up to the page designers, but at least in Japan, they have clearly favored using normal text boxes for this. I should state that there is one notable exception: banks, which have four-digit PIN numbers that are fairly foolproof. If you DO use a password field, by all means use the `istyle` attribute, which determines what kind of text is automatically entered. It won't work on 501-series phones, but will save lots of hassle on all newer models.

Chapter 9: cHTML, the Language Used for Creating i-mode Pages

The code at the beginning of this paragraph would create a password box with number entry as default. See previous section for `istyle` attribute values.

The rest of the tags can just as easily be demonstrated as explained. The example in Listing 9-10 has all of the types of input. Figure 9-10 shows the result.

Listing 9-10: All Types of Input

```
<html>
<head>
<title>
Form Collage
</title>
</head>
<body>
<form action="/cgi-bin/submit.cgi" method=post>
<input type="hidden" name="recipient" value="mail@eimode.com">
Tell me your &#63720#<br>
&#63879<input type=text accesskey="1" name="note" size="10"
    maxlength="20"><br>
Your password<br>
&#63880<input type=password name="password" istyle="4" size="6"
    maxlength="4" accesskey="2"><br>
Call me &#63881<input type=checkbox name="callme" value="yes"
    checked accesskey="3"><br>
Nik is cool<br>
&#63882Yes<input type=radio name="cool" value="yes" checked
    accesskey="4">
&#63883No<input type=radio name="cool" value="no"
    accesskey="5"><br>
&#63884<input type=submit  name="submit" value="Send Details"
    accesskey="6"><br>
&#63885<input type=reset name="reset" value="No Way!"
    accesskey="7"><br>
</form>
</body>
</html>
```

I used the emoji number symbols and the `accesskey` attribute. I did this more as a demonstration than anything because most sites do not make such heavy use of these. I have to say, though, that after playing around with this on my phone, I wish they did use it more! They look a bit goofy, but do make it much easier because there is no using the arrow keys to scroll down, and then the OK button to select it—just dial in the number.

Though it is not visible on this page, if you put a hidden field in, it induces a line break, unless, as in this form, it comes at the top, right after the `<form>` tag.

Figure 9-10: Using lots of different input fields and accesskeys to give you a good idea of what they look like in i-mode.

I used the `post` method for this form. The `get` method, which appends the data to the end of the URL, stands a chance of the data getting dumped because the maximum length of a URL is 256 bytes.

<select></select>

This defines a selection menu, or a pull-down menu. Within these tags are a number of options, defined by the `<option>` tag, explained in the next item.

This tag takes the following attributes:

- `name`   The name the data is sent to the script under. Should be something the script receiving the form data recognizes.

    `name="favorite pizza"`

- `size`   This is the number of rows to initially display. After clicking, a user can view all options. Minimum is 1, and maximum is 8.

    `size="1"`

- `multiple`[2]   This attribute allows more than one row to be chosen. It does not take any values.

<option>

This tag is used before each option in a `<select>` menu list. This tag takes the following attributes:

- `value`   This value is one of the values to be included in the select list, and is the value sent to the form script. Maximum length is 42 bytes.

    `value="pepperoni"`

- selected   If included, this option value becomes the default value. If the multiple attribute is used with the select tag, more than one option may be selected.

The `select` and `option` tags are basically the same as HTML. There is one attribute, `multiple`, which is part of the cHTML 2.0 implementation, so it won't work on 501-series phones. Listing 9-11 creates an example of a simple form using select and option. Figure 9-11 walks you through selections in the form.

Listing 9-11: A Pull-down Menu, Using a <select> List

```
<html>
<head>
<title>Geek Survey</title>
</head>
<body>
<div align=center>
<form action="/cgi-bin/submit.cgi" method="post">
Favorite Pizza<br>
<select name="favorite pizza" size="1" multiple>
<option value="pepperoni" selected>pepperoni
<option value="sausage">sausage
<option value="three">artichoke
<option value="canadian bacon">canada bacon
</select><br>
Favorite<br>
Caffeine Source
<br><select name="caffeine" size="1">
<option value="mellow">Mellow Yellow
<option value="mountain">Mountain Dew
<option value="dietcherrycoke">Diet Cherry Coke
<option value="coffee">Coffee
</select>
<br>
<input type="submit" value="Be Counted!" name="send data">
</form>
</div>
</body>
</html>
```

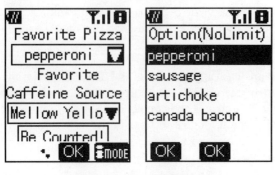

Figure 9-11: The main menu and the two menus that appear when you select the pull-down menus.

In terms of how much information you can get on just one screen, these select pull-down menus cannot be beat. You will notice that the pizza menu, which used the `multiple` attribute, has a slightly different option menu than that of the caffeine menu. The left OK button selects multiple choices, whereas the middle one takes you back to the main menu. I would put this method of data input way ahead of the radio buttons and check boxes seen in the previous example, simply because they only take up one line. If you keep the words in the menu very short, in fact, it dynamically sizes horizontally, so it is possible to get the label and the menu on one line. A bit tricky, but give it a try!

\<textarea\>

This is another kind of form field, a text box. This is used within the `<form>` tags. This tag takes the following attributes:

- `name` The name of the field the data is transmitted to the server as. Should be something that the script recognizes.

  `name="dreams"`

- `rows` Number of lines of text. The minimum is 1, and the maximum is 8. Because clicking the box to enter data takes a user to a separate data entry window, the number of rows displayed on the screen is not so important.

```
rows="3"
```

- cols   Number of characters of width. The minimum is 1, and the maximum is determined by screen width. Fourteen is a useful maximum, but a maximum greater than screen width will be cut down to the maximum allowed on that particular handset anyway, so don't sweat this number too much.

```
cols="14"
```

- istyle   This attribute sets the default input method. In this case, it is the Roman alphabet.

```
istyle="3"
```
[2]

The last form-related tag is the `<textarea>` tag. This tag is pretty much exactly the same as its HTML form. There is one new attribute in cHTML 2.0, the `istyle` attribute. This is a useful addition to cHTML 2.0 that allows a developer to define one of four options for input: Hiragana (1), Katakana (2), Alphabet (3), or Numeric (4). Hiragana and Katakana are two Japanese alphabets that are used for native Japanese words and foreign loan words, respectively. All i-mode-capable phones have the capability to input these four sorts of characters, so it makes sense to allow a developer to define which type of input is required, which also saves a user from having to remember to adjust his input method. This is something that will probably be much less necessary when i-mode rolls out in Europe and the U.S., and will probably consist of only two options, letters and numbers; or uppercase, lowercase, and numbers.

The `textarea` tag allows a maximum of 512 characters to be input. Considering the input method, this should be more than enough! Listing 9-12 shows a simple example of using the `textarea` tag:

Listing 9-12: Using the <textarea> Tag to Make User Input Fields

```
<html>
<head>
<title>Textarea</title>
</head>
<body>
Dream Mail<BR>
<form action="/cgi-bin/submit.cgi" method="get">
<textarea name="dreams" cols="14" rows ="3" istyle="3">
Please tell us your dreams.</textarea>
<input type="submit" value="Send!" name="send data">
<input type="reset" value="Don't" name="no">
</form>
</body>
</html>
```

Figure 9-12 shows what you would see.

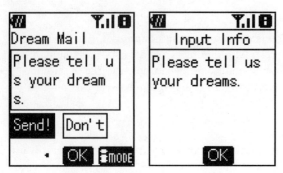

Figure 9-12: The form and the window that comes up if the text entry field is selected.

<h1></h1><h2></h2><h3></h3><h4></h4>

The heading tags, which typically assign successive levels of headings, are available in cHTML, but don't do very much. Probably on future models, which may have font styles, such as bold and italic, these tags will be displayed somewhat differently. At present, the code in Listing 9-13 would result in Figure 9-13.

Listing 9-13: Using Heading Tags to Delineate Rank

```
<html>
<head>
<title>Org Chart</title>
</head>
<body>
<h1>Honcho</h1><br>
<h2>Col. Santiago</h2><br>
<h3>Lt. Mitchell</h3><br>
<h4>Sgt. Bilko</h4><br>
<h5>Pfc. Pile</h5>
</body>
</html>
```

The heading tags act almost exactly like the <p> tag at present, and I would recommend against frequently using either the <h1> tag or the <p> tag because they both waste screen space.

Figure 9-13: Though heading tags were used, there is no difference in how the ranks are displayed.

\<head\>\</head\>

This tag does exactly the same thing that it does in HTML The only available tags for use in the head are `<title></title>` and `<meta>`, which go between the head tags. See below for how both of these tags are used.

\<hr\>

The horizontal rule is a valuable tag in cHTML because it gives you a "free" graphical element. This tag takes the following attributes:

- `align` The alignment of the rule, either `left`, `center`, or `right`.

    `align="right"`

- `size` The thickness of the rule. A number of pixels; minimum of 1, maximum of 12.

    `size="3"`

- `width` A number of pixels, or a percentage. Pixels greater than the width of a screen are ignored.

    `width="33%"`

This is a really useful element to have on a page, and, like emoji, it doesn't "cost" anything in terms of memory. It is probably used much more in cHTML than in HTML, and its attributes are very definitely used more. Listing 9-14 shows how to use the `<hr>` tag to save space, and Figure 9-14 shows the results.

Listing 9-14: Using the Horizontal Rule with Its Align

```
<html>
<head>
<title>
Horizontal Rule
</title>
</head>
<body>
<center>This Ceiling</center>
<hr size="5">
<div align="right">said</div>
<hr align="right" size="3" width="33%">
<div align="center">her boss</div>
 <hr align="center" size="3" width="33%">
Is Not Glass!
</body>
</html>
```

Figure 9-14: Using the horizontal rules graphically.

\<html>\</html>

This most basic of tags doesn't have any attributes. Also, just to clarify, despite writing cHTML, nearly all (if not all) the people who write pages for i-mode name their pages exactly like HTML: `mypage.html`. The first version of Wap Profit's development tool required the name to change to `mypage.chtml`. One of the beauties of using cHTML is that when you open a file in an HTML editor, it looks at the HTML extension and knows what to do, displaying tags and attributes in different colors. All the tools used for making regular HTML pages can be used in exactly the same way for making cHTML pages, provided you stick to the supported tags and attributes.

<img>

The image tag is used to insert an image into a document. Be careful! Only GIF images are supported on all i-mode phones. This tag takes the following attributes:

- `src` This attribute takes the URL of the image source.

    `src="img2.gif"`

- `align` This attribute takes the values `left`, `right`, `top`, `middle`, and `bottom`.

    `align="left"`

- `width` This attribute takes either a value of a number of pixels or a percentage of current width.

    `width="30%"`

- `height` This attribute takes either a value of a number of pixels or a percentage of current height.

    `height="60%"`

- `hspace` This attribute takes a number of pixels. Watch out! If the number of pixels exceeds half the screen's width, the attribute is ignored.

    `hspace="15"`

- `vspace` This attribute takes a number of pixels.

    `vspace="10"`

- `alt` This attribute takes whatever you want a user to see if he has graphics turned off or if a graphic doesn't load for some reason.

    `alt="Llama Communications"`

Most of these attributes are used in exactly the same way in cHTML as in HTML. The way they are displayed, however, although not fundamentally different, should be understood.

First, take a look at exactly how the align attributes display (see Listing 9-15).

Listing 9-15: Using the align Attribute with Images

```
<html>
<head>
<title>Img</title>
</head>
<body>
<img src="9-15-2.gif" align="left" alt="LLama Communications">
eimode
```

Continued

Listing 9-15 *(Continued)*

```
<br>i-mode
<br clear="all">news in English
<img src="9-15-1.gif " width="100" height="30">
</body>
</html>
```

Listing 9-15 displays as shown in Figure 9-15.

Figure 9-15: Using align to place images on the page.

The `align` tag used with the first graphic was "left". Without this, the text to the side would get broken into one part to the side of the graphic, and the other below. If you want to have a better idea of how this looks, I really recommend that you take a look at it using one of the emulators. The problem with the previous example is that the graphics are color, which don't translate well to black-and-white in this case. This just serves as a reminder, though, that not all mobile phones in Japan are color yet. So, when i-mode rolls out in Europe and the U.S., most handsets probably won't be color, at least at first. Colors, too, can be very tricky. Some handsets display some colors well; on others, the same colors are almost invisible and washed-out. Even the best emulator can't emulate this, and testing colors on real handsets becomes quite necessary. If you must use colors in your graphics, then, use a very basic palette with darker colors. Using yellow is a bad idea because it will simply not show up on most models. Using pale colors of any shade is also not a good idea. And even if a few models support more than 256 colors, they are in the minority. So keep your palette down to 256 colors (fewer if you can).

Next, play with the `height`, `width`, `hspace`, and `vspace` attributes. You probably will not use many more than three or four graphics on your screen at a time. Try the code in Listing 9-16 and see what comes up (shown in Figure 9-16).

Listing 9-16: Using align, height, and width Attributes with Images

```
<html>
<head>
<title>Img3</title>
```

```
</head>
<body>
<img src="9-16.gif" align="right" width="50" height="30">
<div align="left">eimode</div>
<br clear="all">
<img src="9-15-2.gif" width="40" height="40" hspace="10">
<img src="9-16.gif" align="left" width="60%" heigth="80%">
News
<br clear="all">
<body>
</html>
```

Figure 9-16: The size and alignment of images and text in Listing 9-16 look like this.

<marquee></marquee>[2]

This tag is used to move text across the screen like the stock tickers on CNBC or CNNfi. This tag takes the following attributes:

- `direction` Values for this attribute can be `left` or `right`. Left means it is going left from right; and vice versa for right (it is going from right to left).

  `direction="left"`

- `behavior` Possible values for this attribute are: `scroll`, `slide`, or `alternate`. The default is `scroll`.

  `behavior="scroll"`

- `loop` This value is the number of times the marquee loops. Possible values are 1–16.

  `loop="10"`

This tag is a little difficult to demonstrate in a book because it moves across the screen. The use of the tag is quite clear: Put a text string of up to 64 characters (bytes) between the two tags, and watch them dance across the screen. The marquee can be only one line, and if the line is longer than the screen, it is not broken up when the action stops. So, anything that is off the screen is invisible, and the only way to view it is to reload the page. The action of scroll is to move continuously across the screen until its maximum number of loops has been reached. It then stops with the text off the screen. The slide behavior scrolls across the screen until the last character in the string is showing, and then stops. If you use a loop with slide, once it has stopped, the text disappears and slides in from off-screen all over again. The alternate behavior scrolls across the screen, stops when the last character is on the screen, and then scrolls in the other direction. On my phone, this begins as right to left. The loop simply performs the function of having the string move back and forth across the screen. You should use the slide or alternate tags if you want your text string to be visible after its movement has stopped.

\<menu>\<li>\</menu>

This is another tag that is displayed in exactly the same way as the unnumbered list (`<ul>`) tag. It may be of use when migrating to XHTML because it is a part of the XHTML specification. You may be better off using `<ul>`, just to keep things clear.

\<meta>[2]

This tag in cHTML 2.0 can be used only to indicate character encoding. At this point, the only possible option is Shift_JIS because it is the only character set supported by the i-mode handsets sold in Japan. It is fair to say, however, that other character sets will be supported in the future, and it will be used to distinguish character encoding. The meta tag has only one attribute: charset, which takes these forms:

```
charset=SHIFT_JIS.
<meta http-equiv="Content-Type" content="text/html ; charset=SHIFT_JIS">
```

\<object>\</object>[3]

This tag defines the location of a piece of outside code. In i-mode it is only presently used to declare Java objects. This tag takes the following attribute, declare id, as follows:

```
declare id="Sample" data="sample.jam" type="application/x-jam"
```

The `id` property is what the object can be referred to by an `<a ijam>` tag, the data property is the URL of the ADF (.jam) file, and the type is what sort of file this is. Only x-jam is supported at this point.

This tag and how to use it will be covered more completely when we come to a discussion of writing Java midlets for i-mode phones in Chapter 12.

<ol><li></ol>

This is an ordered list, which takes list items defined by the `<li>` tag and orders them numerically or alphabetically. This tag takes the following attributes:

- `type` Values can be A for uppercase alphabet, a for lowercase alphabet, and 1 for numbers.

    `type="A"`[2]

- `start` A number, maximum of 26, with a type of A or a. Not necessary unless you want to start the list at something besides A, a, or 1.

    `start="2"`[2]

Just on a style note, I recommend using either `type="a"`, or `type="A"` to distinguish the list from a list of access-key accessible links, which tend to be numbered. Not doing so risks having a user think she can access one of the list items by an access key. Listing 9-17 demonstrates these tags.

Listing 9-17: Using an Ordered List

```
<html><head><title>OL</title></head><body>
Major Tasks<BR>
<ol type="a">
<li>Save world
<li>Walk the dog
<li>Sell soul
<li>Go jogging
<li>Study Cajun
</ol>
</body></html>
```

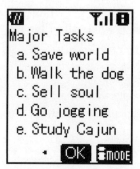

Figure 9-17: The list items are automatically put in alphabetical order.

You will notice that items indent one space in this browser list. In other browsers, there is no indent. You need to be careful of this difference, especially if you are trying to fit something on one line. Rather than 14 characters (normal screen width of 16, minus two for the item letter or number, plus one more for the trailing period), you may only have 13, or in the case of one emulator I tried, only 12.

<p></p>

This tag is a basic one, but not needed very much because many other tags also start a new line, and the `<div>` tag gives you more flexibility concerning alignment. It inserts a blank line, as in standard HTML, which in the case of a telephone handset, could be space better used otherwise. Use of `<div>` tags, together with `<br>`, gives better results. This tag takes the `align` attribute as follows, with one of the values `left`, `right`, or `center`:

```
align= "left | right | center"
```

<plaintext></plaintext>

This is not a part of HTML 4, and is not something you will probably make much use of. It was originally a part of HTML 2, but it wasn't a tag that someone writing a page actually used, but rather one that an HTTP server used when it encountered text that appeared to not be formatted in HTML. This told the browser to not bother looking for tags; and to treat anything included as preformatted text, and just display it. Listing 9-18 creates a plain text example, as shown in Figure 9-18.

Listing 9-18: Using <plaintext>, but Don't!

```
<html>
<head>
<title>Plaintext</title>
```

```
</head>
<body>
Before the tag
<PLAINTEXT>
after the tag
</PLAINTEXT>
after close tag
</body>
</html>
```

Figure 9-18: The <plaintext> tag tells the browser to stop processing HTML.

What it was really meant for and why you shouldn't really be using it is because, invisible to the eye, the server is using it to tell the little browser in your handset to chill out when it can't find any tags. So, if I had a file named `notag.html`, with only this text string inside "`this is what it was originally meant to do`", it would be displayed as shown in Figure 9-19:

Figure 9-19: The HTTP server uses <plaintext> when it encounters text not formatted in HTML.

<pre></pre>

This tag shows text preformatted, including line breaks and spaces. Quite useful for certain things, I am sure, but I can't think of what they might be on a mobile phone handset. Anyway, the cHTML in Listing 9-19 produces the screen shown in Figure 9-20:

Listing 9-19: Using Preformatted Text

```
<html>
<head>
<title>Pre</title>
</head>
<body>
<pre>
   things get
shown as they
      Look!
No    Formatting
</pre>
</body>
</html>
```

Figure 9-20: Line breaks and spaces are maintained.

<title></title>

I have been using this tag in every single example, and you may well have been saying to yourself, "Where in the heck is the title on the screen? If there ain't one, why in the heck is this clown using the tag?" Both fair questions. I could give a snotty answer like this: "Because it is simply good HTML practice," but I hate those kinds of answers. There is actually a practical reason to include the tag. The <title> tag is basically used only when a user bookmarks a site or saves it as a screen memo. In that case, the site appears in the bookmark or screen memo window under its title. Because you presumably want people to remember you or what it is that you have on your site, it is a good idea to use this tag.

<ul><li></ul>

This tag, which is the same as standard HTML, creates an unnumbered or unordered list. The bullets display differently on different models of handset. On Fujitsu-made handsets, the bullets are quite small; on the Panasonic-made 209is model I use for these screen shots, they are huge.

The following to-do list (see Listing 9-20) displays as shown in Figure 9-21:

Listing 9-20: An Unnumbered To-Do List

```
<html>
<head>
<title>UL</title>
</head>
<body>
Major Tasks<BR>
<ul>
<li>Clean my room
<li>Kiss my wife
<li>Write my book
<li>Go jogging
<li>Learn to code
</ul>
</body>
</html>
```

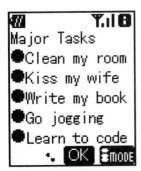

Figure 9-21: The unnumbered list uses bullets for each item. So much to do!

Color chart

Figure 9-22 shows the color chart for assigning various colors in the color attributes of text, background, and other elements for your color screens. You actually have 256 colors on most handsets, but the way in which all of those colors are displayed varies substantially among handsets. The colors seem to give good and fairly consistent results regardless of handset model. If you are doing a graphic, I would highly recommend using colors from this palette wherever possible.

Black #000000	Gray #808080	Maroon #800000	Purple #800080
Green #008000	Olive #808000	Navy #000080	Teal #008080
Silver #C0C0C0	White #FFFFFF	Red #FF0000	Fuchsia #FF00FF
Lime #00FF00	Yellow #FFFF00	Blue #0000FF	Aqua #00FFFF

Figure 9-22: The basic i-mode color palette, with color names and hex codes.

Notes:

[2] Indicates attributes and tags are available only on the second-generation models, such as the 502i-series, 209i-series, and 821i-series.

[3] Indicates attributes and tags are available only on the third-generation models, such as the 503i-series or 210i-series.

Chapter 10

Playing Sounds in i-mode

THIS CHAPTER LOOKS AT USING SOUNDS in i-mode. You learn about the sound format, the tools for working with that format, and how to implement sounds in your application.

Sound Formats in i-mode

i-mode uses a proprietary format called Melody Format for i-mode, or MFi. The files created by this format are called i-melodies by NTT DoCoMo. It is a variation on the MIDI format, and is sometimes called the MLD format because that is its three-letter file extension.

DoCoMo is the only company using this format at this time. The format is not actually NTT DoCoMo's own format; it was developed instead by another company, originally for delivery of karaoke over dialup connections.

The format is extremely compact when compared to MIDI. For example, in Chapter 14, you build a game that uses sound. The original sounds were in MIDI format, and took up 237 bytes. The MFi sound, on the other hand, took up only 68 bytes. At about three-and-a-half times smaller, it is definitely a savings over MIDI in terms of memory.

MIDI is a format used by musicians and others to store musical information, in the form of what note is being played, on what instrument, for what duration, at what tempo, with what intonation, and so on. It is a rich format that is used in a way similar to the way Postscript is used by graphic artists — rather than painting each pixel, in the case of MIDI each second of a sound file represents enough information that a user's computer can reproduce the graphic or sound. This has tremendous space savings, and is naturally looked at when talking about downloading a certain kind of file.

Although MIDI is significantly smaller than Wave files or MP3 files, it is still bigger than it needs to be on a mobile phone. There are some reasons for this. MIDI is not only a format that is meant to be played back, but it is also mixed with other sounds by a studio engineer. To do this without a lot of trouble, headers are used on all parts of the data to identify which part the data belongs to. MFi is purely a playback format, and does not need the same amount of identifying data. On most current computer sound cards, MIDI is supported as a matter of course. This is done using a processor that converts the MIDI data into Wave data and then outputs it. Similarly, all current NTT DoCoMo mobile phones have an MFi chip. Many of these sound chips (although not all) are made by Yamaha.

Despite using the same MFi chip, however, the sounds on different models of phones are reproduced quite differently because the software and settings that control the chip are programmed by each handset manufacturer. You end up in the curious situation in which Yamaha, who makes the chips, has to maintain different versions of the ringing tones on its ringing tone service for each model of handset. This is something that Yamaha can do, but we cannot.

There are several utilities publicly available that emulate how actual phones reproduce sound. They are in Japanese, only, however, which makes them somewhat inaccessible to the non-Japanese developer.

Creating MFi files

Although there are no emulators except the Japanese ones, there are tools you can use to create MFi files. Most people, including Yamaha, first create their sounds in whichever format is most comfortable and then use a conversion (or in the case of Yamaha, different conversions for each model of handset) to MFi. The only really foolproof method that is compatible is to create the file in MIDI.

Professional and amateur musicians alike have come to use the MIDI standard to connect electronic instruments to computers. My inexpensive consumer-oriented keyboard has a MIDI port, and MIDI controllers are about $80 for an inexpensive one. The MIDI controller sits between a MIDI device and a computer, and most also come packaged with software allowing you to edit your sounds and perform effects on them. The most common MIDI input method, by far, is certainly the keyboard, but there are many other instruments that are also capable of this. Most are rather expensive, however, and out of the price range of peasants like us (at least until our first IPO).

You can do the same thing with a variety of software applications, including the Music MasterWorks program. Download it from www.musicmasterworks.com. Figure 10-1 shows what this program looks like. These applications generally imitate either a piano or (in the case of Music MasterWorks) a staff upon which you can place notes. These are far less expensive than a keyboard and MIDI controller, and for your purposes, they are just as good.

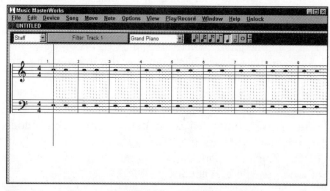

Figure 10-1: The opening screen on MusicMasterWorks.

Using this or another similar utility, follow these steps:

1. Play around with the sound utility or record your own tones, and see if you can get something that sounds good. Figure 10-2 shows my own feeble attempt at musicality.

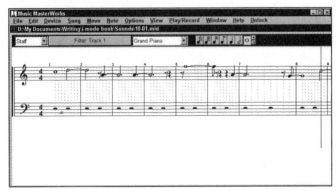

Figure 10-2: After you have added notes to the staff, save the file as a MIDI file.

2. Save this file as a MIDI file.
3. Now, you need to use a conversion program to convert this to MFi. The only one I can find that does this without much trouble and without much Japanese involved is called Chaku Melo Convertor. Download it from `http://hp.vector.co.jp/authors/VA005084/503i.lzh`. When you decompress it, it goes into a folder called 503I, as you can see in Figure 10-3.

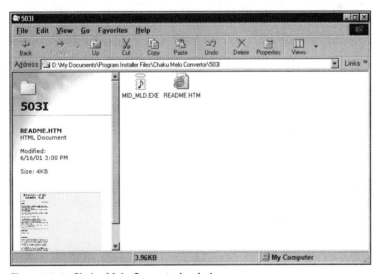

Figure 10-3: Chaku Melo Convertor's window.

4. Select the file to convert.

5. You can either double-click on the MID_MLD.EXE icon, or drop a MIDI file on top of the icon. Double-click to bring up the window.

6. Select your file, as shown in Figure 10-4.

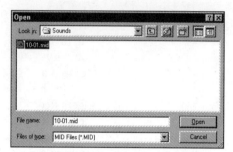

Figure 10-4: Select a MIDI file to convert.

Opening the file automatically creates an MFi file.

After you press the Open button, the message shown in Figure 10-5 comes up.

Figure 10-5: A message tells you that if you want to send a mail with this melody, you can simply paste from the Clipboard to your mail software, and the melody will be sent.

The new file is created.

There should be a new file of the same name, but with the .mld extension in the same folder the original was in. Nothing comes up to tell you this, and the little program doesn't really linger around after it is done. But it does do its job quite well.

7. Upload your MFi file to a server.

You need to upload the resulting MFi file to a Web server, or put it in the resource folder if you plan to use it in an i-Appli. To allow users to download the ringing tone from your Web server, they only need to know the URL to it, or to be able to click a link to that URL, and voila! The downloaded ringing sound is saved to their phone's memory automatically.

A couple of things to mention: The last message about saving the file on the Clipboard for mail is important. There are actually two formats for sounds; one is called i-melody, and one is called mail melody. When talking about e-mailing melodies, that is mail melody. Mail melodies are contained in the body of the e-mail, so they cannot be in binary format. Melodies in the binary format can be downloaded, but not sent to your friends: It is i-mail melodies that are sent.

DoCoMo's decision to separate the two kinds of ringing tones has to do with copyright protection. JASRAC, the Japan Association for Singers Recording Artists and Composers, charges fairly low royalties on the use of the intellectual property of its members for use in ring tones. If users were able to send these copyrighted songs to friends, copying would be rampant, and it would have to charge more to those who did purchase the songs. So files with .mld on the end go through the HTTP pipe, and they are put directly into an area of the system for ringing tones. There is no way to get them out of there once they are in, except to erase them. Mail, on the other hand, goes into the user's mailbox, and stays there until it is forwarded to someone else or deleted.

Other ways of creating music

The method I used in the previous section to create a sound file for i-mode is a basic one that didn't depend very much on language, hardware, or learning a new description language.

There is a format, which was proposed by a South African, called Music Markup Language, or MML. This is not to be confused with NTT DoCoMo competitor J-Phone's own markup language, Mobile Markup Language, also abbreviated MML.

You can find the specifications for it at www.mmlxml.org/.

This is basically a language to textually tell a sound browser what to produce, in the same way that HTML tells a visual browser what to draw to the page. I have not heard of many uses of this outside of mobile phone ringing tones, but it has become a sort of intermediary language between various ringing tone formats. The Chaku Melo Convertor converted only between MIDI and MLD formats, so it probably used a straight conversion. But what if you wanted to convert from MLD to SMAF, Yamaha's new standard? There are programs, in Japanese, which do all sorts of conversions by turning the source into MML and then converting from MML to the target format. This way, each format needs only two conversions.

The problem with going very far into creating ring tones using MML is that the tools are completely in Japanese and behave in a very buggy manner when installed on an English system, even if you are using NJWin.

For more information, if you read Japanese, go to google.co.jp and look up MLD and chaku (in kanji) and melo (in katakana); or if you don't read Japanese, go to my Web site: http://sound.eimode.com/.

Again, if you are going to be developing for the Japanese market, you will probably need to install a Japanese version of Windows in order to be able to use all of the development tools properly. Our simple creation and conversion above does the job in the meantime.

Chapter 11

Programming in cHTML: A Tutorial

IN THIS CHAPTER, YOU CREATE WEB PAGES in cHTML, the variation on HTML used on i-mode mobile phones. The first project is to take a current Web page and to pare down the essential elements of the page to fit on a small mobile phone screen. The second project is to design a new project from the ground up.

Modifying an Existing HTML Document for i-mode

The eimode Web site, of which I am Webmaster, has a problem: Despite proclaiming to offer all the i-mode news in English that anyone needs, it doesn't actually possess an i-mode version. Shame! It is up to you in this section to build the new eimode site, which will be viewable on i-mode phones. Figure 11-1 shows the Web page you want to adapt.

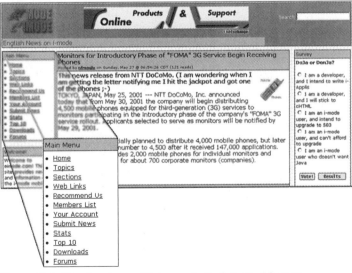

Figure 11-1: The eimode.com Web page in its full-sized format.

Figure 11-1 is a PHP document, not an HTML one. This is a matter of semantics because what comes out for the Web browser to see is pure HTML, and that is what you will work with. This site uses a very good free news and information content management package called phpNuke, which produces pages with a wide variety of database-driven content that is put in easy-to-use modules at user-specified places on the page. The HTML code for this page is on the CD, sample code 11-1.

The HTML for this page poses problems when porting this site to cHTML, though, because the preceding page relies heavily on tables, which are unavailable in cHTML. Aside from that, there is simply too much material for one page designed to be viewed on a mobile phone browser. So you look at the page, and alas! Something pops out at you—the index menu on the top left. This menu is just about right for a mobile phone. Listing 11-1 shows the code, taken from the whole page, for that part of the page.

Listing 11-1: The code for what is shown in Figure 11-1. On the CD ROM, the full text of the page is listed.

```
<table border=0 cellspacing=0 cellpadding=2 width=100%>
<tr><td valign=top width=150>
<table border=0 cellspacing=0 cellpadding=0 width=100% bgcolor=000000>
<tr><td>
<table width=100% border=0 cellspacing=1 cellpadding=3>
<tr><td colspan=1 bgcolor=CCCCCC>
<font size=2>Main Menu</td></tr><tr><td bgcolor=FFFFFF>
<font size=2><li><a href=index.php>Home</a>
<br>
<li><a href=topics.php>Topics</a>
<br>
<li><a href=sections.php>Sections</a>
<br>
<li><a href=links.php>Web Links</a>
<br>
<li><a href=friend.php>Recommend Us</a>
<br>
<li><a href=memberslist.php>Members List</a>
<br>
<li><a href=user.php>Your Account</a>
<br>
<li><a href=submit.php>Submit News</a>
<br>
<li><a href=stats.php>Stats</a>
<br>
<li><a href=top.php>Top 10</a>
<br>
```

```
<li><a href=download.php>Downloads</a>
<br>
<li><a href=forum.php>Forums</a>
</td></tr></table></td></tr></table><br>
```

Removing unneeded tags

The code in Listing 11-1 is basically a table containing a list of links. The table is used to frame the list to add borders and to position the header and elements, rather than to format the actual list. That is done using list tags. Because those are also available in cHTML, try the easy thing, which is to simply cut the list part of this code out and put it in a new document, shown in Listing 11-2.

Listing 11-2: The same index list, without the table elements

```
<html>
<head><title>eimode Main</title></head>
<body>
<li><a href=topics.php>Topics</a>
<br>
<li><a href=sections.php>Sections</a>
<br>
<li><a href=links.php>Web Links</a>
<br>
<li><a href=friend.php>Recommend Us</a>
<br>
<li><a href=memberslist.php>Members List</a>
<br>
<li><a href=user.php>Your Account</a>
<br>
<li><a href=submit.php>Submit News</a>
<br>
<li><a href=stats.php>Stats</a>
<br>
<li><a href=top.php>Top 10</a>
<br>
<li><a href=download.php>Downloads</a>
<br>
<li><a href=forum.php>Forums</a>
</body></html>
```

Now, there are a few things about this that just don't look right. First, of course, is that there are list items, but no list! You have two choices, an unnumbered list, as in the original, or a numbered list. I choose to go with the numbered list. I will explain that choice in a minute. There are also unneeded `<br>` tags. In i-mode, and actually in most other implementations of HTML I have seen, the `<br>` tags are repetitive when used in a list. I am not actually sure why they are there, but get rid of them.

And, finally, why did I choose a numbered list? So that you can use the `accesskey` attribute of `<a>`. There is a minor problem with this, though, which is that there are more than ten items. If all the items were necessary, this could be gotten around, but this is actually a good time to look at what is actually in the list and whether that is appropriate for the mobile phone platform. Stats and downloads are also not particularly helpful on this platform, though you could do what many sites do to attract users: offer ringing tone downloads. Maybe later. You can also rearrange things a little to reflect their estimated order of importance to a person accessing the site on a mobile phone. Listing 11-3 shows the code for the simpler i-mode menu shown in Figure 11-2.

Listing 11-3: The list is reduced and re-ordered to fit the mobile phone platform

```
<html>
<head><title>eimode Main</title></head>
<body>
<ol>
<li><a href=topics.php accesskey="1">Topics</a>
<li><a href=sections.php accesskey="2">Sections</a>
<li><a href=forum.php accesskey="3">Forums</a>
<li><a href=links.php accesskey="4">Web Links</a>
<li><a href=submit.php accesskey="5">Submit News</a>
<li><a href=top.php accesskey="6">Top 10</a>
<li><a href=user.php accesskey="7">Your Account</a>
<li><a href=friend.php accesskey="8">Recommend Us</a>
<li><a href=memberslist.php accesskey="9">Members List</a>
</ol>
</body></html>
```

Figure 11-2: The results of the pared-down menu.

Trimming graphics to size

Now, notice that something seems to be lacking in this picture, and that is the picture. Who are you? No, this is not a philosophical question, but one that a user would be asking after viewing the page with the simplified menu in Figure 11-2. You need to let them know who you are.

To do this using as little memory as possible, which was discussed in Chapter 8, it is perfectly acceptable to use simple text, or text with emoji symbols. But it turns out that eimode is a registered trademark in the U.S., and you wish to use that trademark.

So, you need a graphic that looks like the one on the full-sized Web page, but in a wee version that won't take up lots of memory or take too long to download. My tool of choice is Photoshop, but actually that is probably overkill in this case. The original graphic was not very big, only about 5k. There are a couple of things about it, though, that don't make it ideal in its present form. Drop shadows, although they look nice on a Web page, just add clutter to an image on a small screen. The actual dimensions of the image, too, are obviously too big for a mobile phone screen.

Luckily, the original image was created in Adobe Illustrator, and removing the drop shadows is quite easy, as is changing the size. Then, when I go to export it, I realize that if it is 100 pixels wide, which is about right for a mobile phone screen, it will be 50 pixels tall. Now, there are many sites that use a startup graphic that takes up the whole first screen and requires users to scroll down to get to the menu. Fair enough, but I want users to do as little scrolling as possible, so I decide to arrange the two typographic elements side by side, rather than on top of one another. By making the graphic more wide than tall, I save space: The 100 pixel graphic goes from being 50 pixels tall to 13 pixels tall.

The resulting image is tiny, about 300 bytes. There is only one good way to test it, and that is to put it into the page and have a look. The resulting code looks like Listing 11-4:

Listing 11-4: The index page with the gif logo

```
<html>
<head><title>eimode Main</title></head>
<body>
<img src="11-3.gif" width="100%"height="13">
<ol>
<li><a href=topics.php accesskey="1">Topics</a>
<li><a href=sections.php accesskey="2">Sections</a>
<li><a href=forum.php accesskey="3">Forums</a>
<li><a href=links.php accesskey="4">Web Links</a>
<li><a href=submit.php accesskey="5">Submit News</a>
<li><a href=top.php accesskey="6">Top 10</a>
<li><a href=user.php accesskey="7">Your Account</a>
<li><a href=friend.php accesskey="8">Recommend Us</a>
<li><a href=memberslist.php accesskey="9">Members List</a>
</ol>
</body></html>
```

And the results look like those shown in Figure 11-3.

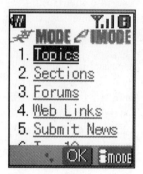

Figure 11-3: The page with the logo added.

Great! It looks good, and the colors (which you can't see, but are in the original) seem to look all right. Everything is hunky dory. Not quite. I used x-9's i-MIMIC emulator for Figure 11-3. It is quite good at emulating the actions of the phone, but uses Internet Explorer's color palette to display colors. In that palette, my image looks fine. Unfortunately, the actual phones do not use this palette, and the light blue I used in the center of the image is nearly completely washed out when I try it on my handset. The red, which is the part on the outside letters, is sketchy. Obviously, I have made the two cardinal sins of doing graphics for mobile phones — using colors that are too light to clearly show up, and using non-primary colors. Here is what I need to do:

1. Chapter 9 lists the primary colors and their codes. Choose two that are not too light. I choose a pure red (#FF0000) and a pure blue (#0000FF).

2. It's important that when I export the image from Illustrator I turn anti-aliasing off, and that when I resize it I use Nearest Neighbor rather than Bicubic, which creates new colors. Dithered colors look very bad on mobile phones.

3. My resulting graphic has only three colors: red, blue, and white. I could use a transparent GIF, which would eliminate the need for white, but the older handsets don't support transparency, and I want the image to be as compatible as I can.

4. I end up cheating a little because the resulting graphic has lost some definition. I went in using a paint program, and used the eraser tool to clean up the look of the image a little bit. Figure 11-4 shows what it looked like on the emulator, though the whole point of this step was to get it looking right on the real thing.

Figure 11-4: The colors are fixed, and definition is added to the logo.

Listing 11-5 shows the code including the new graphic.

Listing 11-5: The only difference between this code and Listing 11-4 is the new graphic

```
<html>
<head><title>eimode Main</title></head>
<body>
<img src="11-4.gif" width="100%" height="13">
<ol>
<li><a href=topics.php accesskey="1">Topics</a>
<li><a href=sections.php accesskey="2">Sections</a>
<li><a href=forum.php accesskey="3">Forums</a>
<li><a href=links.php accesskey="4">Web Links</a>
<li><a href=submit.php accesskey="5">Submit News</a>
<li><a href=top.php accesskey="6">Top 10</a>
<li><a href=user.php accesskey="7">Your Account</a>
<li><a href=friend.php accesskey="8">Recommend Us</a>
<li><a href=memberslist.php accesskey="9">Members List</a>
</ol>
</body></html>
```

Setting text-wrap options

Now, this page looks okay, but I want to make it a little more organized. I will use horizontal rules to do this. These are "free" in terms of memory, and help break a page into logical units. I want to break this page into three basic units: the heading, the main features of the site, and the maintenance and administrative features of the site.

Chapter 9 discussed using emoji symbol characters to tell a user that an `accesskey` was available. This would be a good time to think about that. It requires getting rid of our list and replacing it with simple line breaks, but it adds a more polished look while making it clearer to users. I also want to include a copyright notice. Listing 11-6 shows our new code; and Figures 11-5, 11-6, and 11-7 show the three screens that result.

Listing 11-6: The index with emoji symbol numbers and accesskey attribute added

```
<html>
<head><title>eimode Main</title></head>
<body>
<img src="11-4.gif" width="100%" height="13">
<hr>
&#63879;<a href=topics.php accesskey="1"> Topics</a><br>
&#63880;<a href=sections.php accesskey="2"> Sections</a><br>
&#63881;<a href=forum.php accesskey="3"> Forums</a><br>
&#63882;<a href=links.php accesskey="4"> Web Links</a><br>
&#63883;<a href=submit.php accesskey="5"> Submit News</a><br>
&#63884;<a href=top.php accesskey="6"> Top 10</a><br>
<hr width="80%">
&#63885;<a href=user.php accesskey="7"> Your Account</a><br>
&#63886;<a href=friend.php accesskey="8"> Recommend Us</a><br>
&#63887;<a href=memberslist.php accesskey="9"> Members List</a><p>
<div align="center">&copy;2001,
   Llama<br>Communications,<br>L.L.C.</div>
</body></html>
```

Figure 11-5: The first page of the finished index.

Figure 11-6: The second page of the finished index.

Figure 11-7: The last page of the finished index.

And that is that. This menu is aesthetically acceptable, and obviously a part of eimode's main site, easily navigable using access keys, and manageable from a user's standpoint.

But what of the pages linked to? Those sites have not changed at all, and the information displayed on them has not been formatted, nor has any consideration been taken of viewing on a mobile phone. Because there are nine menu items linked to this page, each of those nine pages would have to be formatted for viewing on a mobile phone. That is overkill for a simple tutorial, so take just one of the pages. You will notice that all of the links in the above examples are to pages that end with the php extension. These pages are written in PHP. All of my content is stored in a MySQL database, and I need some way to get it out of the database. PHP is the simplest way to do this. The beauty of having the information in the database is that both the i-mode page and the normal page have access to the same information simply by accessing the database. And yet, they can display it quite differently to fit the situation. This is powerful.

Listing 11-7 shows the text of the original PHP script used on the full site.

Listing 11-7: The topic page of the portal site

```php
<?PHP

/*************************************************************************/
/* PHP-NUKE: Web Portal System                                           */
/* ===========================                                           */
/*                                                                       */
/* Copyright (c) 2001 by Francisco Burzi (fburzi@ncc.org.ve)             */
/* http://phpnuke.org                                                    */
/*                                                                       */
/* This program is free software. You can redistribute it and/or modify  */
/* it under the terms of the GNU General Public License as published by  */
/* the Free Software Foundation; either version 2 of the License.        */
/*************************************************************************/

if (!IsSet($mainfile)) { include ('mainfile.php'); }

        include("header.php");
        $result = mysql_query("select topicid, topicname, topicimage, topictext
                from topics order by topicname");
        if (mysql_num_rows($result)==0) {
            echo "<table border=0 bgcolor=000000 cellpadding=2 cellspacing=0
                width=95%>
            <tr><td>
            <table border=0 bgcolor=FFFFFF cellpadding=1 cellspacing=0
                width=100%>";
        }
        if (mysql_num_rows($result) > 0) {

        OpenTable();
        echo "
        <font size=3><b><center>".translate("Current Active
                Topics")."</b><br>".translate("Click to list all articles
                in this topic")."</center><br>
        <center><table border=0 width=100% align=center cellpadding=2><tr>";
            while(list($topicid, $topicname, $topicimage, $topictext) =
                mysql_fetch_array($result)) {
?>
            <td align=center>
            <?php echo "<a href=search.php?query=&topic=$topicid>"; ?><img
                src=<?php echo "$tipath$topicimage"; ?> border=0></a><br>
            <font size=2><b><?php echo "$topictext"; ?>
            </td>
            <?php
// Thanks to John Hoffmann from softlinux.org for the next 5 lines ;)
```

```
            $count++;
            if ($count == 5) {
                echo "</tr></tr>";
                $count = 0;
            }
        }
        echo "</tr></table>";
    }
    CloseTable();
    mysql_free_result($result);
    include("footer.php");
?>
```

Now, the key part of this that you want to have on the page is a simple list of topics. You now get a distinct feeling for the time it is going to take you to make your site completely mobile-friendly. There is still one level beneath this level, using the search engine, which will also have to be pared down to work with mobile phones. You stop with this project after you have finished getting the page above into proper shape, however.

Adding access keys and other touches

Looking at the code in Listing 11-7, you see some `include` commands for `footer.php` and `header.php`. Because they are used to include top and bottom stuff that you don't really need on a mobile phone (such as Link Exchange ads, and so on), you need to get rid of them. You also need to cut out all the table formatting because that is also not needed. Further, you need to cut out other sorts of text formatting because it is not compatible with most phones. You also notice that there aren't any starting or ending tags because they are included in the header and footer includes. Because you will get rid of those, you also need to put in the usual HTML tags at the start and finish.

There is one thing you need to include: the `mainfile.php`. It includes necessary functions for opening the database, and also connects to `config.php`, which contains all the necessary environmental variables for the implementation of this package and for the system. Both of these elements are necessary if you are going to maintain compatibility with the data used by the main Web page. In Listing 11-8, you can see the source for this drastically pared-down version.

Listing 11-8: The topic page, with all of the unnecessary code removed. This is topic.php on the CD ROM.

```
<html>
<head><title>eimode Topics</title></head>
<body>
<?PHP
```

Continued

Listing 11-8 *(Continued)*

```
/**********************************************************************/
/* PHP-NUKE: Web Portal System                                        */
/* ============================                                       */
/*                                                                    */
/* Copyright (c) 2001 by Francisco Burzi (fburzi@ncc.org.ve)          */
/* http://phpnuke.org                                                 */
/* Modifications to original made by Nik Frengle on June 16th, 2001   */
/* This program is free software. You can redistribute it and/or modify */
/* it under the terms of the GNU General Public License as published by */
/* the Free Software Foundation; either version 2 of the License.     */
/**********************************************************************/

include ('mainfile.php');
$result = mysql_query("select topicid, topicname, topicimage, topictext from
    topics order by topicname");
if (mysql_num_rows($result) > 0) {
echo "Current Active Topics<br>Click to list all articles in this topic.<br>";
while(list($topicid, $topicname, $topicimage, $topictext) =
    mysql_fetch_array($result)) {
echo "<a href='search.php?query=&topic=$topicid'>";
 echo "$topictext</a><br>";
                }
        }
        mysql_free_result($result);
?>
</body>
</html>
```

This page turns out the still quite ugly screen shown in Figure 11-8.

Figure 11-8: The pared-down topics page.

Up until this point in the examples, I have used simple `<br>` tags and counted every line of text to figure out where to put them. But because you are asking a database for information, you don't know exactly what you will get beforehand,

and formatting the data beforehand is impossible. If computers weren't made for this kind of repetitive work, lord knows what they are for. As it turns out, there is a very good and simple PHP script that I can include called `textwrap.php`. The script, after it is included, is called by

`"textwrap($text,16)"`

if, for example, you want the text wrapped at 16 characters.

You also want to add indicators of `accesskeys`, as well as the `accesskey` attribute. Adding the `accesskey` emoji symbol characters adds three characters in width, so you want to reduce the text wrap to 13. If the results are really long strings, you might want to think of another way to do it because the text wraps at 13 for not only the first line, but all subsequent lines. As it is, this is a bit difficult, and because the topic names are fairly short, I chose to be lazy. But if there are more than nine topics, you can't use either the `accesskey` or the emoji representing them, so you put blanks in after the number of topics reaches ten.

You also want the bar across the top (below the logo), and you want the copyright at the bottom. The astute reader will cotton on that you should really do this by using `include` files for header and footer, as was done in the script before you altered it for i-mode. Such a header would include the logo, and the footer would include the copyright notice. Because I am trying less to show you how to master PHP than how to format your pages for i-mode, I have tried to keep this relatively self-contained and simple (see Listing 11-9).

Listing 11-9: The topics page with logo, accesskeys, text wrapping, and other enhancements. This is listed as i-topics.php on the CD-ROM.

```
<html>
<head><title>eimode Topics</title></head>
<body>
<img src="ilogo.gif" width="100%" height="13">
<hr>
<?PHP


/********************************************************************/
/* PHP-NUKE: Web Portal System                                      */
/* ===========================                                      */
/*                                                                  */
/* Copyright (c) 2001 by Francisco Burzi (fburzi@ncc.org.ve)        */
/* http://phpnuke.org                                               */
/* Modifications to original made by Nik Frengle on June 16th, 2001 */
/* This program is free software. You can redistribute it and/or modify */
/* it under the terms of the GNU General Public License as published by */
/* the Free Software Foundation; either version 2 of the License.   */
/********************************************************************/
```

Continued

Listing 11-9 *(Continued)*

```
include ('mainfile.php');
include ('textwrap.php');
$result = mysql_query("select topicid, topicname, topicimage, topictext from
    topics order by topicname");
if (mysql_num_rows($result) > 0) {
echo "Click for all<br> topic articles<br>";
while(list($topicid, $topicname, $topicimage, $topictext) =
    mysql_fetch_array($result)) {
if ($count<10) {
$softkey=$count+63879;
$accesskey=$count+1;
}
else {
$softekey="";
$accesskey="";
}
echo "&#$softkey<a href='search.php?query=&topic=$topicid'
    accesskey='$accesskey'>";
$topictext=textwrap($topictext,13);
echo "$topictext</a><br>";
$count++;
            }
        }
        mysql_free_result($result);
?>
<br><div align="center">&copy; 2001, Llama
<br>Communications,<br>L.L.C.</div>
</body>
</html>
```

This code comes out looking like Figures 11-9, 11-10, and 11-11.

Figure 11-9: The first page. The text wrap script has done its job, and the logo with <hr> below looks very snazzy.

Chapter 11: Programming in cHTML: A Tutorial

Figure 11-10: You can see in option 4 that the text wrapping worked just right.

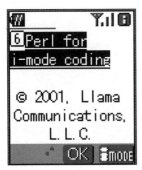

Figure 11-11: The last page, with your footer.

Lessons learned

What have you learned in this small project, besides that it isn't as easy as it sounds? Well, I can think of a few important things you have learned:

- To be very careful of your choice of colors in graphics, and to avoid anti-aliasing at all costs.
- To really consider what will be useful and readable on a mobile phone.
- To really use the tools you have, such as the `textwrap` script and the capability of PHP to put access keys on the first nine selections and to generate the special characters to make a user's experience as easy as you can. There are specific considerations in the case of mobile phone users.

Creating an Online Address Book

Adapting an existing site to the realities of the mobile Internet is one thing, and planning a new site or service from the ground up is something else entirely. One of the things that you want to do before you begin is to think about the things that are going to be useful in a situation in which you may only have your mobile phone and no other device. This makes a mobile phone the perfect device for productivity applications like an address book. In this section, we will develop an address book application that stores its data on a central server, and can be accessed from a mobile phone.

Defining the project

So, assume that you are using those tools that you have been getting comfortable with up until this point — MySQL database, PHP, and cHTML. In this case, the things you need to do to make your goal a reality are fairly straightforward:

- Get the data into a database.
- Get the data out from a mobile phone.
- Make sure those doing the input and output are the ones the data belongs to.

The first objective, getting the data into the database, actually looks to be more difficult than getting it out. Let us look at the problem logically: If, for example, the data is on your PC in Outlook. somehow, you need to upload that data into a Web-connected MySQL database.

Because this tutorial is really about developing for the mobile client, you are going to cheat a little. Assume that you considered all of the above, and decided for the time being to use only a very simple interface, through the mobile phone, to input data. Yes, this is redundant and not exactly ideal. You are going to do something, though, that is an important part of this whole process: You are going to acknowledge the parts that will be most difficult, and put those aside for now. After you have a functioning application, you can make these kinds of improvements much more easily than if you had tried to do everything at once.

So, you have two different needs: one to get the data into the database and one to get it out. You also want to verify who a user is, so that the address book data cannot be used by another user. So, you have two types of data: user data and their address-book data. For this, you use two different tables in our database. We touched briefly on tables before, when you created a user-verification table.

Chapter 8 covered building a user table, with the example allowing users to log in to get their horoscopes.

Creating the data structure

Tables are the storage bins of a database, each with a designated place for specific types of information for that table. In this instance the table would be the collection of names and phone numbers. Each person in the table represents one record and each piece of information in this record, such as a name or phone number, is a field. You can specify the kinds of information, such as text or numbers, that you put into a field. This description fits a very simple version of a database, but it serves our purposes.

Because I hate doing something and not getting any more mileage out of it again, you will use the SQL table and PHP script developed in Chapter 8 to input user data and verify user identity. One thing that I mentioned, but did not go deeply into in Chapter 8, was that on older phones the utn attribute, which passes a phone's unique 11-digit serial number in the user-agent head, does not work. You want to be backwards-compatible with these folks, if for no other reason than a desire to get as much traffic as you can. So, you modify the script to ask for a user's telephone number if it is an older phone, which also happens to be 11 digits, and a password. Because you are assuming the security of a phone's serial number, in the same way that NTT DoCoMo does, you don't need a password from those users. But anyone could access a user's address book if they knew the phone number, which is unacceptable, so you add the password for these users. This is something you will need to add to the SQL table as well: a password field.

Start this way, with the structure of the data that you plan on using. In terms of creating this structure in your SQL database, you do this (as in Chapter 8) by using the Structured Query Language (SQL). The user data and address data are really separate kinds of data, and should be put in different tables. To create these tables, you would use the SQL shown in Listing 11-10.

Listing 11-10: Creating the SQL tables

```
CREATE TABLE users (
   uid int(11) NOT NULL,
   password varchar (20)NOT NULL,
   phone int(11) NOT NULL,
   name varchar(30) DEFAULT '' NOT NULL,
   email varchar(30) DEFAULT '' NOT NULL,
PRIMARY KEY (uid)
);
CREATE TABLE addresses (
   aid int AUTO_INCREMENT NOT NULL PRIMARY KEY,
   uid varchar(11) NOT NULL,
   fname varchar(20)NOT NULL,
   lname varchar(20),
   homestreet1 varchar (40),
   homestreet2 varchar (40),
```

Continued

Listing 11-10 *(Continued)*

```
    homecity varchar(30),
    homestate varchar(30),
    homepost varchar(12),
    homephone varchar (14),
    homefax varchar (14),
    homeemail varchar (30),
    companystreet1 varchar (40),
    companystreet2 varchar (40),
    companycity varchar(30),
    companystate varchar(30),
    companypost varchar(12),
    companyphone varchar (14),
    companyfax varchar (14),
    companyemail varchar (30),
    mobilephone varchar (14),
    mobileemail varchar (30)
);
```

Looking at the user table, you can see that you have eliminated the birthday and horoscope sign field from it, and added a field for a password and for a user's mobile phone number. In the case of users of older phones, this is a bit repetitive because you use their phone number as the `uid` field. You also want to make sure you have the numbers of those using newer phones, and so you need a separate field. You also notice that I have made the password field `NOT NULL`. Although you won't ask for the password in the case of users whose phones support the `utn` attribute, and send their handset's serial number, if you extend the service in future to be accessible from a PC, these users also need a password. It's better to anticipate this eventuality than to have to ask users to do something different in the future.

Looking at the addresses table, you can see that you have two fields at the top that may need a little explanation. Each record needs to have a unique primary field. This is the field used to identify a record. Because two users may have exactly the same person's address in their address books, and because you are storing all addresses of all users in the same table, you need a unique identifier. `AUTO_INCREMENT` tells MySQL to automatically put a number one bigger than the last number in this field. For the first record, the number is 0. I also have a field called `uid`. This field is needed to identify whom the data belongs to. Because records are input sequentially into the table, a particular user does not have his own "area" of the table, and so you need to identify whose records are whose. You use the `uid` to do this. Note that you could actually create a new table for each user and then all records in that table would belong to them. The problem with this is that my virtual hosting provider does not allow unlimited tables, and that this takes up more memory, which could be a *lot* more if the number of users reached 100,000 or even a million!

The rest of the fields in the table should be fairly obvious. I have defined phone numbers as the `INT` type, which means no dashes or parentheses. This is so that they will be compatible with the `phoneto` attribute, and a user will be able to simply click on the number and dial that number.

Directing the flow of user interaction

So, you have the structure, you know what sorts of data you will need to input and be able to output, and all that is left to do is to actually do both input and output. In Chapter 8, I used separate pages to input information and to process that information. You keep to that discipline, but add a step: When a user accesses the index page, you check two things: what his handset's serial number is, and what IP address he is accessing the site from. NTT DoCoMo uses certain IP numbers, and it is actually the combination of the serial number and the IP number that is somewhat secure. Here is the logic of your first page:

- If a user's browser matches, but the IP number doesn't come from NTT DoCoMo, you know this is probably someone using an emulator. You are terribly suspicious, and you send her to your regular Web page.
- If a user accesses the page from a normal Web browser, it takes her to your normal Web page.
- If the IP matches, but there is no information on the model, it is someone using an older or newer model phone, so you take her to the login/signup page that doesn't use the `utn` attribute to get their phone's serial number.
- If the IP matches, but it's a model that cannot use `utn`, you send her to a login/signup page.
- Finally, if the IP is okay, and it is a model that can use `utn`, you take a user to the `utn` signup and login page.

This can fairly easily be done using PHP and server-side redirects, as shown in Listing 11-11.

Listing 11-11: The index.php entry page in our application

```
<?php
$agent=getenv("http_user_agent");
$remoteAddress=getenv("remote_addr");
if (ereg("210.153.84.",$remoteAddress)||ereg("210.136.161.",$remoteAddress))
{
$user="legal";
}
else {
    $user="illegal";
```

Continued

Listing 11-11 *(Continued)*

```
   }
$agentArray= split("/",$agent);
$connection = mysql_connect("localhost", "us12425a", "universal");
mysql_select_db ("db12425a");
if ($agentArray[2]){
$result=mysql_query("SELECT * FROM handsets WHERE model='$agentArray[2]'");
}
if ($agentArray[0]!="DoCoMo"||!$result||$user=="illegal")
{  header("Location: http://www.eimode.com/");
}
elseif ($user=="legal" && mysql_result($result,0,utn)=="Y")
{
header("Location:http://www.eimode.com/address/utnlogin.html");
}
else {
header("Location:http://www.eimode.com/address/login.html");
  }/
?>
```

All right, all right, I admit that wasn't exactly a simple script, but neither was it too terribly difficult. The next section briefly describes what it did, so you can understand what is going on.

First, you want to know who is trying to access our site. You ask for two environmental variables: the `http_user_agent` and the `remote_addr`. The first variable tells you what sort of browser someone is using, and whether her phone supports sending her serial number or not. The second one tells you her IP number. Because NTT DoCoMo requests all go through their IP servers, you can tell whether someone is accessing the site from an i-mode phone or not. This may be more security than you really want because someone using a Web-enabled phone from another company may want to access your site, too. You can easily turn this off by commenting out all of this bit except the `$user="legal";` line. This is the sort of authentication used by official sites that use DoCoMo's billing system, so that they can make sure the customers are subscribed and are really who they say they are. In this case, you are using this because you want your users to feel quite safe in storing their information on your site. Note that in the version of PHP included on the CD-ROM, Boolean values are possible, but on earlier versions of PHP they were not, which is why I assign text values instead.

Next, you split the `$agent` value into all four or five parts, and store them in an array called `$agentArray`. You only use the first value (`$agentArray[0]`), and the third value (`$ua[2]`).

After you have this array, you want to check the model against those in your database, so you first check to see whether there is any value in the `model` field of the array (`$ua[2]`). If there is, you open the `handsets` table in the MySQL database, and search the `model` field for the model of phone trying to access your site.

Next, you go through a series of `if` clauses to shoot your users to the appropriate pages. I am sure there is a more elegant way to do this, but that is one of the other beauties of PHP – inelegant code works pretty much as well as the prettier stuff.

That's it. So, now you have to create the two pages that our users who are actually using mobile phones will get sent to, the sign up/logon page, one for `utn`-enabled phones, the other for older phones. The page that users of browsers other than i-mode browsers get sent to is something I will leave to you. I have to admit one big drawback to all of this is that you can't use an emulator to view the site, because even if iMIMIC actually does send the correct user agent header, it is the wrong IP address.

Building the interface

Really, all of the work that you just finished is just a means of making sure that the right people get to the right page. That is very important, and helps to define the structure of your site.

Now, you need to build the interface that will enable a registered user to input his mobile phone number and password to gain entrance, and that will allow new users to register for the service.

All of your cHTML experience now comes into play, and you need to design an attractive and elegant first page. Remember, this is the page that people see when they first show up. This page gives a user his first impression of your site, and it either encourages him to take the time to input his information, or causes him to jump to another site. Most Japanese sites do this by packing all the needed information onto the first page, giving users a clear picture of what is there.

A couple of sites I have seen by non-Japanese, including the pretty well-known ImaDoko (Where are you now) service, use a graphical first screen to get a user's attention, and stash all of the cluttered text down below. One of the advantages, actually, to the slow speed of i-mode is that it usually takes 5–10 seconds for the first graphic to load, during which time the text contents are visible, even if the graphic pushes the text to the bottom (off the page) when it finally loads it. Using graphics in this manner can improve the look of your page while allowing a user to see the information she may need before the graphic has loaded. You need to make sure not to use the height or width attributes if you want this to happen.

For the non-`utn` pages, you use one-half page for the opening graphic and then use two buttons to take users to internal links to either log in or to input their new user information.

For the `utn`-enabled handset, you also use a half-page graphic, but only one of the links, the new user signup link, takes the user to another spot on the page. The login button, rather than taking him to a sign-in form (because this is done automatically), will have three buttons that take him to the lookup page, the data entry page, or the user management page.

Actually, these pages take him to the `login.php` page, which checks the information that either he or his phone gave. If it checks out, he gets redirected to the appropriate page. If it doesn't, he is either directed to sign up, or to try logging in again.

Listing 11-12 shows the basic cHTML page for non-utn phones.

Listing 11-12: The non-UTN-enabled login. Listed on CD-ROM as login.html.

```
<html>
<head><title>i Address Book</title></head>
<body>
<div align="center">
<img src="address.gif"><br>
Login/Signup<br>
<hr width="60%" size="4"></div>
&#63879; <a href="#login" accesskey="1">Login</a><br>
&#63880; <a href="#newuser" accesskey="2">New Users</a><br>
<br><br><br><br><br><br><br><br>
<a name="login">
<form action="login.php" method="post">
<div align="center">Login<br>
<hr width=60% size="2"></div>
Go:
<select name="gotoPage" size="1">
<option value="lookup.php" selected>Lookup
<option value="input.php">Input Addresses
<option value="user.php">User options
</select><br>
Mobile number:<br>
<input type="text" name="uid" size="11" maxlength="11" istyle="4"><br>
Password:<br>
<input type="text" name="password" size="11" maxlength="15" style="3">
<div align="center"><input type=submit value="Login"></div>
</form>
<br><br><br><br><br><br><br>
<a name="newuser">
<div algin="center">New User Sign-up<br>
<hr width=80% size="2"></div>
Name:<br>
<form action="signup.php" method="post">
<input type="text" name="name" size="14" maxlength="30" istyle="3"><br>
E-mail Address:<br>
<input type="text" name="email" size="14" maxlength="30" istyle="3"><br>
```

```
Mobile Number:<br>
<input type="text" name="phone" size="14" maxlength="11" istyle="4"><br>
Password:<br>
<input type="text" name="password" size="14" maxlength="20" istyle="3"><br>
<div align="center"><input type="submit" value="Sign Up"></div>
</form>
</body>
</html>
```

The output looks like the screen shown in Figure 11-12.

Figure 11-12: The Login/Signup screen.

The form in Figure 11-12 takes him to the login.php page, and takes with it the information he input, either the uid (in the form of the mobile number) and password, or in the case of a new user, all of his information.

I am feeling pretty pleased with myself. Figure 11-12 of this page couldn't be better: I get my very small logo at the top, my login button (which I anticipate being most often used) is automatically ready to push because it is the first link on the page, and my new user button is visible right below it. I have managed to get everything I want on the first screen, and still not have it look too cluttered. A user hitting either of the buttons is taken immediately to his screen because it is actually in the same HTML document, and therefore already loaded. It has the handy emoji buttons indicating the handy access keys.

This couldn't be too much better. Keep in mind that this is not an entertainment product, but rather a utility, so my standards are based on that. There is a whole methodology that big entertainment sites use, and a lot of it is based on customer psychology, which is, I have the feeling, probably somewhat different in Japan than in the U.S. I have to say, though, that as utilities go, this is definitely one of the better designed and implemented pages I have seen. That doesn't say much for the other guys, I guess.

Okay, you are going to do exactly the same thing for users logging in with utn-enabled handsets, except you are going to call it a menu instead because if these users are registered, there is no need to have them input their information, and you can offer them buttons taking them directly (or so they think) to their desired service. Actually, the links are to the `login.php` script, with information on what their final destination should be in the URL. If they are new users, you refer them to the new user login on the login page you just came from. This saves you from having to repeat exactly the same code as you did previously. To make it work correctly, you have to add the `utn` attribute to the `<form>` tag in `login.html`. Listing 11-13 contains the `utn`-enabled user's first page:

Listing 11-13: The menu and sign up page for UTN-enabled handsets. Called utnlogin.html on the CD ROM.

```
<html>
<head><title>i Address Book</title></head>
<body>
<div align="center">
<img src="address.gif"><br>
Menu<br>
<hr width="60%" size="4"></div>
&#63879; <a href="login.php?gotoPage=lookup.php" accesskey="1" utn>
Lookup</a><br>
&#63880; <a href="login.php?gotoPage=input.php" accesskey="2" utn>
Input Data</a><br>
&#63881; <a href="login.php?gotoPage=user.php" accesskey="3" utn>
User options</a><br>
&#63882;  <a href="login.html#newuser" accesskey="4">
New Users</a><br>
</body>
</html>
```

Listing 11-13 looks like the screen shown in Figure 11-13.

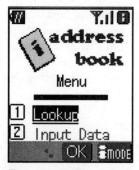

Figure 11-13: The utnlogin.html menu screen.

I think that the code in Listing 11-12 and the screen shown in Figure 11-13 make very clear what this page does, and because you have lots more to do, I won't go through it and explain everything. The only noteworthy thing, really, is the `utn` attributes in the `<a>` tags. So, duly note them and then continue.

Authenticating users

The next piece of code dwarfs all the other bits because it is the security gateway; the proverbial chastity belt used to preserve the virtue of your users. Basically, you want this login code to do the following:

- Check where users are logging on from. Only i-mode users are supposed to have gotten this far, but if they bookmarked the login page, they would have gotten around the first check.

- Check whether these users are using DoCoMo browsers. They should be, but if they aren't, you toss them to another page.

- Check whether the `utn` serial number is available. If it isn't, make sure a password is available. One or the other should have been made available by logging in. If they weren't, send the user to the login page.

- Check the user name against the database. If they aren't registered, send them to the user signup part of the `login.html` page.

- If they have a username, but their password doesn't match, send them back to the login page.

- If they have a matching username and password, send them and those two variables onto the page they want to go to.

These are all important things, and getting them all right can be a headache. Let's look at the product of my migraine (Listing 11-14).

Listing 11-14: The security gateway of our application. Called login.php on the CD ROM.

```
<?php
$utn=getenv("http_user_agent");
$ra=getenv("remote_addr");
if (ereg("210.153.84.",$ra)||ereg("210.136.161.",$ra))
{
$user="legal";
}
else {
    $user="illegal";
    }
$ua= split("/",$utn);
if ($ua[4]&&!$uid){
$uid=substr($ua[4],-11,11);
}
```

Continued

Listing 11-14 *(Continued)*

```
if ($ua[0]!="DoCoMo"||$user=="illegal")
{   header("Location: http://www.eimode.com/");
}
elseif (!$ua[4]&&!$password){
header("Location: http://www.eimode.com/address/");
}
elseif ($password){
    $connection = mysql_connect("localhost", "us12425b", "limited");
    mysql_select_db ("db12425a");
    $result=mysql_query("SELECT * FROM users WHERE uid='$uid'");
    if (mysql_numrows($result)!=0){
     $dbpw=mysql_result($result,0,password);
     $password=crypt($password,$dbpw);
     if ($password==$dbpw){
        session_start( );
        session_register("uid","password");
        header ("Location: http://eimode.com/address/$gotoPage");
        }
    }
    else
        {
            header ("Location: http://eimode.com/address/login.html#login");
        }
}
elseif ($ua[4] && !$password) {
$connection = mysql_connect("localhost", "us12425b", "limited");
    mysql_select_db ("db12425a");
    $result=mysql_query("SELECT * FROM users WHERE uid='$uid'");
if (mysql_numrows($result)!=0){
    $password = (mysql_result($result,0,password));
    session_start( );
    session_register("uid","password");
    header ("Location: http://eimode.com/address/$gotoPage");
}
else {
header ("Location: http://eimode.com/address/login.html#newuser");
}
}
?>
```

Believe it or not, this works! Explaining how it works, however, will be a challenge. Before you even start, though, I want to prepare my defense: From the standpoint of object-oriented programming, this is terrible. I don't declare any functions, include any `includes`, or really reuse any of my code as I should.

I agree, this is a quick and dirty solution, and in the real world of doing any kind of consulting, I would take a fair amount of time cleaning it up and commenting it. However, the quick-and-dirty solution does have a reason: I wanted to show a series of tests in one file, and wanted to give the reader a useful and clear idea of how PHP can work.

Basically, you do what you said you were going to do, which is to go through a series of tests, sending users to the appropriate places depending on the results of the test. The user, of course, wants to be taken directly to where he should be, and he will be taken there if he just plays by the rules. Below is the first test performed on a login, this one to test whether a user is logging on from a DoCoMo phone or not. If not, he is sent to our regular home page. The first tests eliminate people who logged on from outside of i-mode, who logged on from non-DoCoMo browsers, who didn't go through the login page, or who didn't enter information if it was needed.

Listing 11-14.1: The first test: checking for an i-mode browser

```
if ($ua[0]!="DoCoMo"||$user=="illegal"){
{   header("Location: http://www.eimode.com/");
}
```

The next tests look for whether a user has passed a `utn` number or a `$uid`. One comes from the `utnlogin.html` page, and one comes from the `login.html` page. If he has a `utn`, but not a `$uid`, he needs a `$uid`, which you give him by assigning the `utn` value to `$uid`. This is tested against the `uid` field of the users table in the database. If it exists, he is okay, and his password is also pulled out and passed to the next page using the `sessions` function.

Listing 11-14.2: The second test: whether a phone is utn-enabled, and a user did not log on with another user id

```
if ($ua[4]&&!$uid){
    $uid=substr($ua[4],-11,11);}
```

A quick note on sessions: This is a function introduced with PHP 4.0. If you are using 3.16 or one of the other common versions I am sorry — the security, the ease of use, and common sense make an upgrade the smart move. Progress is progress, and the new version is freely available on the included CD-ROM or online, so there really isn't any reason to hold off any longer. Especially if you are doing i-mode applications, sessions-capability alone is reason enough to upgrade, since most applications for i-mode make heavy use of them, in the absence of cookie support. PHP 4 is nearly 100% compatible with scripts from PHP 3, and you will be able to run our examples. The best of both worlds.

If the serial number is not present, you assume that it's a non-utn enabled handset, and you perform some tests to determine whether a $uid and $password have been passed from the login.html script. If they have, you see whether they match a known user in the users table of the database. If they do match, the user is sent to the page she requested after setting the session id. If they don't match, she is pushed to the newuser signup page. If there is no value in either password or uid, she is sent back to the login page.

Wow, that wasn't so hard!

But critics of the preceding code have a reasonable gripe with it (especially if you will make it the central authorization page): It isn't very portable, so it takes little effort to make it work on another server. I did this for a reason, as I said before — the reader can see a clear flow of everything that is happening in the script.

Our next step will be to build three pages that have different relationships with this login script, but which use many of the same variables. This is really the ideal situation in which to use a config.inc file, which is an include file. Basically, it is just a list of the variables, with some amount of tweaking done where such tweaking won't hurt the universality of the use of the data. Everything that you would need to change if you moved the script to another server should also be in this file, so that you can define all alternatives at one time and in a simple way.

Listing 11-15 shows what this config.inc file looks like:

Listing 11-15: The file containing your configuration variables. Called config.inc on the CD-ROM.

```php
<?php
$topDirectory="eimode.com/";
$packageDirectory="address";
$db="db12425a";
$dbhost="localhost";
$dbuser="us12425b";
$dbpw="limited";
$ipcheck="";
$browserCheck="";
$unknownPath="http://eimode.com/";
$language="en";
$agent=getenv("HTTP_USER_AGENT");
$agentArray= split("/",$agent);
$browser=$agentArray[0];
$model=$agentArray[2];
$utn=$agentArray[4];
$remoteAddress=getenv("REMOTE_ADDR");
?>
```

It is quite evident what most of the variables mean, thanks to more care being taken with naming them than previously.

And Listing 11-15 is our newly modular login script, which is somewhat shorter now. It also saves me money on my mobile phone bill, because it is much easier to turn the IP checking and browser checking on and off, as these are defined in the `config.inc` file. That means that I can use a normal browser to test the site, which is convenient, though not totally indicative of the user experience. At this stage, though, you are more interested in the workings of our code than its appearance. Listing 11-16 shows the final version, and is on the CD-ROM in the sample/address/ directory.

Listing 11-16: The login.php Script

```php
<?php
include('config.inc');
if ($ipcheck=="YES")
{
if (!ereg("210.153.84.",$remoteAddress)&&!ereg("210.136.161.",$remoteAddress))
{
    header("Location: $unknownPath");
}


}
elseif ($browsercheck=="YES"){
    if ($browser!="DoCoMo")
{
    header("Location: $unknownPath");
}
}
else
{if ($utn&&!$uid){
$uid=$utn;
}
elseif (!$utn&&!$password){
header("Location: $topDirectory.$packageDirectory");
}
elseif ($password){
    $connection = mysql_connect("$dbhost", "$dbuser", "$dbpw");
    mysql_select_db ("$db");
    $result=mysql_query("SELECT * FROM users WHERE uid='$uid'");
    if (mysql_numrows($result)!=0){
     $storedPw=mysql_result($result,0,password);
     $password=crypt($password,$storedPw);
     if ($password==$storedPw){
        session_start( );
        session_register("uid","password");
```

Continued

Listing 11-16 *(Continued)*

```
            header ("Location: http://$topDirectory$packageDirectory/$gotoPage");
            }
    }
    else
        {
        header ("Location: http://$topDirectory$packageDirectory/login.html#login");
        }
}
elseif ($utn && !$password) {
$connection = mysql_connect("$dbhost", "$dbuser", "$dbpw");
    mysql_select_db ("$db");
    $result=mysql_query("SELECT * FROM users WHERE uid='$uid'");
    if (mysql_numrows($result)!=0){
        $password = (mysql_result($result,0,password));
    session_start( );
    session_register("uid","password");
    header ("Location: http://$topDirectory$packageDirectory/$gotoPage");
}
else {
header ("Location: http://$topDirectory.$packageDirectory/login.html#newuser");
}
}}
?>
```

Managing account setup

The next bit of code you need to create allows new users to log in. This is a pretty straightforward script, which processes the information sent from a form on the `login.html` page. The script checks whether someone else has an account with either the same mobile phone number or serial number. If they do, it warns the user and gives him a link back to the login page. If not, it puts his information into the database, thanks him for joining, and gives him the link to log in.

One of the things that you did with the `login.php` script was to encrypt the password the user input before comparing it to the one in the database. That is because you were assuming that the one in the database was encrypted, which means that you have to encrypt it before entering it into the database.

There are some other things that you can and probably should do, such as checking to make sure the e-mail and phone number are in the correct formats, and that there aren't any empty fields, but those things will just have to wait for the 0.2alpha version. Listing 11-17 shows the script, minus the niceties.

Listing 11-17: The signup.php script

```
<!doctype html public "-//W3C//DTD HTML 4.0 //EN">
<html>
<head>
</head>
<body>
<title>New User Signup</title>
<?php
include ('salt.php');
include ('config.inc');
if ($utn&&!$uid){
    $uid=$utn;
}
elseif (!$uid&&$phone){
    $uid=$phone;
}
$connection = mysql_connect("$dbhost", "$dbuser", "$dbpw");
mysql_select_db ("$db");
$result=mysql_query("SELECT * FROM users WHERE uid='$uid'");
if (mysql_numrows($result)!=0){
    echo "The phone you<br>are calling<br>from, or the<br>mobile number<br>";
    echo "you have input<br>have already<br>been registered.<br>";
    echo "Please <a href='login.html#login'>login</a><br>if this is
        your<br>account.";
        echo "If this is not<br>your account and<br>this is your first<br>
        visit to this site,<br>";
        echo "there may be<br>someone else<br>using your phone<br>or number.
            Please<br>";
    echo "<a href='mailto:Webmaster@eimode.com'>contact us. </a>";
}
elseif (!$uid && !$phone){
        echo "You probably did<br>not access this<br>page through our<br>main
            signup<br>";
        echo "page. Please go <a href='index.php'>there</a><br>now.";
}
else {
     $password=crypt($password,gensalt("des-std"));
    $insertresult=mysql_query("INSERT INTO users(uid,password,phone,name,email)
            VALUES('$uid','$password','$phone','$name','$email')");
    if ($insertresult){
        echo "Thank you $name!<br>You have been<br>registered. Please click
            on<br>";
        echo "<a href='index.php'>here</a>to login.";
    }
```

Continued

Listing 11-17 *(Continued)*

```
    else {
        echo "There was some<br>problem, and you were<br>not registered.<br>";
        echo "Please use the<br>backarrow, and<br>try again.";
    }

}
?>
</body>
</html>
```

Most of this script, you will notice, is text. The logic parts, finding whether there is already an entry for this person or not, and determining whether a user's phone number or serial number should be used for the `uid` field of the database, are straightforward and use the following line:

```
$password=crypt($password,gensalt("des-std"));
```

This line contains one thing that may need explaining: The `crypt` method of encryption is one-way. There is no way to decrypt it except to hack it, and this would probably take a couple of powerful computers a while to do. This makes it reasonably secure. The problem is: How do you use it? I mean, if you need to check the password against the one in the database, which is encrypted and can't be decrypted, what do you do? The answer is that you use exactly the same encryption for the password you want to check as you did to create the encrypted password in the database.

The DES standard encryption scheme is available on UNIX systems. In PHP, using this function allows you to put in a string to encrypt and some "salt." The salt is what the encryption algorithm uses to start the encryption. The same algorithm, given the same string and the same salt, will encrypt a string in exactly the same way every time. On my provider, if you leave out the salt, it seems to randomly choose its way of encrypting and doesn't seem to use salt. This makes the encrypted passwords completely useless because there is no way to know how to replicate them. So, you have included the `salt.php` file, and you call the `gensalt` function to generate the proper salt for DES-standard encryption.

In `login.php`, you include the following line:

```
$password=crypt($password,$storedPw);
```

If you used a salt to create the encryption, putting the encrypted password as the salt when you encrypt the user input to check it will pass its salt on to the `crypt` function.

Although you are being quite security-conscious about passwords and such, using encryption is good protection against someone hacking your database and discovering everyone's passwords. If you transmit a password from a user to the server by normal HTTP, your risk of having his information discovered increases. If possible, use an HTTPS server for the login and signup pages. A combination of considered password and database security with secure server security should be quite secure, although a determined attacker with a large amount of resources at her disposal could launch a denial of service or other attack, which neither of these two measures provide much protection against.

There is no absolutely secure server, but it is your job to protect your users to the best of your ability. Take this seriously! Though it is easier to test when it is turned off, implement the IP checking and the browser checking in the `config.inc` file. This could possibly foil a denial of service attack, depending on its ferocity, and your server's resources. One of the dangers of going overboard, though, is that you make things difficult for the customers. So far, for customers with newer handsets, you have been able to make the process nearly painless. Using IP checking would really be advisable, though, because without it, someone could fairly easily use PHP or another scripting language to imitate the `HTTP_USER_AGENT` of a known user, along with his serial number. Because you don't check for the passwords of these users, the only real security is in using this together with the IP checking. It is much more difficult to imitate an IP number, because this is where the information is sent, and a hacker presumably is hacking to have information sent.

The guts of the application: entering and finding addresses

So, you have a way for new users to sign up, a way to check that they are who they say they are when they log in, a couple of pretty attractive login pages, and a `config.inc` file with commonly used variables. What's next?

Well, this application still doesn't actually do anything! It might be good to actually code those things it was meant to do: enter address information and retrieve it. The good thing about your work up until now is that you have built a fairly secure structure in which to do this, and you have a config file from which you can take the information you need. You have also been passed the session id, username, and encrypted password in the session function, so that you don't have to ask a user to log in at every screen, and yet you can check her information if she tries to manipulate data.

Cool! So basically, you just have to build one page that takes a user's address information and inserts it into the address database, and build another page that lets him search for it. You should probably build the page that puts the information into the database first because without information in there to search for, it is difficult to test the search application. You also can tweak any inconsistencies in the data structure before you actually try to implement it.

So, for inputting information, what do you need to do?

1. You need to check the session ID and to make sure the IP address that he logged in with is the same one he is using. If it isn't, you need to blast the user out of there.

2. You need to provide a screen for users to input the necessary information.

3. When a user submits information, you need to have a way to process it.

4. Before processing the data, you need to check the user's ID and password to make sure they match.

5. You need to insert the data into the addresses table, with the user's ID inserted and properly verified.

6. A user who wants to input more data needs to be taken back to step 2; one who wants to look up data needs to be taken to that page; and one who wants to log out needs that option.

Now, one limitation of using sessions is that you need to be able to pass session variables to the next page. If the next page is a cHTML page, it cannot do anything with this information. Actually, because you are using a PHP function to pass this information, a cHTML page would not even receive it. I point this out because in looking at the previous list of tasks to perform, it looks as if some will use mostly cHTML. As you have seen previously, however, a PHP script can actually include mostly cHTML or HTML with no problem while maintaining its PHP functionality.

The other thing about the tasks is that some of them can be put into functions. Functions are PHP code that is performed when the function is called. They make the code cleaner, especially when you have lots of information to output (as you do in this section), by allowing you to simply output the results of a function.

You put both the printing of the form, and the menu that pops up after the task is complete, into functions. Actually, the menu that pops up afterward can easily be used in other parts of the application, so you put this function into the `config.inc` file.

Listing 11-18 shows the code that you end up with:

Listing 11-18: The first part of the script to input addresses

```php
<?php
    include ('config.inc');
    session_start();
    $currentIP=getenv("REMOTE_ADDR");
    if (!ereg($remoteAddress,$currentIP)){
    session_destroy();
    header ("Location:login.html");
}
    else {

    $connection = mysql_connect("$dbhost", "$dbuser", "$dbpw");
    mysql_select_db ("$db");
    $result=mysql_query("SELECT * FROM users WHERE uid='$uid'");
    if (mysql_numrows($result)!=0){
        if (mysql_result($result,0,password)!=$password){
            session_destroy();
```

```
                header ("Location:login.html");
        }
    }
    else {
            session_destroy();
            header ("Location:login.html");
    }
}
$inserterror="There was some<br>problem
    adding<br>".$fname."'s<br>address to your <br>address book.<br>";
$insertsuccess=$fname."'s address<br>was added to<br>your
    address<br>book!<br>";

if ($return){
    $insertresult=mysql_query("INSERT INTO addresses
        (uid,fname,lname,homestreet1,homestreet2,homecity,homestate,
     homepost,homephone,homefax,homeemail,companystreet1,
     companystreet2,companycity,companystate,companypost,
     companyphone,companyfax,companyemail,mobilephone,mobileemail)
VALUES('$uid','$fname','$lname','$homestreet1','$homestreet2',
     '$homecity','$homestate','$homepost','$homephone','$homefax',
     '$homeemail','$companystreet1','$companystreet2',
     '$companycity','$companystate','$companypost','$companyphone',
     '$companyfax','$companyemail','$mobilephone','$mobileemail')");
    print $header;
    if (!mysql_error()){$message=$insertsuccess;}
    else {$message=$inserterror; print mysql_error()."<BR>";}
    nextMenu($message);
    print $footer;
}
else {
    print $header;
    print "<div align='center'>Input Address</div><p>";
    outputForm();
    print $footer;
}
function outputForm()
{
print '<form name="FormName" action="input.cgi?'.SID.'"
    method="POST">';
print 'First Name<input type="text" name="fname" size="14"
    maxlength="20"><br>';
print 'Last Name<input type="text" name="lname" size="14"
    maxlength="20"><br>';
```

Continued

Listing 11-18 *(Continued)*

```
print 'Home Street<input type="text" name="homestreet1" size="14"
    maxlength="40"><br>';
print 'Home Street2<input type="text" name="homestreet2" size="14"
    maxlength="40"><br>';
print 'Home City<input type="text" name="homecity" size="14"
    maxlength="30"><br>';
print 'Home State<input type="text" name="homestate" size="14"
    maxlength="30"><br>';
print 'Home Postcode<input type="text" name="homepost" size="14"
    maxlength="12" istyle="4"><br>';
print 'Home Phone<input type="text" name="homephone" size="14"
    maxlength="14" istyle="4"><br>';
print 'Home Fax<input type="text" name="homefax" size="14"
    maxlength="14" istyle="4"><br>';
print 'Home E-mail<input type="text" name="homeemail" size="14"
    maxlength="30"><br>';
print 'Company Street1<input type="text" name="companystreet1"
    size="14" maxlength="40"><br>';
print 'Company Street2<input type="text" name="companystreet2"
    size="14" maxlength="40"><br>';
print 'Company City<input type="text" name="companycity" size="14"
    maxlength="30"><br>';
print 'Company State<input type="text" name="companystate" size="14"
    maxlength="30"><br>';
print 'Company Postcode<input type="text" name="companypost"
    size="14" maxlength="12" istyle="4"><br>';
print 'Company Phone<input type="text" name="companyphone" size="14"
    maxlength="14" istyle="4"><br>';
print 'Company Fax<input type="text" name="companyfax" size="14"
    maxlength="14" istyle="4"><br>';
print 'Company E-mail<input type="text" name="companyemail"
    size="14" maxlength="30"> <br>';
print 'Mobile Phone<input type="text" name="mobilephone" size="14"
    maxlength="14" istyle="4"><br>';
print 'Mobile E-mail<input type="text" name="mobileemail" size="14"
    maxlength="30"><br>';
print ' <input type=hidden name="return" value="true">';
print ' <input type="submit" value="Put Address In"></form>';
}
?>
```

This code ends up being rather long, especially with all of the cHTML `print` commands in the `ouputForm` function, and yet it is still fairly simple to read. Although not exactly object-oriented programming, it is at least modular. The first section looks at the information sent in the session, which should be a user id, password, and IP number. If they are all correct, it takes you to the next section, which has a simple test of whether or not the variable $result is present or not. This was a variable present in the form, so you can tell if the form has been submitted to this page, or if this is the first time a user is accessing the page.

If a form was submitted, you insert the results into the addresses table of the database, and print a message with either the error or a success message, and a menu of options for a user.

If this is the first time to the page, the user is given a form to input a new address.

Troubleshooting problems

There are several problems I had making this script work. One regards the use of sessions, which I was not so familiar with. To start a session, the start_session() function is called. That is pretty simple. But to continue a session, you must remember to do a few things:

- First, you must include ?SID on the end of the URL of any page that you link to that you want the session variables to get passed to.

- Next, you must be sure that you have declared the session variables, as I did in the login.php script.

- It is not terribly intuitive, but having started the session in one script doesn't mean anything in another script. So, even if you may have correctly put ?SID on to the end of the URL of a page, after you access this page, you still need to use the start_session() command to be able to access the session data.

Ah, but the potential problems don't end there! The next was a serious security issue. Using sessions is not secure – even if the URL line in Internet Explorer or Netscape does not show it, your session ID is actually being passed with the URL. It is quite possible to find out what the session ID is, hijack it, and get access to a user's data. The secure part is that even if they know what the session ID is, hackers can't change the data in the session. So, you send the IP address that a user started a session from, and check it against where he is now. If they are not the same, you do not give him access. Most of the time, this is because his modem timed out, and he connected using a different IP number. As such, you could see this as an inconvenience for a user, and it is. But it also protects him, so you balance one against the other, and in my book, protection comes out ahead where it doesn't add too much inconvenience.

The last problem was trying to force everything in functions, all the better to demonstrate them. There are basically three parts I thought I could get into functions: outputting the form, processing the form results, and outputting a menu when a user finished. Outputting the form and outputting a menu posed no problems. Processing the form results did. Functions are discrete elements within a PHP script, and do not share variables from the rest of the script unless the variables are passed to it. Passing all the form variables is possible, but the names of those variables also have to be declared in the function declaration. In this case, they once again had to be listed when you told the function which pieces of data you wanted inserted into the designated fields of the table.

There is a simple way around this, which is to load all of the variables into an array and then just pass that array to the function. But there is a cost in terms of clarity of what is happening — the array variables would be called $myarray[0] instead of the more easily understandable $uid, which is also what this is called in the database table. There is a way around this, which is to use the list function to assign names to each of the values in the array. But this means listing up all of the variables again.

All to what purpose? This task is performed only in this script, and only once per page iteration. Making a script that displays the same list of variables more than once is a waste of time, so you took the easiest route, which is to *not* use a function for this task.

Searching and editing the entries

The last two tasks, finding addresses and modifying user information, should be easy. The first parts will look exactly as they did in the Listing 11-18 input script, confirming a user's password and IP number. Hmm... Did I just say, "look exactly as?" That sounds distinctly like something that could be made into a function, and indeed you will start by making a function to do these things, something that I missed in writing the last script.

Take a look at what the config.inc file looks like at this point (see Listing 11-19):

Listing 11-19: The config.inc file

```
<?php
$topDirectory="eimode.com/";
$packageDirectory="address";
$db="databasename";
$dbhost="localhost";
$dbuser="username";
$dbpw="password";
$ipcheck="";
$browserCheck="";
$unknownPath="http://eimode.com/";
$language="en";
$agent=getenv("HTTP_USER_AGENT");
$agentArray= split("/",$agent);
$browser=$agentArray[0];
$model=$agentArray[2];
$utn=substr($agentArray[4],-11,11);
$remoteAddress=substr(getenv("REMOTE_ADDR"),0,11);
```

```
$header="<html><head><title>Addressbook</title></head><body><div align='center'>
    <img src='address.gif'></div><br>";
$footer="<br>&copy; 2001, Llama<br>Communications,<br>L.L.C.</body></html>";
$inserterror="There was some<br>problem adding<br>".$fname."'s<br>address to
your
    <br>address book.<br>";
$insertsuccess=$fname."'s address<br>was added to<br>your address<br>book!<br>";
$searchempty="No names found.<hr>Try again:<br>";
function nextMenu($message)
{
print $message;
print "<br>&#63879; <a href='lookup.cgi?".SID."' accesskey='1'> Look
    up<br>addresses</a><br>";
print "&#63880; <a href='input.cgi?".SID."' accesskey='2'> Add
addresses</a><br>";
print "&#63881; <a href='user.cgi?".SID."' accesskey='3'> Change
    user<br>information</a><br>";
print "&#63882; <a href='logout.cgi?".SID."' accesskey='4'> Log out</a><br>";
}
function
checkSessionUser($remoteAddress,$uid,$password,$dbhost,$dbuser,$dbpw,$db)
{$currentIP=getenv("REMOTE_ADDR");
    if (!ereg($remoteAddress,$currentIP)){
    session_destroy();
    header ("Location:login.html");
}
    else {

    $connection = mysql_connect("$dbhost", "$dbuser", "$dbpw");
    mysql_select_db ("$db");
    $result=mysql_query("SELECT * FROM users WHERE uid='$uid'");
    if (mysql_numrows($result)!=0){
       if (mysql_result($result,0,password)!=$password){
            session_destroy();
            header ("Location:login.html");
        }
    }
    else {
         session_destroy();
         header ("Location:login.html");
    }
}


}

?>
```

This is not what you would call a very short and concise config.inc, and it has gotten away from the original reason you created it, which was to store variables that got used again and again in a variety of scripts. By adding these functions, you have muddied the waters somewhat. Your little project has become a monster!

Fear not! There is an easy way around this, which is to create a new include file to hold the functions. You can call this file library.inc because (like a library) it stores more complete works than the simple variables stored in config.inc. You can also store functions that you created in input.php, and store functions that you will soon create in lookup.php.

Look at look up, and see what you need to do with this script:

- The first task is to ask a user for the first letter or letters of the name of the person whose address she wants to find. You need to make sure that the search is not case-sensitive because names usually start with capital letters, but many cell phones require more key clicks for capital letters than they do for lowercase letters.

- Next, you need to display all matches, and ask the user which one of them she wants to see. There isn't that much room on a mobile phone screen to display every one, so asking a user which address they would like to display makes it easier.

- When a user has told you the name of the person whose address she wants, you need to display it (making sure to make the e-mail, phone, and fax information into links that bring up an e-mail browser, or dial the phone or use a mobile fax to fax something).

That's it. You see the evolution of the code in Listing 11-19, as the scripts, by separating functions from logic, more and more begin to resemble the list of tasks you set previously. This is the reason you have separated logic from function: to keep things straight in your mind, and to be able to clearly see (and have others clearly see) what is going on in the scripts, separately from exactly how that is done. Listing 11-20 shows the script for lookup.php:

Listing 11-20: The lookup.php script

```
<?php
include ('config.inc');
include ('textwrap.php');
include ('library.inc');
session_start();
checkSessionUser($remoteAddress,$uid,$password,$dbhost,$dbuser,$dbpw,$db);

if ($return) {//A form has been sent
print $header;
getnames($searchstring);
print$footer;
```

```
}



elseif ($aid){//A user has chosen a name
    print $header;
    printAddress($aid);
    nextMenu("<hr><br>Where next?");
    print $footer;
}
else {//It is their first time to the script
    print $header."Find Address<p>";
    searchform();
    nextMenu("<hr><br>Menu");
    print $footer;
}


?>
```

This is by far the most elegant script in terms of clarity and straightforwardness, but you do lose a clear understanding of how it's doing its work because you put the functions in the library.inc file.

Really, this code needs no explanation, because it has done exactly what you said you wanted to do in the list you made beforehand. That is not to say that nothing went on in the time between making that list and coming up with usable code — most of the hair-pulling stuff is in functions in the library.inc file. And I should add that only after a really messy-looking code took shape and had been tested and debugged, did I stick the function in the library.inc file.

So, what is in those functions? You call five in the script, three of which are new; and two are checkSessionUser and nextMenu, which you have used before. The first new one is getnames. This script finds names after a user inputs the first letter or letters of the person's name whose address he is looking for (see Listing 11-21).

Listing 11-21: The getnames function, in the library.inc file

```
function getnames($searchstring,$uid)
{
    $searchfor = "^".sql_regcase($searchstring)."*";
    $result = mysql_query("SELECT * FROM addresses WHERE fname
        REGEXP '$searchfor' OR lname REGEXP '$searchfor' HAVING uid
        ='$uid'");
    if (mysql_numrows($result) != 0)
{
$num = mysql_numrows( $result );
```

Continued

Listing 11-21 *(Continued)*

```
$searchfound="Search found ".$num."<br>name(s):";
print $searchfound;
print"<form action='lookup.cgi?".SID."' method='post'>";
for ($i=0;$i<$num;$i++){
    print"<input type='radio' name='aid'
      value='".mysql_result($result,$i,aid)."'>
      ".mysql_result($result,$i,fname)."
      ".mysql_result($result,$i,lname)."<br>";
    }
    print"<div align='center'><input type='submit' value='Get
      Address'></div></form><br>";
    nextMenu("<hr><br>Where now?");
    }
    else {
      nextMenu("No names found.<hr>Try again:<br>");
      }
}
```

You do a little SQL stuff with the string before you put it into the query. The ^ symbol that you append to the front means "the first thing after the beginning of the field." So, if somebody is named Doug, and you put in only an "o," it won't return anything. An asterisk is appended to the end to signify that everything is okay after the initial letter.

In the query, you use the `OR` operator to do a regular expression search for the letter in either the first name field or the last name field. The `HAVING` operator may be one you are not familiar with, even if you use SQL. I say that because I was not familiar with it, but ended up needing it because it wasn't possible to do a regular expression search on the two fields and still test to make sure the returned records were from the user requesting them, using either `AND` or `OR`. Putting this sort of logic in the query is much faster than returning all of the results and then sorting it using PHP. Believe me.

If the query finds results, it prints the beginning of the form and then uses a `for` loop to print the names found with radio buttons next to each one, so that a user can choose which one to display. You only give them the option of displaying one because that is all that comfortably fits on a screen.

```
for ($i=0;$i<$num;$i++){
```

This `for` statement is a beaut! It has everything that you need in one easy package. The first variable passed is what your counting variable starts at (in this case $I). The next is the condition at which the `for` loop ends at is defined, and the third tells what to do to the counting variable to move it toward the end.

The last `else` statement just catches cases in which there are no results returned. You can, by the way, call a function from within a function in PHP, as I do with the `nextMenu` function from within the `getnames` function.

The next function is the `searchform` function. This basically just ouputs the form you will use to search for a name (see Listing 11-22).

Listing 11-22: The searchform function in library.inc on the CD ROM

```
function searchform()
{
print "<form action='lookup.cgi?".SID."' method='post'>";
print "Look for:<input type='text' name='searchstring' size='6' max='10'><br>";
print "(first letters<br>first/last name)<P>";
print "<div align='center'><input type='submit' value='Find Address'></div>";
print "<input type='hidden' name='return' value='TRUE'>";
print "</form>";
}
```

This form isn't terribly elegant, but it does the nitty-gritty work of getting a user to enter a letter to search for. This is nearly pure HTML, with the exception of the `SID` variable appended to the end of the URL of the action. This is the variable that passes on the session information to the next page.

The last function (and the longest) is the one you use to output an address found in the database. A lot of handling is needed to display the data clearly. Among other things, you use the `textwrap` function. To use that, you first create a long string of data consisting of returned fields and punctuation, and send that string to the `textwrap` function.

Listing 11-23 shows the `printaddress` function.

Listing 11-23: The printaddress function in library.inc on the CD ROM

```
function printAddress($aid)
{
   $result =mysql_query("SELECT * FROM   addresses WHERE aid='$aid'");
   $address=mysql_fetch_object($result);
   if ($address->homestreet1){$homeaddress=$address->homestreet1." ".$address-
      >homestreet2." ".$address->homecity." ".$address->homestate." ".$address-
      >homepost."<br>";}
   if ($address->homephone){$homephone="<a href='tel:".$address-
      >homephone."'>Tel. ".$address->homephone."</a><br>";}
   if ($address->homefax){$homefax="<a href='faxto:".$address->homefax."'>Fax
      ".$address->homefax."<br>";}
   if ($address->homeemail){$homeemail="<a href='mailto:".$address-
      >homeemail."'>Email ".$address->homeemail."</a><br>";}
   if ($address->companystreet1){$companyaddress=$address->companystreet1."
      ".$address->companystreet2." ".
   $address->companycity." ".$address->companystate." ".$address-
      >companypost."<br>";}
```

Continued

Listing 11-23 *(Continued)*

```
    if ($address->companyphone){$companyphone="<a href='tel:".$address-
        >companyphone."'>Tel. ".$address->companyphone."</a><br>";}
    if ($address->companyfax){$companyfax="<a href='faxto:".$address-
        >companyfax."'>Fax ".$address->companyfax."</a><br>";}
    if ($address->companyemail){$companyemail="<a href='mailto:".$address-
        >companyemail."'>Email ".$address->companyemail."</a><br>";}
    if ($address->mobilephone){$mobilephone="<a href='tel:".$address-
        >mobilephone."'>Cell ".$address->mobilephone."</a><br>";}
    if ($address->mobileemail){$mobileemail="<a href='mailto:".$address-
        >mobileemail."'>Mobile Email ".$address->mobileemail."</a><br>";}
    $fulladdress=$address->fname." ".$address->lname."<br>Home
        Address:<br>".textwrap($homeaddress,16).$homephone.$homefax.$homeemail.
    "<br>Work Address:<br>".textwrap($companyaddress,16).$companyphone.
        $companyfax.$companyemail."<br>Mobile Contact:<br>".$mobilephone.
        $mobileemail;
    print $fulladdress;
}
```

This function uses lots of `if` clauses to see whether there is any data to display. If so, it creates strings to display. The other really important feature is the links that you include with e-mail addresses, telephone numbers, and fax numbers. This is the sort of linkage between the device and the service that actually would make it useful enough for people to use. You could, and would if you planned to make a real commercial service of this, also include features to allow users to add the address to their in-phone address book, to send the address to their friends, and other little features that are quite easy to program, but which allow the sort of functionality that sell a service to customers for a couple of dollars a month.

So, you are down to your last task (and one of the easiest), which is the user account management task. The following is what you want to do with this task:

- Print the current information and text entry boxes to change that information.

- Check the password again just to be sure, both against the one in the database and the one passed as a session variable.

- If a new value hasn't been assigned by a user, fill it with the current value.

- Replace the old record in the database with the new one.

- Tell the user whether it was successful or not, and link back to the login page if it was — you don't want him continuing with the same session if he has a new password.

Chapter 11: Programming in cHTML: A Tutorial

One thing that you cannot let a user change is his user ID, which is used to store all of his addresses. If he ends up changing his phone number, and doesn't have utn, his user ID and new telephone number will be different. Similarly, if he buys a new phone that is utn-enabled, he will not be able to log in using utn because he needs to input his old phone number. These are issues, however, that are not imperative right now, and you will need to address this later.

The code for this page is shown in Listing 11-24:

Listing 11-24: The user.php code

```
<?php
include ('config.inc');
include ('textwrap.php');
include ('library.inc');
session_start();
checkSessionUser($remoteAddress,$uid,$password,$dbhost,$dbuser,$dbpw,$db);

if ($return){//A user has returned the form
    print $header;

changeUserInfo($name,$email,$phone,$uid,$password,$oldpassword,$newpassword);
    print $footer;
}
else {//It is the first time to the script, and an input form is given
    print $header."Change User Info<p>";
    userform($uid,$password);
    nextMenu("<hr><br>Menu");
    print $footer;
}
```

Again, most of the functionality, as that adjective suggests, lies in the functions, which are in the library.inc file. You have only two new ones in this script, the changeUserInfo function and the userform function. The userform function outputs the form, and the changeUserInfo function processes the information, either updating the database or refusing because of bad password or other problems.

Listing 11-25 shows these two functions:

Listing 11-25: The changeUserInfo and userform functions

```
function changeUserInfo($name,$email,$phone,$uid,$password,
       $oldpassword,$newpassword)
{
$result=mysql_query("SELECT * FROM users WHERE uid='$uid'");
    if (mysql_numrows($result)!=0){
      $storedPw=mysql_result($result,0,password);
      $oldpassword=crypt($oldpassword,$storedPw);
```

Continued

Listing 11-25 *(Continued)*

```
            if ($oldpassword==$storedPw&&$oldpassword==$password){
                if (!$name){$name=mysql_result($result,0,name);}
                if (!$email){$email=mysql_result($result,0,email);}
                if (!$phone){$phone=mysql_result($result,0,phone);}
                if (!$newpassword){$newpassword=$storedPw;}
                $replaceresult=mysql_query("REPLACE INTO users(uid,password,phone,
                   name,email)VALUES('$uid','$newpassword','$phone','$name','$email')");
                if ($replaceresult){
                    print "Your info has<br>been updated.<br>Please click<br><a
                        href='index.php'>here</a>to login again.<br>";
                }
                else {print "There was a <br>problem updating<br>your user info.<br>";
                session_destroy();
                }
            }
        else {nextMenu("Your password<br>did not match.");
        }
    }
    else {
        print "I am sorry<br>but your info<br>couldn't be<br>found. Please<br><a
            href='index.php'>login</a><br>again.";
        session_destroy();
    }
}
function userform($uid,$password)
{
    $result=mysql_query("SELECT * FROM users WHERE uid='$uid'");
    if (mysql_numrows($result)!=0){
        if (mysql_result($result,0,password)==$password){
print "Enter new info<br>to change, leave<br>blank to keep.<br>";
print "<form action='user.cgi".SID."' method='post'>";
print mysql_result($result,0,name)."<br><input type='text' name='name' size='14'
    maxlength='30' istyle='3'><br>";
print mysql_result($result,0,email)."<br><input type='text' name='email'
size='14'
    maxlength='30' istyle='3'><br>";
print mysql_result($result,0,phone)."<br><input type='text' name='phone'
size='14'
    maxlength='11' istyle='4'><br>";
print "You must enter<br>your password<br>to make these<br>changes:<br><input
    type='text' name='oldpassword' size='14' maxlength='20' istyle='3'><br>";
print "New password:<br><input type='text' name='newpassword' size='14'
    maxlength='20' istyle='3'><br>";
```

```
   print '<input type=hidden name="return" value="true">';
   print "<div align='center'><input type='submit' value='Change Info'></div>";
   }
   else {print "There was some<br>problem with<br>your password.<br>Please
      click<br><a href=index.php>here</a>to login<br>again.";
         session_destroy();
   }
}
else {print "There was some<br>problem with<br>your login.<br>Please click<br><a
   href=index.php>here</a>to login<br>again.";
      session_destroy();}
}
```

There isn't really a lot in these functions that you haven't seen before, and what is happening is quite clear. The only thing that is really even questionable is choosing to use the REPLACE function in our SQL query. I do this just to cut down on the accesses to the database, replacing the entire row of data with either the new values or the old ones. At this point, optimization is not that important, but it would be if you had 100,000 users with 20 addresses each. That totals two million addresses, and cutting the number of accesses will help keep you going even when a significant portion of the registered users might be using the service. MySQL is extremely fast, as is PHP when compiled into the Apache server as a module. Both are also extremely stable, and it is more than conceivable that they be used for serving gigabytes, even terabytes of data. But you do have to keep an eye on queries that might break the system. The name search query, actually, is one that could conceivably break the system. It could become necessary to specify only first names or only last names.

This is the part of developing a successful service that is difficult without some experience. You could do what one quite-famous official i-mode site did, which is to buy very expensive solutions from Oracle (eating up a lot of its seed money); or what another soon-to-be official i-mode site has done, which is to use PHP and MySQL and really spend a good effort optimizing them for what they are doing, and not have to depend at all on seed money, except that which they themselves have put in. And forget venture capitalists! Personally, given a choice, I would take the latter option. But it takes a little more time to do this than it does to write a check to Oracle or Microsoft. I wish I had a choice! Most people starting out don't have the money for Oracle or Microsoft's supercharged databases.

Testing the code for various handsets

If you know that you will be serving customers using specific models of mobile phones, having such a database can have other benefits, which in the rush to get the application ready, you really did not address. The textwrap function, for example, could be set to wrap at the screen width particular to a particular phone. Or the logo displayed could be chosen depending on whether a handset supported colors, and how many colors. If you ever added a Java portion to the application, various handsets have shown very tiny incompatibilities, and you could direct those handset users to the particular version suited to their handsets. Or, if you were serving customers across several carriers, you could use this handset database to figure out how you should deliver certain features, such as GPS-one, or i-area.

In the example application, you used several DoCoMo-specific attributes. The main one was the `utn` attribute, which you spent a lot of time making identify a handset, so that login could be as seamless as possible. You also use the emoji characters for number buttons, which do not show up properly on the screen of non-i-mode phones or browsers. Because these really aren't used too much, except in the number lists, you could probably get away with not doing anything. But you could also very easily use a condition of some sort to check the model of handset and output characters that would match the model of handset that a user had. The more customization you do, the better experience you can provide your user. Sure, some of the customization is merely cosmetic, but in the i-mode paradigm, in which you are *selling* your content, looks really do matter.

I have included the handset database in the `address.sql` file, included in the addressbook application in the code examples folder on the CD-ROM. If you build an application, you have the current information (as of writing, that is) on handsets, and if you want to use the database to add some customization features to your own application, it's a good resource to do that with.

The application you have created is fully functional, though not quite ready for primetime. Some of the things that you would probably want to do if you were pursuing this as a business would be more customization, adding the ability to modify addresses, optimizing the SQL that you use so that it would scale reasonably seamlessly, and about 10,000 other things that you would probably only think of when you got feedback from people testing the application.

The basis of this application is a database-script-browser interaction, which is applicable to almost any application and is particularly well-suited to the i-mode platform with its small memory cache, limited browser functionality, and relatively slow connection speeds.

Chapter 12

i-Appli: The i-mode Version of Java

IF JAVA IS SOMETHING YOU HAVE BEEN MEANING TO LEARN, starting with the i-mode version called i-Appli is a good place to start. It is simple enough in its framework to grasp in a fairly short time, and the 10 KB limit on the size of applications means that even if you want to get unmanageable, you can't. This is a perfect situation for a beginner.

For a pro, it may actually be harder because it means you must scale your expectations down to mobile phone size. The J2ME, unlike J2SE or J2EE, is missing many of the basic classes that you take for granted. Learning to work within these limitations in this chapter will give you a better idea of how to apply your current Java knowledge to this new platform.

Getting to Know the i-Appli Java API

The first thing to know is that i-Appli is based on the J2ME or Java 2 Micro Edition and its Connected Limited Device Configuration (CLDC), but not the Mobile Information Device Profile (MIDP) defined by Sun, instead using its own API. The J2ME consists of the KVM (Kilobyte Virtual Machine) and basic Java class libraries. The CLDC is a configuration of the J2ME for mobile devices. The MIDP is actually a manufacturer-specific API, customized by Sun for each device, depending on its functionality. All implementations share a basic set of MIDP classes.

NTT DoCoMo's implementation outside the MIDP framework flies in the face of the "write once, run anywhere" motto of Java, and it breaks any direct attempts to port an application developed using the MIDP API directly to an i-Appli. The first attempts at programming, before i-Appli phones were actually released, were doomed to failure because there were no emulators available to emulate the way that the class libraries embedded in phones actually worked. Building one's own emulator based on the API was beyond the sort of thing most developers wanted to do just to start messing around.

Luckily, at this point DoCoMo made public a couple of excellent tools, which make developing on this platform much easier. Porting programs from J2ME directly will still break them, but I have translated the API into English (see Appendix C).

The DoJa API in Appendix C is referred to quite often in this chapter and in Chapter 13. Please have a look at it, and if you have the MIDP API, compare them. Many of the methods and classes are similar or the same, though they may be named differently.

i-Appli and the DoJa API are very much in keeping with what NTT DoCoMo did with cHTML, which is to reduce the existing standard down even further, to its bare essentials. Though the MIDP was designed for "mobile" devices, this really includes a broad range of electronic devices, from the Palm, which is the platform that has first been used with MIDP, down to the much smaller platform of the mobile handset. NTT DoCoMo appears much less in love with standards than in getting its developers the specific tools they need to develop for a quite specific platform, the DoCoMo handset. It probably has a very good point because it is hard to imagine porting a 10 KB midlet to use on a Palm device or vice versa, porting a 250 KB midlet from the Palm device to a mobile phone.

Where DoCoMo probably screwed up is in making its classes and methods specifically for use on DoCoMo devices, rather than just saying, "Okay, here is the MIDP; you can use these classes and methods, and you can't use those." Then at least there would have been the option of playing the tiny games on a Palm device. Because it added some specific classes and wanted to allow manufacturers to add other classes, it chose not to do this. Be that as it may, the API provides a reasonable set of tools to be able to develop for this platform, and porting the applications to other CLDC implementations, depending on how close they are to the MIDP, shouldn't be quite as difficult as it seems.

Making Preparations

The very first thing that you need to do right this moment is to get your set of tools together. If you are an experienced developer, please excuse me for a few minutes while I get myself kitted for the battle ahead.

See Appendix A for what is on the included CD-ROM, and how to install it.

1. I suggest that at this time you install NJWin, if you haven't already. This is a little utility that allows you to see Japanese, Chinese, or Korean without having those versions of Windows installed. Actually, if you can't read Japanese, it may not make any difference.

2. Install the Java 2 SDK from the enclosed CD-ROM, if you don't already have it. If you don't have it installed when you try the next step, you will be prompted to install it.

3. Next, follow the following link:

   http://www.nttdocomo.co.jp/i/java/j2mewsdk4doja_release2_2.exe

   Instead of reaching a Web page, this URL leads directly to the download file for the KToolbar. A dialog box should come up in your browser and ask where you want to save the file. The installation wizard screens are in Japanese, but for this installation, you really need to press only the key that says "next" in Japanese. Just hit the Return button because the next key should be the one highlighted. Though this is all in Japanese, the KToolbar, the thing that you just installed, is actually in English, so you are out of the Japanese woods for now. (See Figure 12-1.)

4. Next, you need a text editor. Truly, anything will work, from Notepad to a fancy-schmancy IDE. For now, I recommend something closer to Notepad. I use UltraEdit32, a shareware editor that is included on the CD-ROM. This editor is quite fast, has some nice basic features, and has one important advantage over Notepad, which is that it color-codes your code so it is easy to spot what you are looking for.

Figure 12-1: The KToolbar window.

Okay, assuming that you have KToolbar and UltraEdit32 set up, that is really all that you need. Wow! The simplicity of it! Wait until you see KToolbar in action, and you will wonder what all the fuss is about Java. It doesn't look so hard at all using these kinds of tools! And that is really the point: Using simple tools to create small and fairly simple applications is something most of you can mess about with and enjoy. They enable you to do some cute things on your own, and to possibly share your work and impress others. If you are really adept at this, you can actually do a lot with it, and can dream of being Big in Japan, as the great song from that '80s band Alphaville put it.

Creating i-Applis

Now, to quickly explain the process:

1. You write your Java in your text editor. Hah! Easier said than done, you say.

2. Next, you use the KToolbar to create a package, with a name the same as your executing class (for example, `hello`), by pressing the New button in the KToolbar window.

3. Now you put your Java source into the `src` folder of the `hello` folder that was created in the `J2ME4DoJa/application/` folder.

4. There is a bug in KToolbar or an oversight that causes it to not automatically fill in the AppClass line in the jam file. This causes it to not properly compile. You can solve this by pressing the Settings button in the KToolbar window. Look for the AppClass line, and enter the name of the main class for application in the field to the right. This should probably be named the same as the package, for simplicity.

5. You need to compile the program, so your next step is to open the project in the KToolbar if it is not already open, and press the Compile button in the KToolbar window.

6. Assuming that there are no errors, your last step is to execute the program. Far from killing the poor innocent program, this makes an emulator window pop up and the program run. You cannot simply click a jam file and have it run. You must always use the KToolbar emulator or another emulator, and open the application from within the emulator.

Yes, You Guessed It: Hello World

There seems to be some sort of conspiracy in the world of computer language education to spread this rather strange sentence of "Hello World." I come to computer languages in a rather roundabout way, however, having taught English for eight years, and saying "Hello World" just doesn't seem quite as useful to me as saying, "Hello World, nice to meet you. What can I do for you?"

Using the low-level API

The low-level API has nothing to do with its education or income! It is low level because it is closer to clearly defining all of the things that are happening within it. If you imagine the busy workers in the basement, slaving away, the low-level API is on the first floor – not very much above the workers. There is no middle management to sort out which job goes where, and so on. Which means that in this API, nearly everything is specified by the programmers. The next few sections show you examples of some of the main classes in the low-level API and basic coding constructs that you will use to paint your miniature canvas.

CREATING A BASIC PACKAGE

The low-level API is also called the canvas form of a Java midlet because it uses the `Canvas` class to display information. You see how this works in the simple Hello World program shown in Listing 12-1:

Listing 12-1: Hello.java

```java
package hello;
import com.nttdocomo.ui.*;

public class Hello extends IApplication {
    MainCanvas gc;

    public void start() {
        gc = new MainCanvas();
        Display.setCurrent(gc);
    }
}

class MainCanvas extends Canvas {
    public void paint(Graphics g) {
        g.lock();
        g.setColor(Graphics.getColorOfName(Graphics.WHITE));
        g.fillRect(0, 0, getWidth(), getHeight());
        g.setColor(Graphics.getColorOfName(Graphics.RED));
        g.drawString("Hello World", 18, 10);
        g.drawString("Nice to meet you", 1, 30);
        g.drawString("What can I do", 14, 50);
        g.drawString("do for you?", 20, 70);
        g.unlock(true);
    }
}
```

Figure 12-2 shows the resulting page in an emulator after compiling the code in the KToolbar.

There are some things about the code in Listing 12-1 that I want to explain and caution you about. First is the first line. Without the first line, the code still compiles fine, but may not run correctly from the KToolbar, and you may get a `class not found` exception. This line tells the compiler what to call the package. If it doesn't have it, it may not execute correctly.

Figure 12-2: The first application!

The next line is the `import` command, which imports all the classes of com.nttdocomo.ui (the asterisk is called a wildcard operator, which includes everything in the com.nttdocomo.ui class). You actually use only three of the classes (the `Graphics` class, `Iapplication` class, and `Canvas` class), but there is no penalty for importing all the classes, and it is quicker than specifying each one.

The next line is declaring this class and telling the compiler that this is an application. Anything that executes is an extension of `Iapplication`, and therefore must have the following line:

```
YourClass extends Iapplication {...}
```

So, is your class both an extension of another class and a class on its own? Yes. `Iapplication` is called a superclass because other classes are derived from it. In iAppli development, all applications derive from this single superclass.

The next line declares the `MainCanvas` class:

```
MainCanvas gc;
```

Taking a look at the `Maincanvas` class, which does the actual drawing onscreen, you see the `setColor`, `fillRect`, and `drawString` methods of the `Graphics` class being called. Though other flavors of Java have `fillOval` and `drawOval` methods, and the MIDP has the `fillarc` and `drawarc` methods, the J2ME4DoJa does not have these methods. But even without these as native methods, the low-level API provides you the basic tools you need to draw circles or ovals, though it is a lot more trouble than if it simply included the methods in the API. In this case, however, the methods are sufficient to find the width and height of the screen, and fill the screen with a white box. You then change the color that the `Graphics` class uses to draw to the screen, and use the `drawString` method to write your string directly, and exactly, at the specified coordinates onscreen in the new color (red). The `Graphics` class also allows you to set the font you want (`setFont`), draw lines (`drawLine`), and do many other things with the screen.

You notice that this is a lot of work to write four simple lines of text. Of course, if this is all you wanted to do, you could much more easily use cHTML or the high-level API, which is discussed later on in the High Level API section. The `Canvas` class is quite useful for precisely positioning text and graphics onscreen. This is important in any sort of graphical game or application, and is something that cannot be done using cHTML. In addition, because you can position the text on the screen, you can also change that position, thereby moving your text and graphics. This can be done in a few different ways, but all involve a loop.

USING LOOPS TO FILL THE SCREEN

In Java, there are three kinds of loops: `for` loops, `while` loops, and `do-while` loops. So, using the simplest of the three, the `for` loop, you can encapsulate the screen draws in your hello world example in a loop, and play with the x or y coordinates that tell the `Graphics` class where on the screen to draw the strings. By increasing incrementally, you create animation. Listing 12-2 shows what the code would look like after you try this:

Listing 12-2: Using Loops to Move Text Around the Canvas

```
import com.nttdocomo.ui.*;

public class Hello extends IApplication{
private static Thread runner;
      MainCanvas gc;

      public void start() {
            gc = new MainCanvas();
            Display.setCurrent(gc);
      }
public class MainCanvas extends Canvas implements Runnable {

      public MainCanvas() {
}

      public void run() {
            while (true) {
                  try {
                        Thread.sleep (2000);
                  }
                  catch (Exception e) {}
                  repaint();
            }
}
```

Continued

Listing 12-2 *(Continued)*

```
public void paint(Graphics g) {
    int i;
    int x;
    int y;
    for ( i = 1; i<40;i++){

g.setColor(Graphics.getColorOfName(Graphics.WHITE));
        g.fillRect(0, 0, getWidth(), getHeight());


g.setColor(Graphics.getColorOfName(Graphics.RED));
        x=-40+i;
        y=10 +i;
        g.drawString("Hello World", x, y);
        y=30+i;
        g.drawString("Nice to meet you", x, y);
        x=40-i;
        y=50-i;
        g.drawString("What can I do", x, y);
        y=70-i;
        g.drawString("do for you?", x, y);
        }
    }
}
```

Figures 12-3 and 12-4 show the two steps in the process.

Figure 12-3: The graphic starts in various positions.

Figure 12-4: It then moves into alignment.

There are a few things that are different about your first example and the previous code. The first is that you have done something strange with your MainCanvas class. What you have done is to allow threads to call particular instances of the class. The reason you did this is to use the capability of a thread to sleep for a given length of time, thereby enabling precise timing of your animation. This capability of threads, as well as many others, makes them invaluable, especially when working with some widely varied processors, as is the case with i-Appli-enabled mobile phones. On Sony's SO503i, for example, screen draws and processing are very speedy — up to four times faster, in fact, than on Panasonic's P503i. If you did Listing 12-2 without threads, the speed that it ran at would depend on the processor, which is not acceptable if you try to animate something smoothly.

The next thing you notice is the declaration of the variables. In Java, unlike in PHP, for example, you cannot use a new variable without first declaring it to be of type whatever (in this case int, which is integer). The variable i is consequently used in the for statement, and the x and y variables are used in drawString method calls.

One last, important thing is that you took the locks off the graphics. Listing 12-1 had this set of lines:

```
public void paint(Graphics g) {
            g.lock();
```

The lock is designed to hide the screen until all the processing has been done. It is the capability to view the processing, however, that actually creates the animation, and so locking the screen is not what you want to do.

This modified Hello World example is exactly the kind of thing you saw when Java applets first made their debut. It is not terribly useful, but it is cool to be able to animate your little mobile phone screens a little. I didn't use the example to show you how to do simple animation, but to point out the possibilities of the Canvas class. Your example is, in fact, quite crude. But think about being able to allow a user to control the movement of a graphic, rather than text, and how you might use that in your application. The previous example is just the first step in that direction.

Using the Panel class or the high-level API

Learning to use the low-level API first may seem to be a bit backward, and in some ways it is. The reason I did this was to give you a sense of the sort of control that you have using the low-level API, and conversely, the amount of detail you need to include in a class based on the `Canvas` class. Using the high-level API, as in the previous analogy, is like having lots of middle managers between what is happening to the bit players (excuse the pun) in the basement and the application on the top floor. Rather than having to specifically tell the peons exactly what to do, the application can give more general instructions. This means less work for the application, but also less control in appearance and less specificity in what can be done. The reason I started with the `Canvas` class was to give you a clear idea of what you are giving up and what you are gaining in using the `Panel` class.

DISPLAYING AND ALIGNING TEXT IN A LABEL

So, get started. Boring as it is, you will do pretty much the same thing with the `Panel` class as you did with the `Canvas` class, which is to do a simple message (although this time, the message will be a little more creative than Hello World).

Actually, the code for doing this is very short, as follows (see Listing 12-3):

Listing 12-3: A Simple Panel to Display a Message

```java
import com.nttdocomo.ui.*;

public class HelloWorld extends IApplication {
        public void start() {
                Panel panel = new Panel();
                Label label = new Label("Hi Ma!");
                panel.add(label);
                Display.setCurrent(panel);
        }
}
```

Wow, that was simple! You create a new panel, create a label, put the label in the panel, and display the panel. Very straightforward, and very easy. Figure 12-5 shows what this looks like:

Figure 12-5: A simple display using the Panel class.

Notice that although it is easy, it doesn't look great. You can actually determine the alignment of the label by inserting the numbers 0 (left), 1 (center), or 2 (right); or by specifying a position, like this:

```
Label centerLabel = new Label("Hi Ma!", Label.CENTER);
centerLabel.setSize (96,15);
```

The first line could also be the following, which would look like Figure 12-6:

```
Label centerLabel = new Label("Hi Ma!", 1);
```

Figure 12-6: This is what happens when you use alignment.

There is no difference, except that Label.CENTER is more intuitive. So, for people who have a hard time remembering which codes mean what, that would probably be a good choice.

This would align the text in the middle (in the middle of the label whose size you have to define). If you don't define it, the label is, by default, only as long as the text it contains, and using the alignment attribute does nothing because it aligns the text *within* the label.

DISPLAYING TEXT WITH THE TEXTBOX CLASS

You also have the Textbox class, in which you can display text. As its name suggests, this class is quite similar to the HTML form element of the same name. Many of the classes used with the Panel class, in fact, are a lot like form elements. One difference is that the textbox class can be used to display text, not to only accept user input.

Listing 12-4 shows a quick example of how the Textbox class works, and Figure 12-7 shows the result.

Listing 12-4: Using the Textbox Class in the Panel

```
import com.nttdocomo.ui.*;

public class TextBoxExample extends IApplication {
    TextBox textBox;

    public void start() {
        Panel panel = new Panel();

        textBox = new TextBox(
            "Hey ma, I'd like to write a letter, but I am
                busy now", 20, 5,
            TextBox.DISPLAY_ANY);

        panel.add(textBox);

        Display.setCurrent(panel);
    }
}
```

Figure 12-7: A simple example of the Textbox class.

There are several fields of this class that can be set. You can see four of them in this example. The actual text to display is the first, second is the number of columns, third is the number of rows, and last is the input mode.

The first three of these fields are quite clear, I think. The last, the input mode, requires a little explanation. In the previous example, you have it set to DISPLAY_ANY. You can also set this field to DISPLAY_PASSWORD, ALPHA, KANA, or NUMBER. Setting DISPLAY_PASSWORD means that anything entered in the box is displayed as asterisks. KANA is the Japanese input method.

To allow or disallow a user to edit whatever text is in the box, the setEditable method can be used. For example:

```
textBox.setEditable(false);
```

This only displays text, and does not allow a user to edit it. Please see the API in Appendix C for the other methods.

OFFERING CHOICES IN A LIST BOX

The next kind of Panel object is the listBox object. Again, this resembles the HTML form element in many ways. Listing 12-5 shows an example:

Listing 12-5: Using the listBox in the Panel

```
import com.nttdocomo.ui.*;

public class ListBoxExample extends IApplication {
    ListBox listBox;

    public void start(){
        Panel panel = new Panel();

        listBox = new ListBox(ListBox.CHOICE);
        listBox.append("Pizza");
        listBox.append("Cola");
        listBox.append("Dim Sum");
        listBox.append("Sushi");

        panel.add(listBox);

        Display.setCurrent(panel);
    }
}
```

This brings up a list box of type CHOICE, as shown in Figures 12-8 and 12-9.

Figure 12-8: The first item, Pizza, appears until you hit the select button.

Figure 12-9: The Japanese in the lower-left corner says Clear, and pressing it causes the list box to retract.

There are five other types of list boxes, and changing the bold in the following code to CHECK_BOX, MULTIPLE_SELECT, NUMBERED_LIST, RADIO_BUTTON, or SINGLE_SELECT creates other list boxes:

```
listbox = new ListBox(ListBox.CHOICE);
```

Figures 12-10 through 12-14 show each type of list box.

Chapter 12: i-Appli: The i-mode Version of Java 225

Figure 12-10: Using a CHECK_BOX. More than one item may be checked.

Figure 12-11: Using a MULTIPLE_SELECT.

Figure 12-12: Using a NUMBERED_LIST. An item is selected using a handset's number keys.

Figure 12-13: Using a RADIO_BUTTON.
Only one item can be selected.

Figure 12-14: Using a SINGLE_SELECT.

I have only covered how these buttons look and what they allow a user to select, but there are many methods available in the Textbox class that determine how data is requested from users, how input is received, and more. Please see the API for more information and details on how to implement these objects.

USING THE BUTTON OBJECT

The next object available from the Panel class is the Button object. Just like it sounds, it writes a button to the panel. Listing 12-6 shows a very simple script to do exactly that:

Listing 12-6: Using the Button Object in a Panel

```
import com.nttdocomo.ui.*;

public class ButtonExample1 extends IApplication {
    Button button1,button2;
```

Chapter 12: i-Appli: The i-mode Version of Java

```
        public void start() {
                button1 = new Button("Button 1");
                button2 = new Button("Button 2");

                Panel panel = new Panel();

                panel.add(button1);
                panel.add(button2);

                Display.setCurrent(panel);
        }
}
```

This displays two buttons on the screen named, appropriately enough, Button 1 and Button 2. Figure 12-15 shows what the screen looks like:

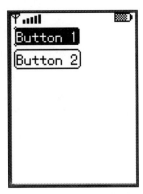

Figure 12-15: Using the button object in the Panel class.

Consult the API in Appendix C for the methods associated with this object. It is pretty straightforward, and there isn't really anything that is going to significantly change the way it looks to the user, but the methods let you do the things you need to make this class useful in a program. You can use the `ComponentListener` class to detect whether a button has been pushed or not, and can then associate that with an action.

CREATING A SCROLLING MESSAGE

The next object available in the `Panel` class is the `Ticker` object. This object basically is like marquee in cHTML or HTML, pushing a ticker-like line of text from right to left across the screen. It is a straightforward class with only one method, which is the text to scroll across the screen.

Listing 12-7 shows an example:

Listing 12-7: Using the Ticker Object to Display a Scrolling Message In a Panel

```
import com.nttdocomo.ui.*;

public class TickerExample extends IApplication {
    public void start(){
        Panel panel = new Panel();
        Ticker ticker = new Ticker("Better read this fast
            before...");
        panel.add(ticker);

        Display.setCurrent(panel);
    }
}
```

This scrolling is pretty slow on the emulators that go with the J2ME4DoJa emulator and the i-Jade P503i emulator, so I assume that this is probably a standard speed of scrolling, and using threads with a sleep method is not neccesary.

Figures 12-16 and 12-17 show the screen shots from the DoJa emulator and the i-Jade, just for a bit of variety:

Figure 12-16: The DoJa emulator's rendering of the Ticker object.

Chapter 12: i-Appli: The i-mode Version of Java

Figure 12-17: The same text and object in i-Jade's P503i emulator.

USING IMAGE LABELS

The next class object for use in the `Panel` class is the `ImageLabel` object. Basically, this is similar to an `img` tag in HTML. It displays a non-moving image in-line on the screen. Listing 12-8 shows some simple code to demonstrate this, and Figure 12-18 shows the result.

Listing 12-8: Using the imageLabel Object in a Panel

```
import com.nttdocomo.ui.*;
import com.nttdocomo.io.*;

public class ImageLabelExample extends IApplication {
    public void start() {
        MediaImage mediaImage =

        MediaManager.getImage("resource:///address.gif");
        try {
            mediaImage.use();
        } catch (ConnectionException ce) {}
        Image image = mediaImage.getImage();
        ImageLabel imageLabel = new ImageLabel(image);

        Panel panel = new Panel();

        panel.add(imageLabel);

        Display.setCurrent(panel);
    }
}
```

Figure 12-18: Placing an image in the panel with ImageLabel.

There are no placement methods in the class itself, but this class, like `ListBox`, `Button`, `Label`, `TextBox`, `Ticker`, and `VisualPresenter`, inherits methods from the `com.nttdocomo.ui.Component` super class, which allows you to set a background color, the location, the size, and the foreground color. Please see the API for more details.

 You should keep in mind when you use graphics that the graphics are included in the 10 KB size limit of midlets. Your graphic in this example is the same one you used in Chapter 11 to demonstrate building an address book, so it is already optimized to its smallest possible size. The whole file size is still 1118 bytes, though. Remember, you have only 10 KB! I used up over 10 percent of it just by putting a graphic on the screen. You can save this space by using the `com.nttdocomo.io.HttpConnection` class to download your graphics each time you need them, rather than compiling them into your `.jar` file, but then you have to worry about how long it takes them to download.

CREATING SIMPLE ANIMATIONS

I reiterate this limitation in anticipation of the next class object, which can be used with the `Panel` class: the `VisualPresenter` class. This is the class you use to display animated gifs, currently the only kind of moving files supported by the DoJa classes. If you are not familiar with gif animations, they basically include a number of gif files piled one on top of the other, with some rules about how fast they can be played thrown in for good measure. You are much better off, if your animation is more than a couple of frames, to use the `Canvas` class and the `MediaManager.getImage` method to get an image, and use the x, y coordinates to move the image around. This works for a lot of the things people use animated gifs for, though not all.

Chapter 12: i-Appli: The i-mode Version of Java

I would be remiss for not at least showing you how to use the `VisualPresenter`, though (see Listing 12-9):

Listing 12-9: Using the VisualPresenter to Show an Animated gif

```
import com.nttdocomo.ui.*;
import com.nttdocomo.io.*;

public class VisualPresenterExample extends IApplication {
      VisualPresenter visualPresenter;

      public void start(){
            Panel panel = new Panel();

            MediaImage mediaImage =
                  MediaManager.getImage(
                        "resource:///addressanimation.gif");
            try{
                  mediaImage.use();
            } catch (ConnectionException ce) {}

            visualPresenter = new VisualPresenter();
            visualPresenter.setImage(mediaImage);

            panel.add(visualPresenter);

            Display.setCurrent(panel);
            visualPresenter.play();
      }
}
```

This class first opens a panel, gets a gif animation, tries to see whether there is a problem with the animation, and if so, stops the program. If there is no problem, the `VisualPresenter` class is called, and the image is set. Then, the `VisualPresenter` is added to the panel, displayed, and the playback started.

I should mention one thing about the way this performs looping. Though you are able to define how many times an animation loops when you build an animated gif, for some reason `VisualPresenter` treats even those looping animations with finite looping as infinite loops.

The resulting data, which it isn't exactly possible to display in its animated glory, looks something like Figure 12-19:

Figure 12-19: Using VisualPresenter to display an animated gif.

By adding just a couple of lines of code, you can make this a little more useful, by allowing you to stop this infinite loop:

```
panel.setSoftKeyListener(this);
panel.setSoftLabel(Frame.SOFT_KEY_1, "Stop");
.
.
        public void softKeyPressed(int no) {
                visualPresenter.stop();
        }
        public void softKeyReleased(int no) {}
}
```

This sets the softkey listener and label, and then tells the VisualPresenter to stop playing if it detects that the softkey has been pressed. It looks basically the same, with the addition of the softkey labeled Stop (see Figure 12-20).

Figure 12-20: Using VisualPresenter with a softkey to display an animated gif.

INCLUDING AUDIO FILES

Along with playing video, you can also use audio in your applications. The audio can be played back using the `AudioPresenter` class. Listing 12-10 shows an example, though I encourage you to try it yourself because there is no way to demonstrate how it sounds without actually running the midlet.

Listing 12-10: Using the AudioPresenter Class to Output Sounds

```
import com.nttdocomo.ui.*;
import com.nttdocomo.io.*;

public class AudioPresenterExample extends IApplication {
    AudioPresenter audioPresenter;

    public void start() {
        audioPresenter = AudioPresenter.getAudioPresenter();
        MediaSound mediaSound =
                MediaManager.getSound("resource:///sound.mid");
        try{
            mediaSound.use();
        } catch (ConnectionException ce) {}
        audioPresenter.setSound(mediaSound);
        audioPresenter.play();
    }
}
```

There is one very important thing to remember. In Listing 12-10, you used `sound.mid`. The emulator cannot play native i-mode sound file-formatted files that end with `.mld`. But if you want this to actually be playable on an i-mode phone, you have to convert the sound from MIDI to MFi format, and change the name of the resource before you build the application.

For information on how to create and convert sound files to i-mode's specific format, see Chapter 10.

ADDING DIALOG BOXES

From the `Panel` class, you also have dialog box objects available. There are several that can be used:

    DIALOG_ERROR
    DIALOG_INFO
    DIALOG_WARNING
    DIALOG_YESNO
    DIALOG_YESNOCANCEL

The appearance of these dialog boxes is completely dependent on the model of phone, and is generally of a style consistent with the user interface on the phone. Please see Appendix C for specifics of how each dialog box is implemented on various handsets.

The error, info, and warning dialog boxes simply display a header, a message, and a button at the bottom to exit the dialog box. The code in Listing 12-11 works for all three kinds of boxes, though you will, of course, need to change the type of dialog from `Dialog.DIALOG_ERROR` to `Dialog.DIALOG_INFO` or `Dialog.DIALOG_WARNING`.

Listing 12-11: Using Dialog Objects in the Panel

```
import com.nttdocomo.ui.*;

public class ErrorDialogExample extends IApplication {
        public void start() {
                Dialog dialog = new Dialog(Dialog.DIALOG_ERROR,
                    "ERROR");
                dialog.setText("Error: Something bad happened");
                dialog.show();
                terminate();
        }
}
```

Figure 12-21 shows the error dialog box. Notice the button in Japanese at the bottom. Because the way the dialog boxes display is completely dependent on how this class is implemented on a particular handset, there is no way around this, if you want to use dialog boxes.

Figure 12-21: An example of an error dialog box.

The other two types of dialog boxes ask a user for some input, displaying buttons to take that input. Again, the way these buttons are displayed depends on the handset, and you should see Appendix C for specifics.

Listing 12-12 contains an example of the DIALOG_YESNOCANCEL type:

Listing 12-12: A YesNoCancel Dialog Object in the Panel

```
import com.nttdocomo.ui.*;

public class YesNoCancelDialogExample extends IApplication {
        public void start() {
                Dialog dialog = new Dialog(Dialog.DIALOG_YESNOCANCEL,
                        "YesNoCancel");
                dialog.setText("You want to do something?");

                int result = dialog.show();

                Dialog dialog2 = new Dialog(Dialog.DIALOG_INFO,
                        "Result");
                switch (result) {
                        case Dialog.BUTTON_YES:
                                dialog2.setText("You said yes, you
                                        proactive honey!");
                                dialog2.show();
                                break;
                        case Dialog.BUTTON_NO:
                                dialog2.setText("You said no, you
                                        dullard!");
                                dialog2.show();
                                break;
                        case Dialog.BUTTON_CANCEL:
                                dialog2.setText("Are you a member of
                                        GenX?");
                                dialog2.show();
                                break;
                }
                terminate();
        }
}
```

Again, the way the three buttons actually display can be found in Appendix C. Figure 12-22 shows how the code appears on the emulator, which is not actually the way they would display on any particular model:

Figure 12-22: The YesNoCancel dialog box.

This chapter touched on the basics of building applications in both the high-level API (also called the `Panel`) and the low-level API (also called the `Canvas`). It is not expected that this brief summary of methods will actually enable you to build an application, but is intended instead to famiarize you with the environment in which you are working.

This book really is meant to be an overview of all things i-mode, and if you are going to be writing Java code, it is expected that you have at least some experience with that. A clever person can teach himself a lot by looking at examples, emulating those examples, looking up parts he doesn't understand, changing the examples, and experimenting. If you are that sort of person, there is probably enough material in the previous examples to allow you to teach yourself. If you would prefer more structure in your learning of Java, I suggest reading *Java 2 For Dummies* by Bill Burd or one of Hungry Minds' other fine titles.

Chapter 13

Programming i-Applis: A Tutorial

IN THIS CHAPTER, YOU DEVELOP A NEW APPLICATION from the beginning. This process will take you through all of the following stages:

- Thinking about what you want the application to do
- Deciding the way in which you achieve the application
- Building the back end
- Declaring the variables and building the main method
- Building the interface
- Connecting to the network to enable client-server functionality
- Retrieving the data from the server
- Formatting your information
- Getting the flow of interfaces and user interaction matched
- Finishing and testing the application

These steps will show a basic application develop from idea to fully functional implementation.

What You Want the Application to Do

You have had time to ingest a fair amount of information about Java since Chapter 12, but it is now time to start thinking about what you are doing with this knowledge. Because many of you are the Gen-X types that will probably never answer this question to your own satisfaction, you end up doing that which comes most naturally — improving that thingamabob you did before. Yes, it is back to the old address book.

Let me apprise you of what exactly it is you are going to do with your i-Appli.

I start this process with "Wouldn't it be cool if?" questions, from the user's point of view.

Wouldn't it be cool if

- you didn't have to go through five menus before getting to an address?
- you could choose which information you wanted displayed on the screen?
- you could have nifty address-grabber animations make you feel happy every day?
- the price of gasoline would go down to below a dollar again?

So, you have your parameters of what you want to have happen with this midlet. The first one, which I would argue is the most useful of the lot and the one most likely to attract users to your service, is luckily also the easiest. By allowing a user to store the client on their own handset, you eliminate connecting, going through the i-mode menu to My Menu, and selecting the application from My Menu page. You eliminate about three pages right there.

Think about an application that does the bare minimum as far as racking up connection charges, too. It just gets the information as efficiently as possible, and displays that information to your user. In this process, one thing you won't have problems with is figuring out whether a handset is the 503 series or not because they won't be able to download the midlet if it isn't. If the user agent doesn't give you the information you need to verify a user and to verify their model, you shut them out.

The second and third parameters from the preceding list are of a customization nature, and the animation that you created in Chapter 12 is frivolous for the sort of application you are developing. I say this with full knowledge that as an American adult male, having an animation hand me an address is not exactly a killer app, while to a young Japanese female, this would make it a killer application. Don't believe me? Just look at the "Post Pet" phenomenon – millions of users have a cute bear or other animated character go to a visual postbox, and check to see if there is any e-mail there. Anyway, we will leave that to the creative folks at our one-man operation (that means me, much later in the millennium).

I will leave the price of gasoline to OPEC's kind offices.

When challenged, laziness seems the better part of valor. Allow a user to determine what information he wants displayed. If he never uses anything except phone numbers, let him display only phone numbers. If he uses only e-mail, that is all he needs to display. By eliminating unwanted information, you earn your bread and cut some of the clutter that the user feels at having to scroll through information for which he has no use.

So, this gives you the basic outline of a midlet! How this midlet will do its work is another question. There are a few things to keep in mind. The first is that if you look at your application written in PHP, with all of the files, functions, and so on, it is much larger than your 10 KB size limit. Another thing to keep in mind is your overwhelming antipathy to coding any sort of function twice when you should be able to reuse what you did the first time.

Building the Back End

Your environment, the Connected Limited Device Profile (CLDP), obviously does not have the functions that PHP does for connecting to your MySQL database. Although you have your PHP functions sitting in a library, ready to be reused, they don't work very well for your i-Appli. So you need to build a simple script that repeats much of what you did previously in Chapter 10. You also need a way to connect to your MySQL database, and you are as lazy as any self-respecting programmer should be. The answer is obvious! You use PHP on the server and Java on the client; and you lessen your work, ensure some compatibility between your cHTML version of your address book, and get to do lots with the application without having it take up lots of memory. Voila!

The code for the PHP script that will do most of the communicating with the midlet appears in Listing 13-1.

Listing 13-1: The addressappli.php Code

```
<?php
header ("Content-type: text/plain");
include ('config.inc');
include ('textwrap.php');
$connection = mysql_connect("$dbhost", "$dbuser", "$dbpw");
    mysql_select_db ("$db");
 if ($state==2){
     $searchfor = "^".sql_regcase($searchstring).".*";
     $result=mysql_query("SELECT * FROM addresses WHERE fname REGEXP
'$searchfor'
        OR lname REGEXP '$searchfor' HAVING uid ='$uid'");
     if (mysql_numrows($result) != 0){
$num = mysql_numrows( $result );
for ($i=0;$i<$num;$i++){
echo mysql_result($result,$i,fname)."\n";
}
} else {print "No results found.\n";}
}
elseif ($state==3){
    switch ($category){

    case Home:
    $result=mysql_query("SELECT * FROM addresses WHERE fname= '$searchstring'
AND
        uid ='$uid'");
     if (mysql_numrows($result) != 0){
    $address=mysql_fetch_object($result);
```

Continued

Listing 13-1 *(Continued)*

```
        if ($address->homestreet1){$homeaddress=$address->homestreet1." ".$address-
            >homestreet2." ".
        $address->homecity." ".$address->homestate." ".$address->homepost."\n";}
        if ($address->homephone){$homephone="T:".$address->homephone."\n";}
        if ($address->homefax){$homefax="F:".$address->homefax."\n";}
        if ($address->homeemail){$homeemail="E:".$address->homeemail."\n";}
        $fulladdress=$address->fname." ".$address-
            >lname."\n".textwrap($homeaddress,14,'\n').$homephone.$homefax.$homeemail;
        print $fulladdress;}
        else {print "There was no data.\n";}
        break;

        case Work:
            $result=mysql_query("SELECT * FROM addresses WHERE fname= '$searchstring'
AND
            uid ='$uid'");
        if (mysql_numrows($result) != 0){
        $address=mysql_fetch_object($result);
        if ($address->companystreet1){$companyaddress=$address->companystreet1."
            ".$address->companystreet2." ".
        $address->companycity." ".$address->companystate." ".$address-
            >companypost."\n";}
            if ($address->companyphone){$companyphone="T::".$address-
>           companyphone."\n";}
        if ($address->companyfax){$companyfax="F:".$address->companyfax."\n";}
        if ($address->companyemail){$companyemail="E:".$address->companyemail."\n";}
        $fulladdress=$address->fname." ".$address->lname."\n".textwrap
            ($companyaddress,14,'\n').$companyphone.$companyfax.$companyemail;
        print $fulladdress;}
        else {print "There was no data.\n";}
        break;

        case Mobile:
            $result=mysql_query("SELECT * FROM addresses WHERE fname= '$searchstring'
            AND uid ='$uid'");
        if (mysql_numrows($result) != 0){
        $address=mysql_fetch_object($result);
        if ($address->mobilephone){$mobilephone="M:".$address->mobilephone."\n";}
        if ($address->mobileemail){$mobileemail=$address->mobileemail."\n";}
        $fulladdress=$address->fname." ".$address-
            >lname."\n".$mobilephone.$mobilemail;
        print $fulladdress;}
        else {print "There was no data.\n";}
        break;
```

```
}
}
else {print "A problem occurred with the i-Appli. Please download it again.\n";}

?>
```

So, what exactly is it that this code does? Reasonable question. First, it gets the `config.inc` file, which is used to get the `HTTP_USER_AGENT` variable, separated into reasonably named components. The `textwrap` function is used just as its name would suggest; it wraps text.

Next, you connect to the mySQL server using the `mysql_connect` command, which is the same server that your cHTML-based application connected to. The only reason you have this script is to communicate between the i-Appli and your database, so the connection will be needed in whichever situation the script is used.

The `addressappli.php` script will get two variables sent to it, which tell it what the i-Appli wants from it. The `state` variable tells the script which state the i-Appli was in when the information was sent, which in turn lets the PHP script respond with the correct information. The `category` variable lets the script know, in the case of a request from the i-Appli with a state of 3 (which means the request was sent from the panel used to display an address) what information to display. There are only three categories: `Home`, `Work`, or `Mobile`. It would be a simple task to modify the script to group the information in whatever manner one wanted, and would only require the i-Appli to add a menu option for this grouping.

The other `state`, state of 2, is used to ask a user whose address they would like to display. If you looked at Chapter 11, this is the same format you used with your cHTML-based application.

To see the cHTML-based usage of PHP for the address book lookup, see the section entitled "Searching and editing the entries" in Chapter 11.

Be sure to add `addressappli.php` to the same folder in which you have the address application created in Chapter 11. It shares some of include files with this application.

Building your main method

This seems to be a perfect case of where to use panels. You can use three different panels, one for the startup screen, which also has an input for the first letter of a name. The next panel is the lookup panel, which displays names starting with that letter, and the last one is the display panel, displaying the address. Listing 13-2 shows how this looks in Java:

Listing 13-2: Declaring your Panels, Softkey Listeners, and Component Listeners

```
public void start() {
 main_panel = new Panel();
 search_panel = new Panel();
 display_panel = new Panel();

 main_panel.setSoftKeyListener(this);
 main_panel.setComponentListener(this);
 main_panel.setSoftLabel(main_panel.SOFT_KEY_1, "Find");
 main_panel.setSoftLabel(main_panel.SOFT_KEY_2, "Quit");
 main_panel.setLayoutManager (null);
 mainInit();

 search_panel.setSoftKeyListener(this);
 search_panel.setComponentListener(this);
 search_panel.setSoftLabel(search_panel.SOFT_KEY_1, "Select");
 search_panel.setSoftLabel(search_panel.SOFT_KEY_2, "Back");
 search_panel.setLayoutManager (null);
 searchInit();

 display_panel.setSoftKeyListener(this);
 display_panel.setComponentListener(this);
 display_panel.setSoftLabel(search_panel.SOFT_KEY_1, "Back");
 display_panel.setSoftLabel(search_panel.SOFT_KEY_2, "Quit");
 display_panel.setLayoutManager (null);
 displayInit();

 Display.setCurrent( main_panel );
 state = MAIN;
}
```

This code basically declares and sets your panels and softkeys, as well as the component listeners, which listen to text being changed in text boxes, option lists being clicked on, and so on. You use the `state` variable to let the program know which state it is in. You also use it, as you saw in your PHP script, to tell the script which data to send.

Be sure to add this file to the same directory as your address application because they share some of the same `include` files.

Building the Interface

The previous code declares all sorts of things, but it still hasn't actually shown the user anything. Figure 13-1 shows you where you should be headed.

Figure 13-1: The starting screen.

Showing the user a screen like Figure 13-1 is done by the one part that doesn't exactly make sense in Listing 13-2: the calls to the `mainInit()`, `searchInit()`, and `displayInit()` methods. These are user-defined methods, which I use to initialize all sorts of things in each panel, such as the menus, color of the background, text boxes, and so on. Listing 13-3 shows the guts of these panels in their `Init` methods: These are the elements of the user interface.

Listing 13-3: Three Methods that Declare the Elements of the User Interface

```
void mainInit(){
 main_panel.setBackground(Graphics.getColorOfName
 ( Graphics.WHITE ) );

 MediaImage mediaImage =
     MediaManager.getImage("resource:///address.gif");
 try {
  mediaImage.use();
  }
 catch (ConnectionException ce) {ce.printStackTrace();}
 Image image = mediaImage.getImage();
 int imagelength=(main_panel.getWidth()-image.getWidth())/2;
 ImageLabel logo = new ImageLabel(image);
 logo.setLocation (imagelength,0);

 Label label1 = new Label("First/last name:");
 label1.setLocation (0,40);

 textbox1 = new TextBox("",8,1,TextBox.DISPLAY_ANY);
 textbox1.setLocation (0,55);

 main_panel.add(logo);
 main_panel.add(label1);
 main_panel.add(textbox1);
 }
```

Continued

Listing 13-3 *(Continued)*

```
void searchInit() {
  search_panel.setBackground(Graphics.getColorOfName(Graphics.
    WHITE));
  sp_label = new Label("Choose a name");
  int  labellength=(search_panel.getWidth()sp_label.getWidth())/2;
  sp_label.setLocation (labellength,0);
  listBox = new ListBox(ListBox.RADIO_BUTTON);
  listBox.setLocation (0,20);

  categoryBox = new ListBox(ListBox.CHOICE,3);
  categoryBox.append("Home");
  categoryBox.append("Work");
  categoryBox.append("Mobile");  categoryBox.select(0);
  categoryBox.setLocation (0,60);

  search_panel.add(sp_label);
  search_panel.add(listBox);
  search_panel.add(categoryBox);
}

void displayInit(){

display_panel.setBackground(Graphics.getColorOfName(Graphics .WHITE)
    );
  addressdisplay = new TextBox("",14,20,TextBox.DISPLAY_ANY);
  addressdisplay.setLocation (0,2);

  display_panel.add(addressdisplay);

}
```

This is the nuts and bolts of what is displayed in the panels. The `labellength` variable in the `searchInit` method has a useful suggestion on how to do centering because this isn't one of the two options using the `setLocation` property, which takes only `right` or `left`.

Besides that, for each component of the panel, you need to do the following:

1. Declare the component and define its properties: `sp_label = new Label("Choose a name");`.

2. Set its location — `sp_label.setLocation (labellength,0);`.

3. Add it to your panel — `search_panel.add(sp_label)`.

This is a necessary process. Setting the location is necessary, unless you want all the components stacked on top of one another. Though you have not done it in this application, positioning components correctly really requires a good understanding of the properties of the particular handset on which the application runs. Spacing lines of text 15 pixels apart, as I have mostly done in this application, is a guess at best, and the chances are that it will look bad on some models of phones. This is the sort of thing that can't be emulated, and that really does seem to call for actual testing on i-mode handsets. After you do that, it is a good idea to write a routine to display the components in a way that fit each handset.

This leads to an important fact about i-Appli: In its security-conscious zeal to protect users, NTT DoCoMo has rendered much of i-Appli less usable than even cHTML. How so? In cHTML, as you saw in Chapter 11, you can get both a handset's serial number and the model number. Ironically, the easiest way to get a handset's properties is through an external method such as HTTP_USER_AGENT. There is a method of system, which is System.getProperty("microedition.platform"). This is standard Java, except the value you pass to this method. You are not given a list of these values in the API. And without this value, you will not get the expected response. How many other properties are available? One can get what all of the possible values are by using the getProperty() empty, but this is a pain. It also turns out, for instance, that with this particular property, handset makers have implemented it in a non-standard way. Following is a short list of what is returned:

F503i	f50x
D503i	D503i
SO503i	SO503i
N503i	pdc
P503i	Panasonic P503i
P503iS	Panasonic P503iS

It is fair to assume that handset makers will clean this up, or that DoCoMo will dictate that they clean it up. For now, however, the HTTP_USER_AGENT property of the header is probably more useful because you can use this in your PHP script to look up a particular handset's characteristics, as you saw in Chapter 11. But how to get information such as serial numbers and handset models from a script to a midlet? You can ask a user to input the information, but that adds more complexity than most users want to deal with, and they would probably abandon your service. That is not good. I used a technique that has the PHP script actually writing the jam (ADF) file. The jam file is the file that describes what is in the actual application, or jar file. It is also called an application descriptor file (ADF). By generating it dynamically in PHP, I can write to the AppParam field, which allows the jam file to pass variables and their values to the midlet. This allows me to make a custom application for each user, with information specific to them, including username and handset type. The code looks like Listing 13-4:

Listing 13-4: The PHP Code that Writes the ADF (.jam) File, address.php

```
<?php
include ('config.inc');
header ("Content-type:application/x-jam");
echo "PackageURL = address.jar\n";
echo "AppSize = 3569\n";
echo "AppName = address\n";
echo "AppVer =\n";
echo "AppParam = uid=".$utn." handset=".$model."\n";
echo "AppClass = address.address\n";
echo "SPsize =\n";
echo "UseNetwork = http\n";
echo "LastModified = ".date("D, d M Y H:i:s")."\n";
echo "KvmVer =\n ";
?>
```

Basically, this listing tells the KVM that it is a `jam` file in the header and then proceeds to output the contents of your `jam` file. You have customized two fields: the `AppParam` field and the `LastModified` field. The `AppParam` field was discussed previously as a way to pass values from the script to the midlet. The `LastModified` field doesn't actually need customization because it needs to change only when you update your application. It is used to tell the KVM whether any current versions stored on the handset are older than the version being sent. I just changed this because when I debugged my application, it saved me having to answer the question of whether or not I wanted to install the same application again. The `date` function in PHP can also take a `timestamp` value, so I can also tell it to look at the date the script was uploaded to the server, which automatically returns a new date only when the script has been uploaded anew. This falls into the category of tweaks, though, and you can ignore those for now.

So far, this is pretty simple. To get to the previous script, all you really need to do is to link to it, which you can do from your `utnlogin.html` page. Following is the new line you add to that page to achieve this:

```
<a href="address.php" accesskey="4" utn>Get i-appli</a><br>
```

Simple! All that is left is to ensure that there is a `jar` file with the bytecode in it that is where the `jam` file says it is — `address.jar`. This requires actually coming up with an application. Oh, bother! Yes, yes, scripting is pretty easygoing, as is declaring simple variables in Java, but now you need to get to the guts of this sucker. Before you get to that, look at a simple diagram of what a user needs to do when he downloads the application, as shown in Figure 13-2.

Chapter 13: Programming i-Applis: A Tutorial   247

Downloading an i-Appli

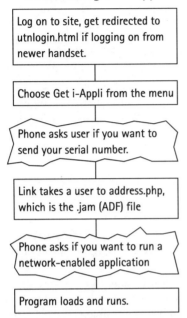

Figure 13-2: The flow of downloading the application.

Figure 13-3 shows what a user must do to use the application.

Obtaining an i-Appli Address

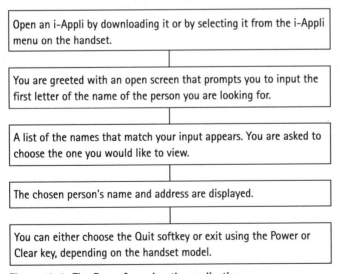

Figure 13-3: The flow of running the application.

So, you have most of your housekeeping done, and now it is time to fill the house with your little children. Would that the procreation of your i-Appli were quite as much fun as that of your children! This sexy little app ought to at least get your juices flowing, though, and drive you to higher and higher heights, and I better let up on the idioms here before you go somewhere you don't want to be going.

Basically, the flowcharts give you an outline of what you want to do in the application. Most of the work in Figure 13-2 has already been done: The PHP scripts take care of the downloading end of things. The flow of actually running the application shown in Figure 13-3 is where the real work of this i-Appli gets done. The next section takes you through coding the functions outlined in Figure 13-3.

Using the Network in the i-Appli

Your first bit of code is that which will get you connected to the network. This is something that seemed so daunting to programmers at first that the first i-Applis were mostly not connected to the network. The first book on writing i-Applis purely taught standalone i-Appli programming. Why is this so difficult? Well . . . because it is. It probably had lots to do with the dearth of good information on the subject. From the mailing lists and the questions and problems that people talk about in making their i-Applis work, this seems to be the number one problem. Take a look at some common (and easily remedied) mistakes:

- Forgetting to set the `UseNetwork` parameter in the `jar` file to `http`. You go nowhere unless you do this first. The parameter should look like this:

    `UseNetwork = http`

- Using a URL to connect to your PHP script that is not exactly the same as the one that the i-Appli came from. Even if it works (that is, you leave off the `www` at the beginning of the address), you get a connection exception. Rather than explicitly stating the URL, use the `getSourceURL()` method, which gives you the right URL.

- There are three methods that you can use to connect to an HTTP server, and they all require different coding to work. The `GET` method and the `HEAD` method require only an input reader because the data is defined in the URL. But URL encoding the data is important! And the header method doesn't actually return any values. The `POST`, on the other hand, uses an input and output stream to work, and it does not work if the data is encoded using URL encoding.

The three methods all have their points, but I suggest that you choose one and stick with it. Only when you are thoroughly familiar with the ins and outs of it, think about another. Because there is a 255-byte limit on URL length, using either the `HEAD` or `GET` method involves the risk of your URL and your data being passed in the URL of getting parsed, and data consequently being lost. I have never

Chapter 13: Programming i-Applis: A Tutorial

personally run into this problem, though I have heard warnings about it since I started doing Web pages in 1994. Go figure. But the 255-byte limit is a real one, and why bother? Just start using the POST method, and you definitely won't need either of the other two. Listing 13-5 shows the first part of my code for the getnames() method. Much trial-and-error went into getting it to behave correctly.

Listing 13-5: The First Part of the getnames() Method

```
public void getnames () {
HttpConnection hc;
InputStreamReader in;
OutputStream out;
String param[] = getArgs() ;
String buf = "";
String parameters =   "state="+state+"&searchstring="+searchstring+"&category="
  +category+"&"+param[0];
String url = getSourceURL()+"addressappli.php";
try {
  hc = (HttpConnection) Connector.open (url, Connector.READ_WRITE, true);
  hc.setRequestMethod(HttpConnection.POST);
  hc.setRequestProperty("Content-Type", "application/x-www-form-urlencoded");

  out=hc.openOutputStream();
  out.write (parameters.getBytes());
  out.close();
  out=null;

  hc.connect();
  in = new InputStreamReader(hc.openInputStream());
```

So, what does it do? First, you define your variables. Because this script has exclusive access to your network connection, you declare your variables as local variables, available only in this method. The parameters variable is basically all the data you send to the server. The param[0] value is the user id that was written to the AppParam field and returned using the getArgs() method. In your AppParam field, you have two arguments, although in this part of the script you only use the first.

Notice that I used the getSourceURL() for the URL string. At first, I did this to be absolutely sure I had the right URL in there, but it turns out that it is also pretty helpful when testing because it means you don't have to keep changing a hard-coded URL all the time, depending on what platform is serving the midlet.

Though this is just a snippet of the code, and you can't see the end of the try method, it doesn't actually do anything with the caught exception.

Inside of the try method, you first open a connection with this line:

```
hc = (HttpConnection) Connector.open (url, Connector.READ_WRITE, true);
```

Because you are using the `POST` method, you must then open an output stream to write your data to the PHP script. This is because you don't pass the parameters in the URL, and it is different from the `GET` and `HEAD` methods.

After you have closed your `OutputStream` and declared it `null` just to be safe, you open an `InputStreamReader`. This is something strongly suggested by DoCoMo because the Java midlet uses Unicode internally, but handsets use SJIS for input and output. That means that technically every input and output needs to be converted by using the `reader` method. I didn't do this when writing because I know which variables will be passed, and in the basic ASCII set there is no difference between Unicode and SJIS. If, however, I expected to write kanji (or any other language encoding for that matter), I would have used the `OutputStreamWriter` method. This method ensures that the output is properly coded, something that a simple `OutputStream` doesn't do.

What the incoming data contains, however, I don't know and have no control over, so I use the `InputStreamReader()` just to be safe.

There is an important method lacking in the CLDC: the `readLine()` method. I am not sure why this was left out, but it adds extra work for you. In your PHP script, data is read out in lines with \n line breaks. All you really want to do is get the names, separated by line breaks and then get the addresses, separated by line breaks. All pretty simple with `readLine()`, and a bit of a pain without it. The `InputStreamReader` method reads the input byte by byte, which means that you have to read all the bytes of the line into a variable to read a line, stop when you find the \n character, write the line to the display, go back and read the next line, and repeat the process until no more data can be read (see Listing 13-6).

Listing 13-6: The Second Half of the getnames Method

```
in = new InputStreamReader(hc.openInputStream());
 String s=" ";
 if (in!=null){
  int n=1;
  if (state==GETNAMES){
   while((s=readLine(in)) != null) {
    listBox.append(s);
   }
  }
  if (state==DISPLAY){
   StringBuffer sb= new StringBuffer();
   while((s=readLine(in)) != null) {
    sb.append(s);
   }
   buf=sb.toString();
   addressdisplay.setText(buf);
  }
 }
 else if (state==DISPLAY){
```

```
  addressdisplay.setText("No names to display");
}
else if (state==GETNAMES){
  sp_label.setText("No names found");
  }
in.close();
hc.close();
```

Okay, what is going on here? The variable s is initialized initially with a space and then the `InputStreamReader` is checked to see if it contains anything. If it does, the state is checked. This is because depending on the state, the information received gets sent to different panels. If the state is equal to GETNAMES, you send the input to the readLine method. Listing 13-7 shows the method created for this workaround:

Listing 13-7: The readLine Method

```
public String readLine (InputStreamReader in){
  String s="";
  int dat=0;
  try {
    while ((dat=in.read()) != '\n' && dat !=-1){
      if (dat !='\f' && dat !='\r')
      s=s+ (char) dat;
    }
  }  catch (Exception e) {}
  if (dat ==-1)
  {return null;}
  else
  {return s+'\n';}
}
```

Coding your own method is a good idea, and it serves the object-oriented dictates of Java, which say that you should separate functions into methods. In actual fact, a really good job of coding would also separate your network connection from your data input handling.

So, anyway, back to the main story. You basically use readLine with a while loop to read in all of the data received, sending it line by line to whichever panel it needs to go to.

That was simple! Yeah, but the devil is in the details, so pay close attention to some of those: The GETNAMES state takes each line and appends it as a list item. You can also read all of them into an array and append the whole array. I don't know which is better, but I am a bit of a simpleton, and a lazy one at that, and I avoid those complicated arrays if I can. In the DISPLAY, on the other hand, you read the entire set of data into a string buffer and append it to the text box when you get to the end. Again, this is easy, and easy is good.

Hey, that's it! Except for some messages if there is nothing contained in the database, you are finished. Wow, piece of cake. No problem!

SoftkeyListeners, Component Listeners, and their Actions

Softkeys are the two keys, generally at the top left and right of the handset, which can be given values to display at the bottom of the screen, unlike other keys. Components are the text boxes, lists, and buttons that make up the components of the high-level API's `Panel` class. A listener must be set for softkeys, components, and keys – though you didn't use the last. When one of these listeners detects an action, it sends that action to the `softKeyPressed()` method or `ComponentAction()` method, which initiates an action based on what key was pressed or component changed.

I have set a `softKeyListener` and `ComponentListener` for all three panels. If data from components or softkeys are to be available to these panels, they must have listeners set. The compiler allows you to compile the program without them, however, so beware! Even if it allows this, the emulator then crashes, and you get messages that are cryptic about problems with threads. Okay, I admit it. I did forget these at first. Hey, to err is human. I am very human.

Listing 13-8 shows the actions methods for the listeners that you already set in the code in your main class. Listeners are set in the main method, and when they hear something happen, they send it to either the `softKeyPressed` or `ComponentAction` method.

Listing 13-8: The componentAction Method

```
public void componentAction(Component c, int type, int param) {
  if(c == textbox1) {
   tbsearchstring = textbox1.getText();
  }
  else if (c==listBox){
   int searchindex = listBox.getSelectedIndex();
   lbsearchstring = listBox.getItem(searchindex);
  }
  else if (c==categoryBox){
   int searchindex2 = categoryBox.getSelectedIndex();
   category = categoryBox.getItem(searchindex2);
  }

}

public void softKeyPressed(int key) {
 switch (key){
```

```
case Frame.SOFT_KEY_1:
 if (state==MAIN||state==DISPLAY){
  state = GETNAMES;
  listBox.removeAll();
  searchstring=tbsearchstring;
  getnames();
  Display.setCurrent(search_panel);
 }
 else if (state==GETNAMES){
  state = DISPLAY;
  searchstring=lbsearchstring;
  getnames();
  Display.setCurrent(display_panel);
 }
break;
case Frame.SOFT_KEY_2:
 if (state==MAIN||state==DISPLAY){
  terminate();}
 else if (state==GETNAMES){
  state = MAIN;
  Display.setCurrent(main_panel);
 }
break;
}
}

public void softKeyReleased(int key) {}
```

All kinds of fun here. Look at the ComponentAction method first. Basically, you have three components (textbox1, listBox, or categoryBox) that might be acted on. If the ComponentAction method is called, you need to know which of them was called in order to get its value correctly. You do this with my die-hard habit of if clauses. Figure 13-4 shows what happens on a handset when the textbox component is selected, and Figure 13-5 shows the input data.

Figure 13-4: Preparing to enter the first letter of a first or last name to search.

When the `textBox1` component is selected to enter data, the screen changes to a data entry screen. This is handled automatically, and is handled somewhat differently by each handset. In this case, the bottom left has an input method selection, the middle has a finished selection, and the right has a backspace option. Though the application is in English, this is part of the handset's settings, and can't be changed from the i-Appli, so it is in Japanese.

Figure 13-5: The first letter is entered.

When you get values from lists, you first must determine which item, which is an integer, is checked. Then, you get the following item:

```
int searchindex = listBox.getSelectedIndex();
lbsearchstring = listBox.getItem(searchindex);
```

In the case of the text box, you simply get the text. The reason that I use different variables to assign the input to is that when you use the back key from the display panel, you want to be able to see your selections again. But if you assigned the input to `searchstring`, it would contain the name that you had selected before, rather than the letter that you first input. That would defeat the purpose of going back.

That is pretty much it for your `ComponentAction` method. I suppose you could also automate it so that when someone put in a letter, it would automatically start the search without having to hit any softkeys, but that isn't necessary, and giving the user the choice of sending it is a good idea, I think.

Okay, the softkeys. They are really the basis of your "action" in this script. They are your means of changing panels. I should point out something important (which I didn't fully appreciate before using the DoJa profile): The panels just sort of sit there, their properties may be appended to or changed, and can be called whenever they are needed, all of the components in whatever state they were left in, unless you tell them otherwise. In a sense you are building a little card, made up of components and the data one enters into those components. Both the component and the data are sitting there until the panel is told to display, using the `Display.setCurrent()` method.

So, what does your `softKeyPressed()` method actually do?

First, you use case to check which key was pressed: softkey1 (left) or softkey2 (right). Wow, he can use if, he can use switch, this is a two-trick wonder dog! Back to your case and back to that old standby if clause, which you use to check what the state is. The sofKeyPressed() method is thrown whenever a softkey is pressed, and there is no inherit distinction between someone pressing a softkey that says Quit and one that says Back. The state variable's current value is used to make this distinction.

Figure 13-6 shows a screen in which the first letter of a name has been entered into the text box. The softkey Find needs to be pushed to send the data to the server and move to the next screen. Figure 13-7 shows the screen you are taken to.

Figure 13-6: After the center selection that means "finished," is selected, you are returned to the search screen, and the text box has your entry in it.

Figure 13-7: You choose the name to look up. In this case, there is only one option.

Review for a moment what the softkeys do when pressed in all three panels:

Main Panel:
Find   Quit

Search Panel:
Select   Back

Display Panel:
Back   Quit

This is like one of those brainteasers — there are three screens with two buttons on each screen. You want to use the fewest number of tests that you have to, so how would you do it? Look at your panels and how you have used the softkeys and the reasonably minimal four tests performed, and I think this is probably the minimum of what can be done. Just my guess because there are no answers in the back of the book.

So, using a combination of case and if clauses, I take each softkey input and do something with it. What?

In the case of softkey1 being pressed, I first change the state so that when it gets sent to the server using the getnames() method, the correct state will be there. Next, I assign either the data entered into the text box or list box to the variable searchstring. Again, the getnames() method uses it to send the query to the server, so you want to make sure it is available. Then, you call the getnames() method. Finally, you display your panel that getnames will have modified, writing text or a list box to it.

In the case of softkey2, two of the three panels use this as a Quit button, so in case either of those buttons got pressed, you simply terminate the application. In the case of the search panel, it is used as a Back button, which takes you back to the main panel.

That's it! Basically your application is nearly done. Notice the following line:

```
public void softKeyReleased(int key) {}
```

Basically, the softKeyListener needs to be able to send either a key pressed event or a key released event somewhere. You need the previous line to ensure that when a softkey is released, the event has somewhere to go. Without it, an error will be thrown.

Figure 13-8 shows an option list that comes up in the select screen, and 13-9 shows the data in the components waiting to be sent on by a softkey event.

Figure 13-8: The option list for the location comes up when selected and you select work.

Figure 13-9: The options are all selected, just waiting for the Select key to be pressed, taking you to your address.

The Complete Source

Bringing all of this together is a relatively simple thing, and Listing 13-9 shows your product. It is 4,747 characters long, and comes out to a 3.5 KB `jar` file, which also includes about 360 bytes of a logo graphic. That means that you only used a third of the resources you could have. Think about adding twice again as much functionality to this i-Appli, and it starts to get exciting. You didn't use the scratchpad at all, which is a way that some developers have found of getting around the memory limitations.

For information on compiling your source code, see Chapter 12, which takes you through this process from beginning to end.

Listing 13-9: The Complete Listing for address.java

```
package address;
import com.nttdocomo.ui.*;
import com.nttdocomo.io.*;
import com.nttdocomo.net.*;
import javax.microedition.io.*;
import java.io.*;
import java.util.*;

public class address extends IApplication implements
    SoftKeyListener,ComponentListener {
 TextBox textbox1;
 ListBox listBox;
 ListBox categoryBox;
```

Continued

Listing 13-9 *(Continued)*

```java
  TextBox addressdisplay;
  int state;
  Label sp_label;
  Panel main_panel;
  Panel search_panel;
  Panel display_panel;
  int MAIN = 1;
  int GETNAMES = 2;
  int DISPLAY =3;
  String searchstring;
  String lbsearchstring;
  String tbsearchstring;
  String category="Home";

  public void start() {
   main_panel = new Panel();
   search_panel = new Panel();
   display_panel = new Panel();

   main_panel.setSoftKeyListener(this);
   main_panel.setComponentListener(this);
   main_panel.setSoftLabel(main_panel.SOFT_KEY_1,      "Find");
   main_panel.setSoftLabel(main_panel.SOFT_KEY_2,      "Quit");
   main_panel.setLayoutManager ( null );
   mainInit();

   search_panel.setSoftKeyListener(this);
   search_panel.setComponentListener(this);
   search_panel.setSoftLabel(search_panel.SOFT_KEY_1,    "Select");
   search_panel.setSoftLabel(search_panel.SOFT_KEY_2,    "Back");
   search_panel.setLayoutManager (null );
   searchInit();

   display_panel.setSoftKeyListener(this);
   display_panel.setComponentListener(this);
   display_panel.setSoftLabel(search_panel.SOFT_KEY_1,   "Back");
   display_panel.setSoftLabel(search_panel.SOFT_KEY_2,   "Quit");
   display_panel.setLayoutManager (null );
   displayInit();

   Display.setCurrent( main_panel );
   state = MAIN;
  }
```

```java
public void componentAction(Component c, int type, int     param) {
 if(c == textbox1) {
  tbsearchstring = textbox1.getText();
 }
 else if (c==listBox){
  int searchindex = listBox.getSelectedIndex();
  lbsearchstring = listBox.getItem(searchindex);
 }
 else if (c==categoryBox){
  int searchindex2 =        categoryBox.getSelectedIndex();
  category = categoryBox.getItem(searchindex2);
 }

}

public void softKeyPressed(int key) {
 switch (key){

  case Frame.SOFT_KEY_1:
  if (state==MAIN||state==DISPLAY){
   state = GETNAMES;
   listBox.removeAll();
   searchstring=tbsearchstring;
   getnames();
   Display.setCurrent(search_panel);
  }
  else if (state==GETNAMES){
  state = DISPLAY;
  searchstring=lbsearchstring;
  getnames();
  Display.setCurrent(display_panel);
  }
  break;
  case Frame.SOFT_KEY_2:
  if (state==MAIN||state==DISPLAY){
  terminate();}
  else if (state==GETNAMES){
  state = MAIN;
  Display.setCurrent(main_panel);
  }
  break;
 }
}
```

Continued

Listing 13-9 *(Continued)*

```java
 public void softKeyReleased(int key) {}

 public void getnames () {
  HttpConnection hc;
  InputStreamReader in;
  OutputStream out;
  String param[] = getArgs() ;
  String buf = "";
  String parameters = "state =" + state+"         &searchstring= "
    +searchstring + "&category="    +category+ "&"+param[0];
  String url = getSourceURL()+"addressappli.php";

  try {

   hc = (HttpConnection) Connector.open(url,
     Connector.READ_WRITE,true);
   hc.setRequestMethod(HttpConnection.POST);
   hc.setRequestProperty("Content-Type",    "application/x-www-
     form-urlencoded");

   out=hc.openOutputStream();
   out.write (parameters.getBytes());
   out.close();
   out=null;

   hc.connect();

   in = new InputStreamReader      (hc.openInputStream());
   String s=" ";
   if (in!=null){
    int n=1;
    if (state==GETNAMES){
     while((s=readLine(in)) != null) {
     listBox.append(s);
     }
    }
    if (state==DISPLAY){
     StringBuffer sb= new       StringBuffer();
     while((s=readLine(in)) != null) {
      sb.append(s);
     }
     buf=sb.toString();
     addressdisplay.setText(buf);
    }
```

```
      }
      else if (state==DISPLAY){
       addressdisplay.setText("No names to       display");
      }
      else if (state==GETNAMES){
       sp_label.setText("No names found");
       }
      in.close();
      hc.close();
     }
     catch (Exception e) {e.printStackTrace();}
    }


    void mainInit(){
     main_panel.setBackground(Graphics.getColorOfName(
       Graphics.WHITE ) );
     MediaImage mediaImage =
     MediaManager.getImage("resource:///address.gif");
     try {
      mediaImage.use();
     } catch (ConnectionException ce)       {ce.printStackTrace();}
     Image image = mediaImage.getImage();
     int imagelength=(main_panel.getWidth()-       image.getWidth())/2;
     ImageLabel logo = new ImageLabel(image);
     logo.setLocation (imagelength,0);

     Label label1 = new Label("First/last name:");
     label1.setLocation (0,40);

     textbox1 = new TextBox("",8,1,TextBox.DISPLAY_ANY);
     textbox1.setLocation (0,55);

     main_panel.add(logo);
     main_panel.add(label1);
     main_panel.add(textbox1);
     }

    void searchInit() {
     search_panel.setBackground(Graphics.getColorOfName(Graphics.
        WHITE));
     sp_label = new Label("Choose a name");
     int labellength=(search_panel.getWidth()-
      sp_label.getWidth())/2;
     sp_label.setLocation (labellength,0);
```

Continued

Listing 13-9 *(Continued)*

```
  listBox = new ListBox(ListBox.RADIO_BUTTON);
  listBox.setLocation (0,20);

  categoryBox = new ListBox(ListBox.CHOICE,3);
  categoryBox.append("Home");
  categoryBox.append("Work");
  categoryBox.append("Mobile");
  categoryBox.select(0);
  categoryBox.setLocation (0,60);

  search_panel.add(sp_label);
  search_panel.add(listBox);
  search_panel.add(categoryBox);
 }
 void displayInit(){
  display_panel.setBackground(Graphics.getColorOfName(Graphics
    .WHITE));
  addressdisplay = new  TextBox ("", 14, 20,
    TextBox.DISPLAY_ANY);
  addressdisplay.setLocation (0,2);

  display_panel.add(addressdisplay);

 }
 public String readLine (InputStreamReader in){
  String s="";
  int dat=0;
  try {
   while ((dat=in.read()) != '\n' && dat !=-1) {
    if (dat !='\f' && dat !='\r')
    s=s+ (char) dat;
   }
  }  catch (Exception e) {}
  if (dat ==-1)
  {return null;}
  else
  {return s+'\n';}
 }
}
```

Testing the Application

In reality, testing is done little by little as you are developing the application. Each method, as you code it, should be tested. This is not always easy to do because many methods do not work separately from others. Using `System.out.println()` to output values of variables may be necessary. Commenting out suspect code may be necessary.

A couple of really basic things to keep in mind:

- Every time you change the code, be sure to recompile.

- Make sure your resources are available. Code will compile without them, but won't run without them. This means sounds and graphics.

- If your application uses a network connection, be sure to declare that in the `jam` file. And be sure that network resources are available when you test, including the mySQL server.

- If you use network connections, be sure that your `POST`, `GET`, or `HEAD` method is formatted correctly.

- Do not forget that each component needs an action assigned to it in the `componentAction()` class. The action can be `null`, but unless it has one, you may get a thread error.

- Make sure that your `address.php` file, which writes the `jam` file for you, has the correct size of the `.jar` file. A normal `jam` file will change this automatically when compiling, but in your PHP file you need to do this manually. Without the correct size, the application won't load.

After you have done all these things, and your application is working, you use an emulator to test it. Figures 13-10 shows the final screen, the address found in the database.

Figure 13-10: The name is displayed in the title, and the address shows in the text box.

A Book Paradigm

Now that you have shown what you can do using a Java client and a back end database server, I'll discuss what else you could have done and other paradigms about how you might do it.

One of the things that Nikkei has done with its site is to allow a user to download a midlet that contains the day's news. The advantage of this paradigm is that after the midlet has been downloaded, it does not need to communicate with the server because all the information is actually contained in the midlet itself or in the scratchpad. This is good for people who might use the application in the subway because the reception is not good underground (in Japan, many subway stations have mini-antennae, so it is actually possible sometimes to get reception).

Your application, too, could use this method. Rather than simply downloading an already-made midlet, you can use a server-side application to get a user's addresses, build the application with the addresses built in, and allow a user to download a true address book, in the sense of a book being a container of information (and a quite portable one at that).

This would require that the code automatically be preverified and then compiled into bytecode. This would be problematic using PHP, and I do not know how difficult it would be using server-side Java. It is a project for a very good programmer, and one who is more proficient than I am at server-side Java.

An alternative is that the midlet downloads all the information to the scratchpad, where it sits and is retrieved by the midlet when needed. This is quite a reasonable way of doing it, but begs the very real question of whether the data would fit in the 5 KB maximum used in the scratchpad. If there were, as in my case, about 120 names in the address book, very little information from each entry could be stored before you reached the limit — about 40 characters worth. This is clearly not enough for someone who wants to look up an address. For phone numbers only, it might be okay, but there are probably at least as many people with bigger address books than mine as with smaller ones.

Building a client-server, data-driven application using Java for i-Appli is a particularly good use of Java. It is a win-win situation, in which the server application is made more useful because of the added flexibility of the client, and in which the Java client is given only the information needed, when it is needed, because of its capability to connect at any time to the server.

This address book application is a clear example of an application that uses backend data with a front-end Java client to provide a user access to data stored in a database. This sort of application is especially attractive to businesses that want to empower their workers with the ability to conveniently and seamlessly view corporate data whenever and wherever they need it.

Chapter 14

Creating an i-Appli Game

IN CHAPTER 13, YOU USED THE HIGH-LEVEL API, or the `Panel` class, to code an address book that connects to a server to retrieve addresses. In this chapter, you use the low-level API, or the `Canvas` class, to create a game.

Creating the Project

In order to create this game, you first need to have an idea of what the game will do. For a very good reason, many of the first i-Applis resemble things done in the past as primitive computer games or video games — it is easier to come up with these sorts of ideas, and not particularly frowned upon in Japan. This project, too, will revive childhood memories.

The game that I decided to create is one that I played when I was a kid. It was a handheld thing that looked a lot like a calculator. Notes played in a random sequence, and little red LEDs next to the chicklet keys lit up to indicate that was the key to get that note. If you were able to correctly repeat the notes in the correct order, you cleared that screen. That is really about all that I remember about the game, but it has planted the idea of what I would like to do in this chapter.

Your program has one advantage that the game I remember didn't have — it has a color screen. So, rather than having little lights next to the keys light up, which mobile phones don't have, this game uses color squares on the screen that "light up." Another potential advantage is that the sounds are almost certainly better on the phone handsets than they were on the game I remember.

With a basic idea of what I want to do with the game, now is a good time to draw a diagram showing the flow of play. Figure 14-1 shows the flowchart of how I want play to proceed in this game:

266 Part III: Developing i-mode Applications

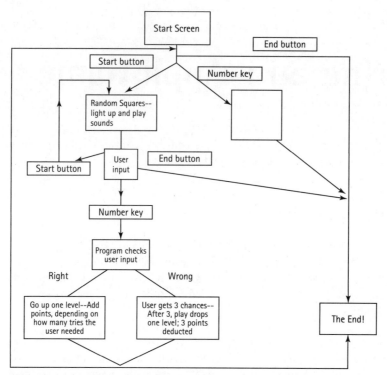

Figure 14-1: A flowchart of how play will proceed in the game.

Drawing on the screen

So, now that you know how the game will flow, you just need to actually make a game that does this. Because you know the number of keys on a keypad, either nine or ten, you can figure out how many squares to draw. Don't use ten because it is difficult to draw ten squares. It's easier to use nine squares, and don't forget our motto — easy is best.

So, your first assignment is to draw the nine squares. You want the game to look reasonably pleasing, which means that an awkwardly placed square is out. So, check the screen's width, figure out how wide and tall each square needs to be, and then paint the squares onto the screen. Listing 14-1 shows the parts of the code for this action.

Listing 14-1: The Main Class and Parts Used for Drawing the Initial Squares

```
import com.nttdocomo.ui.*;
import java.util.*;

public class square extends IApplication {
    private static Main main;
    private static Thread runner;
```

```
  Panel main_panel;

public void start() {
 main = new Main();
 Display.setCurrent( main );
 runner = new Thread( main );
 runner.start();
 }

public class Main extends Canvas implements Runnable {
 MediaSound ms[]= new MediaSound[10];
 AudioPresenter ap[] = new AudioPresenter[10];
 int key;
 int colornumber;
 int x = (Display.getWidth()-4)/ 3;
 int y = (Display.getHeight() -24) / 3;
 .
 .
g.lock();.
g.setColor(g.getColorOfName(g.YELLOW));
g.fillRect(0,0,Display.getWidth(),Display.getHeight();
for (int i=1;i<10;i++){
 g.setColor(g.getColorOfName(startcolor[i]));
 squarePosition(i);
 g.fillRect(startx,starty,x,y);
}
g.setColor(g.getColorOfName(g.BLACK));
g.drawString("Level "+h,2,(y*3)+20);
g.drawString("Score "+score,Display.getWidth()-
font.stringWidth("Score "+score) -2,(y*3)+20);
g.unlock(true);
```

Besides the main class declaration and loading the sounds, which have their own sections later in the chapter, the declarations for the x and y coordinates bear some review. Actually, "coordinates" is not quite accurate – they are the size of the box. The x variable is derived simply by dividing the width of the display minus 4 by 3. For example, you could use this formula:

```
int x = (Display.getWidth()-4)/ 3;
```

The 4 is the two-pixel buffer you want to set on either side of the display. The y variable is determined by dividing by 3 and subtracting 24, as follows:

```
int y = (Display.getHeight() -24) / 3;
```

As you can see, after the `for` loop, 20 pixels are used as a place to display the level, and four more are used for the two-pixel buffer at top and bottom. The program fills the screen with a yellow rectangle, which has the same height and width of the screen and serves as the background. Then, the script begins the `for` loop, which is used to draw all nine colored squares.

Listing 14-1 also sets the color to the `startcolor` array index. It loads the array at the beginning with the colors that you want to use. Next, it calls the `squarePosition()` method, which looks like Listing 14-2 and sets x and y depending on what number position you use.

Listing 14-2: The squarePosition() Method, which Sets the startx and starty Variables

```
public void squarePosition(int key){
 switch (key){
  case 1:
   startx=0;
   starty=0;
   break;
  case 2:
   startx=x;
   starty=0;
   break;
  case 3:
   startx=x*2;
   starty=0;
   break;
  case 4:
   key=4;
   startx=0;
   starty=y;
   break;
  case 5:
   startx=x;
   starty=y;
   break;
  case 6:
   startx=x*2;
   starty=y;
   break;
  case 7:
   startx=0;
   starty=y*2;
   break;
  case 8:
   startx=x;
   starty=y*2;
   break;
```

```
    case 9:
      startx=x*2;
      starty=y*2;
      break;
    default:
      break;
  }
}
```

This switch is quite simple: simply set the `startx` and `starty` variables. It does this for each of the nine rectangles, using the `i` variable of the `for` loop to get the starting coordinates from the `squarePosition()` method. After the coordinates have been set, this script uses the `fillRect()` method to draw to the canvas.

When the `for` loop is finished, the program sets the color to black, and draws the level and score at the bottom of the screen. The score uses some math to calculate how far to the left to draw the string. It uses the `Font.stringWidth()` property to figure out how long the string is, takes two more pixels off of that for the frame, and subtracts this number from the total width of the screen. This result gives you the x coordinate's location where to start writing the string.

Processing events

Now that you have drawn the main screen, the action needs to begin. How do you do this? The easiest way, and the way most comfortable for a player, is to start the action when the users say that they want to start, by hitting a key. Key presses, media events, and timer events are processed in the `processEvents` method. This part of the code is where the action takes place.

Okay, so the user hits a key that starts the game. The event is sent to the `processEvent` method. Once in the method, you should check what sort of event it is. Each event has a binary value, and a `KEY_PRESSED_EVENT` has a value of 0. By using a switch, you can direct the flow of action between the various events that you have going on. This game actually uses four different sorts of events: key pressed events, key released events, media events, and timer expired events.

As the program starts, you need to separate the action within the various events, especially the key events, by which key is being pressed. For this, you can use `if` clauses. In Listing 14-3, you can see the code for the key pressed event:

Listing 14-3: The processEvent Method with a Switch to Separate Events

```
public void processEvent (int type, int key){
  switch (type){
    case 0:
    if (key<10){
      if (state!=4) state=2;
      squarePosition(key);
      currentcolor=6;
```

Continued

Listing 14-3 *(Continued)*

```
   ap[key].play();
   draw=yellowbox;
   repaint();
  }

  if (key==Display.KEY_SOFT2){
   terminate();
  }

  if (key==Display.KEY_SOFT1){
   if (tries<3){
    try{
    for (int i=0;i<=h;i++){
    Random rand = new Random();
    pbkey[i]=Math.abs(rand.nextInt()%9)+1;
    Thread.sleep(30);
    }
    state=4;
    j=0;
    tries=tries+1;
    for (int i=0;i<=h;i++){
    squarePosition(pbkey[i]);
    draw=yellowbox;
    ap[pbkey[i]].play();
    repaint();
    Thread.sleep(300);
    currentcolor=startcolor[pbkey[i]];
    draw=boxcolor;
    squarePosition(pbkey[i]);
    repaint();
    Thread.sleep(300);
    }}catch (Exception e){}
   }
   else {
    if (h>1) h--;
    if (score>3)score=score-3;
    else score=0;
    draw=0;
    repaint();
    tries=0;
   }


 }
  break;
```

The only key events you really need to address are the softkey and the number pad keys. Others will be ignored. The second parameter in the `processEvent` method, what you assign to the key variable, passes the event's number. The number pad keys have the same value as the key pressed. The number you press is important, but you deal with this dynamically inside the `if` clause, which separates the number keys from other types of keys.

Next, consider what you want to have happen when a user hits a number key: You want the square that corresponds to the key to light up and then the music to play. Pretty simple. The trick is that you also need to check whether the user is attempting to imitate a playback that has just occurred, and if so, to capture his input to compare it to that of the one played back. Additionally, you need to make sure that if the users have attempted this level more than three times, their score is reduced by three. As a result, if they aren't already on level 1, they are moved down a level. If the user's input equals the number of notes in the playback, you check each note to see whether it is correct. If it is, you move the user up a level, and add points to the score. If the input doesn't match, you break out of the `case` construct, which also ends this particular event. A player who wants to try again will press the Start button again, and the number of tries it is taking him to correctly repeat increases by one.

Any events that are started need to be finished. So, when you start the note playing by a key pressed event, it makes sense to end the event when the key is released, which is `case 1` in the switch. I talk about the key-released events later in this section, but I want to warn you ahead of time that everything with a beginning must also have an ending.

In the case of a user entering number keys, ending an event is not a big deal because releasing the key is an easy way to know that the event should be stopped. But what about one key press initiating a series of other events, which must be caught? That is what pressing the softkey 1 button labeled "Start" does.

In Listing 14-3, the first `if` clause is catching a player's keypad entries. The next clause catches softkey 2, which ends the game; and the next catches softkey 1. This clause is crucial to the application, and bears a close look.

The program is first using the clause to set the state and test for whether the number of tries exceeds three. You test this here, in addition to the number entry area, because it is very possible that a player just hits the Start button again and again. There wouldn't be any number entry in that case, and the user wouldn't get taken down a level. Double-checking would also catch a case in which, for example, a user is on level 3, which plays four notes. If the user were to input only three and then hit the Start button, the action would be caught by adding this test at the beginning of the Start key.

After you perform these tests, you generate the random sequence to play back. Be careful! Random doesn't really mean random! The number is arrived at from the number of clock cycles that have expired since the program started, or something like that. As a result, you use the `thread.sleep` method — to put the clock forward a little. The pattern is still not totally random, and you still have lots of repeats; but without the sleep method, it simply repeats the same numbers. Actually having repeats seems to make it slightly more musical. You can experiment with the length (in milliseconds) of the sleep and get different sorts of patterns, depending on how long you make it.

Using a `for` loop to populate the array with random numbers between 1 and 9 is really all this `for` loop does. The `else` clause catches when users are over the allowed number of tries, and punishes them by moving them back a level and deducting points, as well as resetting several other counters.

After populating the array, you start the action of playback. Because you use two different `thread.sleep` method calls, it makes sense to just stick the whole loop in a `try` statement, and it probably saves you a few bytes of memory as well. You play back the randomly generated tune using another `for` loop. You set the first sort of box to draw (a yellow box) and the coordinates for the position of the box. You use the `redraw` method, which uses the generated coordinates to draw a yellow box. Then, it starts the playback of the sound, and sleeps for 300 milliseconds. The sound does not seem to be affected by putting the thread to sleep — if you start the sound before the sleep is called, it will play correctly. So, the yellow box appears at about the same time as the sound plays, and stays around for about one-third of a second.

Listing 14-4 shows the code for what is drawn in the `Graphics` method.

Listing 14-4: Drawing the Yellow Box over the Current Box

```
case 2: //In case you want to draw a yellow box
  g.lock();
  g.setColor(g.getColorOfName(g.YELLOW));
  g.fillRect(0,(3*y)+2,(x*3)+4,22);
  g.fillRect(startx,starty,x,y);
  g.setColor(g.getColorOfName(g.BLACK));
  g.drawString("Level "+h,2,(y*3)+20);
  g.drawString("Score "+score,getWidth()- font.stringWidth("Score "+score)-2,(y*3)+20);
  g.unlock(true);
break;
```

You then find what color the current box is by using the `startcolor` array. You set the draw mode to `boxcolor`, which draws a box of the starting color over the yellow box. This makes it appear that the box lights up yellow just one-third of a second, returning to its original color. You then use another `thread.sleep` method because you want some time between when one box plays and another starts. Because the program does not have a sleep timer at the very beginning of the loop, you need to have one here at the end. It doesn't really matter, but you need one between the end of one box event and the beginning of another. This is not exactly intuitive, and it took me more time than I care to admit for it to sink in. Think about it and you will probably get it, or maybe you are a heck of a lot smarter than I am, and get it without much thought. In Listing 14-5, you draw the original color back over the yellow box.

Listing 14-5: Drawing the Original Color Back to the Current Box

```
case 1://In case you want to return the box to its original color
g.lock();
g.setColor(g.getColorOfName(g.YELLOW));
g.fillRect(0,(3*y)+2,(x*3)+4,22);
below the game grid   g.setColor(g.getColorOfName(currentcolor));
g.fillRect(startx,starty,x,y);
g.setColor(g.getColorOfName(g.BLACK));
g.drawString("Level "+h,2,(y*3)+20);
font.stringWidth("Score "+score)-2,(y*3)+20);
g.unlock(true);
break;
```

And that is it! Cool! That really is pretty much it. You should probably look carefully at the KEY_RELEASED_EVENT, which is case 1, and what it does. As I said before, if a number key is pressed by a user, changing the color to yellow, and starting its sound to be played, returning it to its previous state would happen when the key is released. So that is the first thing you do. After that, you need to do some housekeeping.

This script doesn't do any of the checks on a user's input previously mentioned on the KEY_PRESSED_EVENT. I have no special reason for this, except that I knew that I would be using the KEY_RELEASED_EVENT anyway, and saw it as a good time to do this sort of work. So, after turning the sound off and returning the square to its original color, you hit the guts of the part that actually makes this a game. See Listing 14-6.

Listing 14-6: Using the KEY_RELEASED_EVENT Case in the processEvent() Method

```
case 1: //key released event
 if (key<10){
  currentcolor=startcolor[key];
  ap[key].stop();
  draw=boxcolor;
  repaint ();
  if (state==2&&tries<4){
   if (j<h){
    j++;
```

Continued

Listing 14-6 *(Continued)*

```
    uip[j]=key;
  } }
  if (state==4 && j==0){
   state=2;
   uip[0]=key;
  }
  if (j==h){
   int i=0;
   while (i<=h)
   {
    if (i==h&&uip[i]==pbkey[i]){
     h++;
     score=score+(4-tries);
     j=0;
     state=5;
     tries=0;
     break;
    }
    if (uip[i]==pbkey[i]){
     i++;
     continue;
    }
    else{
     j=0;
     state=5;
     if (tries>3){
      if (h>1) h--;
      if (score>3)
      score=score-3;
      else score=0;
      tries=0;
      draw=0;
      repaint();
     }
     break;
    }
   }
  }
 }
break;
```

Okay, after the obvious measures of turning off the sound, setting the color back to the square's original color, setting the draw mode to draw the starting color, and finally redrawing the screen, what is this listing doing? When I say that this portion is the guts of the game part of this application, I mean that this is the part that keeps track of the level, the score, and whether a user's input matches the randomly generated tunes.

First, you check to see whether the `state` is 2 or not. A `state` of 2 means that another key has already been pushed, and you have checked to see that the user is actually responding to a playback event. You add the current keystroke into the `uip` array.

If the `state` is 4, it means that the playback event has just ended. The only time you want to add user input to the `uip` array is when the user is actually responding to something. So you set the `state` to 2, and add the key input to the first register in the `uip` array.

After checking the state, you need to check to see if a user's input has reached the number of keys played back yet. Until that point, you just add her input to the `uip` array, but when her input reaches that number (h), you need to check the user's input against what is randomly generated and played back.

You use a simple `if` clause to check to see if the maximum number of input has been reached yet. If it has, you go into a `while` loop, checking each place in both the `uip` and the `pbkey` arrays against one another. If a mistake is encountered, the user input register, j, is reset. Also, the `state` is set to 5, which means a user has to push the Start button to get anything to happen. Furthermore, you test the `tries` variable to see if it has reached a value of 3 yet. If it has, a user must go down one level, and loses points. Finally, the screen is repainted to reflect the new situation.

Incorporating sounds

I promised that I would take a careful look at how sounds are used in this application. They are an essential part of the game – without them, there would be no game – and clearly understanding how to use them is essential.

Listing 14-7 shows the code for loading the sounds into the program. These sounds are quite short, and are used a number of times in the program, so loading them externally is not such a good idea. They are really an integral part of the game, and having them readily available within the game, without having to download them, is important.

Listing 14-7: The Main Class Loads the Sounds Before Running

```
public class Main extends Canvas implements Runnable {
  MediaSound ms[]= new MediaSound[10];
  AudioPresenter ap[] = new AudioPresenter[10];

  public Main() {
    try{
```

Continued

Listing 14-7 *(Continued)*

```
    for (int noteNum=0;noteNum<10; ++noteNum){
     String resloc = "resource:///"+noteNum+".mid";
     ms[noteNum]=MediaManager.getSound(resloc);
     ms[noteNum].use();
     ap[noteNum] =          AudioPresenter.getAudioPresenter();
     ap[noteNum].setSound(ms[noteNum]);
    }
   }catch (Exception e){}
   setSoftLabel(Frame.SOFT_KEY_1, "Start");
   setSoftLabel(Frame.SOFT_KEY_2, "End");
 }
```

You use two constructors, the `MediaSound` constructor, which you cast as an array because it will hold all of the notes, and the `AudioPresenter` constructor, which is also an array.

`AudioPresenter` contains a `MediaSound` object. `MediaSound` uses the `getSound()` method to retrieve the sound, in this case from the resource folder at compiling time. I named my files with simple numbers and an extension to make it easy to remember what they are called.

You should notice one extremely important thing about the name in this code example — it ends in the .mid extension. This is a standard MIDI file. Standard MIDI files don't, however, play in the i-Appli environment (as discussed in Chapter 10. The KToolbar from Sun is not able to use the i-Appli sound format, MLD. So, I must have two sets of sound files (in two different formats) in my resource folder if I want to use the emulator in the KToolbar to test the application. Don't forget, if you compile with .mid, the sound file doesn't run on i-Appli! You have to change the sound file's format before you can make it work with a real i-Appli phone. The KToolbar compiler compiles MLD files into the application with no problem. The Chaku Melo Convertor, included on the CD-ROM, can be used to convert MIDI files to MLD files, as discussed in Chapter 10.

If you use the i-Jade emulator, included on the CD-ROM, you will not have this problem because it plays native MLD files with no problems. Actually, I used the i-Jade emulator more often than the KToolbar, especially when I wanted to test right before I downloaded the files to the actual phone I used to do real-world tests.

For information on where to find and how to use the KToolbar and how to use a file converter, turn to Chapter 10. Remember that the CD-ROM contains a sample emulator, but the toolbar must be downloaded.

Speaking of problems with sounds on the emulators, both the KToolbar emulator and the i-Jade emulator use your computer's own MIDI board or the MIDI functions on your sound board. Depending on how good your sound card is, it will sound great, so-so, or really terrible. But none of this has much to do with how sound actually sounds on a real-world phone.

In this game, you use simple piano tones of the various notes. The note, at the very least, will be distinguishable as a note. The fact that it is a piano should also be distinguishable. But most of the note's other qualities are pretty highly dependent on how the MIDI chip was integrated by the phone manufacturer into the phone. Ringing tone services actually have a different version of each ringing tone optimized for each model of phone. You don't have the resources for that, but you do need to be aware that you will hear a lot of variance in how sounds sound.

Watching the Game in Action

Seeing the game in action is not really possible within the confines of a book. If you have come this far, I urge you to go to the CD-ROM, install the i-Jade emulator, and run the program.

Following, I have gone step-by-step through the steps of the game. Because this program doesn't use an opening screen or a closing one, there aren't very many steps.

278 Part III: Developing i-mode Applications

In Figure 14-2, you see the screen as it appears when first loaded.

Figure 14-2: The game when it first loads.

Chapter 14: Creating an i-Appli Game   279

In Figure 14-3, you see the screen as it appears when you press the Start button. On level 1, two different squares light up, playing tones at the same time.

Figure 14-3: Pressing the Start button causes the squares to light up.

In Figure 14-4, you see the score change after you correctly input the sequence.

Figure 14-4: Success! The level goes up to 2, and the score increases to 3.

That is really it. In a sense, this is a never-ending game because there is no score at which you win the game. Advancing to as high a level as possible could be the point, or accumulating as many points as possible.

Planning Possible Improvements

You can always improve upon what you have done, and this game is no exception. The following lists some suggestions of what to do with this game:

♦ In Chapter 13, you used a connection to the network to download addresses. How might it be possible to use the network connection to make this game more interesting? Hint: high scores.

- By using different screens, which would mean using different classes in the paradigm of the low-level API, how can you make the flow of the game more attractive? Hint: Start screen.

- How can you let a user know more clearly what the rules are? Hint: Rule screen.

- Is there a way that you could vary the piano tones with guitar tones to make the game more aurally interesting?

All these questions are good ones, and I am sure astute readers will have either more questions of their own, or have answers to these. The way you get to a killer game is to start with a pretty darned good one. What I tried to do in this example was to make a good game. If I want to take the game to the next level, and I definitely do, I have to start asking myself questions such as those I asked previously.

Besides these possible substantive changes, you can look at the whole question of presentation. My game, as I have designed it, is quite functional, but it is not especially attractive. I don't use any rounded edges, the squares are just squares, one right next to the other and stacked on top of one another. I don't use any font attributes to make the text at the bottom attractive, nor do I use a particularly interesting color for the font — black.

Keep in mind that the low-level API gives you total control over how your screen looks. Your canvas is 10,000 square pixels or thereabouts (literally). If you want to make a game that is not only fun, but also possibly sellable, you need to be a DaVinci with this canvas!

Besides the methods of the `Graphics` class I have used in this game, there are many more, and if you want to make an attractive-looking game, you need to familiarize yourself with them.

See the i-mode Java API in Appendix C for full details of the various methods of the `Graphics()` class, as well as all the other classes.

One of the things you definitely run into is the memory limitation. Although this game with the sounds compiled in it takes up barely 6K, the remaining four could easily be eaten up by some of the possible improvements I listed.

And last, but certainly not least, is having your friends play the game. They should probably find ways to crash it, cheat, and otherwise abuse it in ways that you yourself had not considered. Obviously, friends are important assets for more than testing out games for you. Their feedback is important, though watching them and how they play the game can be equally important. If they become engrossed by the game, you know they like it. If they put it down after ten seconds (even if they say "Great game!"), you can probably figure that it wasn't especially interesting to them.

Complete Code Listing

Listing 14-8 includes the complete listing of the code for this game i-Appli.

Listing 14-8: Complete Code for the Game i-Appli

```java
import com.nttdocomo.ui.*;
import java.util.*;


public class square extends IApplication {
 private static Main main;
 private static Thread runner;
 Panel main_panel;
 int key, currentcolor, counter;
 int tries=0;
 int boxcolor=1;
 int yellowbox=2;
 int initialize=3;
 int draw=3;
 int score=0;
 int x = (Display.getWidth() -4) / 3;
 int y = (Display.getHeight() -24) / 3;
 int j=0;
 int h=1;
 int startx=2;
 int starty=2;
 int input=1;
 int state=1;
 int startcolor[]={0,10,13,12,8,4,3,1,2,14};
 int pbkey[]=new int[30];
 int uip[]=new int[30];

public void start() {
 main = new Main();
 Display.setCurrent( main );
 runner = new Thread( main );
 runner.start();
 }

public class Main extends Canvas implements Runnable {
 MediaSound ms[]= new MediaSound[10];
 AudioPresenter ap[] = new AudioPresenter[10];

 public Main() {
 try{
```

Chapter 14: Creating an i-Appli Game

```
    for (int noteNum=0;noteNum<10; ++noteNum){
     String resloc = "resource:///"+noteNum+".mid";
     ms[noteNum]=MediaManager.getSound(resloc);
     ms[noteNum].use();
     ap[noteNum] = AudioPresenter.getAudioPresenter();
     ap[noteNum].setSound(ms[noteNum]);
    }
  }catch (Exception e){}
  setSoftLabel(Frame.SOFT_KEY_1, "Start");
  setSoftLabel(Frame.SOFT_KEY_2, "End");
  }

public void run() {
 while (true) {
  try {
   Thread.sleep(20);
  }
  catch (Exception e) {}
 }
}

public  void paint(Graphics g) { Font font = Font.getDefaultFont();
  switch (draw){
   case 1://In case you want to return the box to its original color
    g.lock();
    g.setColor(g.getColorOfName(g.YELLOW));
    g.fillRect(0,(3*y)+2,(x*3)+4,22);
    below the game grid g.setColor(g.getColorOfName(currentcolor));
    g.fillRect(startx,starty,x,y);
    g.setColor(g.getColorOfName(g.BLACK));
    g.drawString("Level "+h,2,(y*3)+20);
    font.stringWidth("Score "+score)-2,(y*3)+20);
    g.unlock(true);
    break;

   case 2:   //In case you want to draw a yellow box
    g.lock();
    g.setColor(g.getColorOfName(g.YELLOW));
    g.fillRect(0,(3*y)+2,(x*3)+4,22);
    g.fillRect(startx,starty,x,y);
    g.setColor(g.getColorOfName(g.BLACK));
    g.drawString("Level "+h,2,(y*3)+20);
    g.drawString("Score "+score,getWidth()-font.stringWidth("Score
       "+score)-2,(y*3)+20);
```

Continued

Listing 14-8 *(Continued)*

```
    g.unlock(true);
    break;

   default:
    g.lock();
    g.setColor(g.getColorOfName(g.YELLOW));
    g.fillRect(0,0,Display.getWidth(),Display.getHeight());
    for (int i=1;i<10;i++){
     g.setColor(g.getColorOfName(startcolor[i]));
     squarePosition(i);
     g.fillRect(startx,starty,x,y);
    }
    g.setColor(g.getColorOfName(g.BLACK));
    g.drawString("Level "+h,2,(y*3)+20);
    g.drawString("Score "+score,getWidth()-font.stringWidth("Score
       "+score)-2,(y*3)+20);
    g.unlock(true);
    break;
  }

 }
 public void processEvent (int type, int key){
  switch (type){
   case 0:
   if (key<10){
    if (state!=4) state=2;
    squarePosition(key);
    currentcolor=6;
    ap[key].play();
    draw=yellowbox;
    repaint();
   }

   if (key==Display.KEY_SOFT2){
    terminate();
   }

   if (key==Display.KEY_SOFT1){
    if (tries<3){
     try{
     for (int i=0;i<=h;i++){
     Random rand = new Random();
     pbkey[i]=Math.abs(rand.nextInt()%9)+1;
     Thread.sleep(30);
```

```
       }
      state=4;
      j=0;
      tries=tries+1;
      for (int i=0;i<=h;i++){
      squarePosition(pbkey[i]);
      draw=yellowbox;
      ap[pbkey[i]].play();
      repaint();
      Thread.sleep(300);
      currentcolor=startcolor[pbkey[i]];
      draw=boxcolor;
      squarePosition(pbkey[i]);
      repaint();
      Thread.sleep(300);
      }}catch (Exception e){}
    }
    else {
     if (h>1) h--;
     if (score>3)score=score-3;
     else score=0;
     draw=0;
     repaint();
     tries=0;
    }


 }
  break;

  case 1: //key released event
  if (key<10){
   currentcolor=startcolor[key];
   ap[key].stop();
   draw=boxcolor;
   repaint ();
   if (state==2&&tries<4){
    if (j<h){
      j++;
      uip[j]=key;
    }
   }
   if (state==4 && j==0){
   state=2;
   uip[0]=key;
```

Continued

Listing 14-8 *(Continued)*

```
      }
    if (j==h){
      int i=0;
      while (i<=h)
        {
        if (i==h&&uip[i]==pbkey[i]){
          h++;
          score=score+(4-tries);
          j=0;
          state=5;
          tries=0;
          break;
        }
        if (uip[i]==pbkey[i]){
          i++;
          continue;}
        else{
          j=0;
          state=5;
          if (tries==3){
            if (h>1)h--;
            if (score>3)          score=score-3;
            else score=0;
            tries=0;
            draw=0;
            repaint();
          }
          break;
        }
      }
    }
   }
  }
 }
}
public void squarePosition(int key){
  switch (key){
    case 1:
    startx=2;
    starty=2;
    break;
    case 2:
    startx=x+2;
    starty=2;
    break;
```

```
     case 3:
     startx=(x*2)+2;
     starty=2;
     break;
     case 4:
     startx=2;
     starty=y+2;
     break;
     case 5:
     startx=x+2;
     starty=y+2;
     break;
     case 6:
     startx=(x*2)+2;
     starty=y+2;
     break;
     case 7:
     startx=2;
     starty=(y*2)+2;
     break;
     case 8:
     startx=x+2;
     starty=(y*2)+2;
     break;
     case 9:
     startx=(x*2)+2;
     starty=(y*2)+2;
     break;
     default:
     break;
   }
 }
}
}
```

Part IV

Case Studies of i-mode: Implementations and Services

CHAPTER 15
Case Study 1: Walkerplus.com

CHAPTER 16
Case Study 2: Index Corporation

CHAPTER 17
Case Study 3: Nikkei

Chapter 15

Case Study 1: Walkerplus.com

THE CASE STUDY IN THIS CHAPTER PRESENTS THE STORY of a content provider whose content has not realized its full potential. That a company has content does not necessarily mean that it can make a profit on that content. This case study shows a company struggling to make its i-mode business economically successful.

Company History

Walkerplus.com is a young company, formed in April 2000, as a spinoff of the Walker series of magazines published by Kadokawa Publishing. It is 45 percent owned by Kadokawa, with Ticketmaster, JCB, NTT W, Sumitomo Shoji, Toshiba, Zenrin, and MBS as minor shareholders. Figure 15-1 shows the start screen of the Walker-i site.

Figure 15-1: Walker-i; start page of its free content site.

The eight Walker magazines for different cities and regions in Japan are city guides providing dining, shopping, and entertainment information. Walkerplus.com is the online site of the magazines, consisting of a wired Web site, which is supported by advertising, and Walker-i, a wireless Web site that also is sponsored by advertising. Walker-i, available on DoCoMo and EZ Web, is a free service with 220,000 users registered between the two carriers. The service started in December 1999 before the present company (Walkerplus.com, Inc.) existed, according to Tamotsu Gomajiri (Content Business Department General Manager) and company literature.

The copyright of Walker-i is owned jointly by Kadokawa and Walkerplus.com. Most of the material on the Walker-i site is still provided by Kadokawa and the Walker magazines. Of the 20 people who started with the establishment of Walkerplus.com, two or three were responsible for its i-mode and EZ Web content. The company now has about 35 employees, but only a few of them perform i-mode-related work. The digitization of content, data input, and server maintenance has been outsourced to Kadokawa Digix, a subsidiary company of Kadokawa Publishing. Most employees of Walkerplus.com are involved in planning and promotion.

The revenue model for Walker-i consists of an advertisement model. Tamotsu Gomajiri says that this model is not profitable at this point, and the company wants to change the site to a straight pay-for-content model in the future, if possible. In July 2000, Walkerplus.com released My Walker, a premium service for which users pay 250 yen per month. The service currently has about 20,000 customers.

Actually, Gomajiri says that a straight subscription service is the most comfortable revenue model for not only the company's i-mode services, but also for its wired services. The proliferation of free sites makes Gomajiri doubt that this model will be used for the wired services anytime soon. He believes that broadband is one area that may be suited to a straight subscription model, however. What sort of content Walkerplus.com could offer by broadband was a question that Gomajiri could not answer, however, and with Walkerplus.com dependent almost entirely on Kadokawa's content offerings, it is not clear what sort of value-added service might be possible on broadband.

Currently, the earnings on the i-mode site, from subscriptions to My Walker and advertising on Walker-i, are about 5 million yen per month, or US$40,000. The wired site generates revenues of about $400,000, more than 10 times the wireless site. The wired site also is supported by advertising, and the size of advertising contracts for wired versus wireless is substantial, with one contract earning Walkerplus.com as much money as the entire wireless site. Ad revenues, says Gomajiri, have gone down somewhat in the past year or so, but another business of Walkerplus.com, designing and hosting Web sites for restaurants and pubs, has picked up, so the company's revenue from the wired Web side is fairly steady. Walkerplus.com uses D2C (a joint venture between the largest ad agency in Japan, Denso, and NTT DoCoMo) as its agent for i-mode advertising.

If you do the math, you may be asking yourself how a company that makes $40,000 per month in revenues and has only three employees can be losing money.

Accounting is not my forte, but the amount Walkerplus.com pays to Kadokawa and Kadokawa Digix probably is a substantial part of operating costs. This level of revenues doesn't look too bad, and a company not tied to its biggest supplier of content would probably be able to show a profit somehow.

Gomajiri emphasizes the mix of Walker magazines, the Walkerplus.com Web site, and Walker-i mobile contents as comprising the real value of his brand. The demographic of Walker magazines and the Walker Web site is nearly the same. The two services complement one another. Fully 50 percent of registered users on the Internet site also use the magazine, but they use the two services differently. They check the magazine near the beginning of the week; check for the most current information on the Web site at the end of the week; and finally, use the i-mode service on the weekend. Actually, Gomajiri implied that the i-mode site does not fit perfectly with the company's core psychographic audience, meaning an audience of a certain psychological demographic, which consists of people who plan their leisure activities carefully. i-mode is used by the opposite sort of people, such as ones who have a date and no idea of where to go or who completely forgot to plan anything. The highest usage on the Walker-i site is on the weekends.

Analyzing the Walkerplus.com Strategy

The capability to offer a mix of media as Walker-i does, however, does not in itself seem to guarantee revenue. You can look at what Walker is doing in two ways:

- ◆ How the publisher maximizes the use of its available content

- ◆ How the advertiser reaches a specific audience or uses the same outlets for more than one type of ad

The capability to brand over different delivery mediums increases the value of the content on all mediums, thereby luring advertisers. The wired Web site has done this quite effectively, whereas the i-mode Web site is still in its infancy as far as revenues are concerned.

An important aspect of Walkerplus.com's story is the difficulty the company has had in turning revenue from Walker-i, despite having 220,000 registered subscribers. Turning this quite substantial subscriber base into a profit center is a goal that seems to have been approached in a rather laconic way. I was not privy to Kadokawa's thinking in 1999 when it launched the service, but launching a free service at the time probably had a lot to do with the company's perception that although mobile content would be a loss leader, getting in early would be good experience. In this case, the perception dictated the reality. Though it is more difficult to be accepted onto the i-menu as a premium site than as a free site, with Kadokawa's considerable reputation and a stronger effort, offering the site on a subscription basis clearly would have been possible. At the time of its inception, the site was viewed largely as a byproduct of the company's wired Web site, says Gomajiri. Unfortunately, by giving away the content from the beginning, the company made selling it later difficult.

Difficult, but not impossible. The offerings on Walker-i are at this point quite good for the price. The content consists of about 20 weekly listings of restaurants per region or city, along with movie times, events, museums, and other information. Developing new premium services would require an effort that the company apparently does not want to commit to at this point. Deducing the reasons for this noncommittal stance is difficult, but they probably involve a reluctance to sink much money, time, or personnel into a service that was expected to be unprofitable from the start.

Applying the Lessons

So, what are some of the lessons learned?

- ◆ If you have good content, don't give it away! You are giving your business away by providing your product with no clear revenue return. And if you start doing this in the beginning, it is very hard to keep your users if you stop doing it (this is a lesson learned by many dot.com startups, to their detriment).

- ◆ Approach the mobile content not as an experience, but as a business. Sure, it may not be a big business, but dedicating the necessary resources and staff, and approaching it with a measure of seriousness, will ensure that it is a business rather than a charity.

- ◆ Do tie your wired Web content to wireless content. By mobilizing your Web content, you add value to that content from both users' and advertisers' points of view.

- ◆ Don't tie yourself to a specific demographic or psychographic. True, the users of one of your products may fit a certain profile, but one of the beauties of using different outlets for your content is the potential for a wider market. Don't limit yourself!

- ◆ Do your math! If you earn $40,000 per month from a site, why is it not profitable!? The cost of your three employees is about $12,000; your office space (if you prorate) is about $1,000. Assume miscellaneous business expenses of $5,000 or so, and you should still have $22,000 per month. If you want to make a profit, and your supplier of content is charging you more than this, it is time to find a new supplier!

Chapter 16

Case Study 2: Index Corporation

THIS CHAPTER LOOKS AT INDEX, AN ENTERTAINMENT INFORMATION PROVIDER on the official i-mode menu. Index, which listed on the Jasdag market in March 2001, began in 1995 as a travel company. In 1997, Masami Ochiai took over the struggling company. Since that time, aside from a brief stint in the beginning at marketing lightweight videotapes for direct-mail purposes, Index's sole focus has been on developing mobile content.

Company History

A brief chronology of the company's history follows:

- **April 1997:** Acquires exclusive sales rights in Japan for lightweight videocassettes for direct mail from V-Lite Video Corporation, USA.
- **September 1997:** Head office moves to Minami-Aoyama, Tokyo.
- **October 1997:** Enters content business for mobile communication terminals.
- **October 1998:** Launches content service Odekake Denwa-cho (Mobile Yellow Pages) with the start of MOZIO mobile Internet service by Astel Tokyo, a PHS carrier.
- **December 1998:** Launches content service called Yojibaba (which means Four O'Clock Granny) with the start of P-Mail DX text messaging service by DDI Pocket, a PHS carrier.
- **February 1999:** Launches content service Ren-ai no kamisama ("God of Love"), a love horoscope site, with the start of the i-mode service by NTT DoCoMo.
- **April 1999:** Launches content service Private Homepage when EZ Access Service is started by IDO. Acquires exclusive sales rights in Japan on 3D Visual Display System from Optical Products Development Corporation, USA.

- **March 2000:** Connect Productions, Inc. (LA) is made a subsidiary, providing the content service. Kenko Chushin-teki Tsubo Hompo (Health-Centered Acupuncture Point House) is produced for the launch of the J-Sky Web service of J-Phone, a cellular carrier. Mobile Venture Club is jointly established with Open Loop Inc.

- **May 2000:** Connect Corporation, a company that develops Java-application software for mobile terminals, is established through a joint venture with Open Loop, Inc.

- **June 2000:** Reaches agreement with LG Telecom of South Korea on provision of Java content.

- **August 2000:** Makes third-party allocation of new stocks worth 1.1 billion yen, in total, to 13 companies, including Mitsubishi Corporation; Fuji Television Network, Inc.; Asahi National Broadcasting Co, Ltd. (TV Asahi); and Takara Co., Ltd. Head office moves to Meguro-ku, Tokyo.

- **September 2000:** Joins Linktone Mobile Internet Venture Club in China. Connection Corporation makes new stock allocation to third parties.

- **October 2000:** Index Corporation starts to distribute the "God of Love" horoscope service to cellular phones in Taiwan.

- **December 2000:** Along with the start of M-Stage visual, the image distribution service for PHSs of NTT DoCoMo, Index Corporation starts to provide Colosseum Saikyo Densetsu (The Legend of the Strongest Man in the Colosseum).

- **January 2001:** Along with the launch of the 503i series, the mobile phone compatible with DoCoMo's i-mode, Index Corporation starts to provide Mail Anime (mail-animation), the content corresponding to i-Appli (Java). Through a tie-up with Fuji Television Network Inc., Index Corporation provides content for i-mode.

- **March 2001:** Head office moves to Setagaya-ku, Tokyo. Index goes to IPO on the Jasdaq market in Tokyo.

Evolving Mobile Content

The company, which is headed by Masami Ochiai, currently employs a staff of 80. It was one of the first providers to be listed on the i-menu (with its "God of Love" horoscope site), and is unusual among content providers because it had previous experience producing mobile content. It produced this content for Tokyo Telemessage pagers, Astel and DDI Pocket PHS phones, and NTT DoCoMo pagers.

When the company started developing mobile content, so-called "pocket bell" pagers were all the rage among young people. You could see school kids lining up at public telephones to send messages to their friends, using a number-to-character conversion system that still mystifies many, but which these kids mastered as though they were playing the piano. They had to master it: With all their friends lined up behind them, their fellows would start to complain if they used the phone for more than a minute or so.

Index's target user was 14 years old, and in 1997, the price of mobile phones was too expensive for this age group, so pocket bells it was. Because this age group didn't have much money, the pager companies had started to bleed red ink, with NTT DoCoMo, the largest company, losing $320 million in 1998. Though the idea of delivering content to this ubiquitous device was very attractive to Ochiai, the business of pocket bells was in decline.

Kazutoshi Watanabe, who entered the company at Ochiai's bidding in 1998 and currently serves an advisory role as strategy planner, says that the company began to look at developing content for PHS and mobile phones after realizing that its survival depended on such content. The unenviable task of calling on carriers was left to Ogawa, who had a knack for it, says Ochiai. The company, at this point, was extremely small, only ten employees, and still mostly dependent on videotape sales for cash flow.

Yoshimi Ogawa joined Index in 1998 from POV. At POV, she had been in charge of the media department, and had done some work with online content production in that capacity, which put her in a good position when she joined Index. She is currently a board member, and is vice president in charge of operations.

After Ochiai, Ogawa, and Watanabe had made the rounds of all of the mobile phone and PHS carriers, or so they thought, Ms. Ogawa happened on a newspaper story about DDI Pocket's P-mail DX service, slated for rollout in three months. The three of them made an appointment with DDI immediately, and were ready for service in December 1998, when DDI introduced it.

Index's service on DDI Pocket was called "Yoji Baba," which was a communication tool for young people in the form of a character dreamed up by an unknown student that made the rounds at junior and senior high schools. The service was, says Watanabe, sort of like a radio call-in program or variety program using text. People could post messages to one another, read others' messages, contribute anecdotes, and participate in the Yoji Baba community.

The company provided the material for discussion and the system for doing this, with interesting content posted frequently and e-mails posted to the site or answered quickly. The aforementioned radio analogy doesn't make that much sense unless you are familiar with radio in Japan, which has very little to do with music and very much to do with chatty DJs who read lots of readers' letters, interview well-known musicians, and occasionally play music. The entertainment value usually derives less from the music than from how witty or interesting the DJs are.

Yoji Baba, the Four O'Clock Granny, was the ultimate talk show personality, albeit a fictional one. Watanabe says that Yoji Baba contributed to more than 50 percent of DDI's P-Mail DX traffic, and was the number one content service on DDI Pocket, as well as on pagers, in terms of traffic (which at the time was the only measure of revenue generation).

Being number one, though, provided little consolation when Index was still in the red. The company's monthly take was about 2 million yen, or US$16,000 at current rates. This amount came from payments from carriers that used Index's content. The situation was quite desperate at the company, forcing Ochiai to say at one point, "If we can't make a profit, let's get out of this business." Now that Index has made a success of itself, Watanabe attributes that success to the fact that the company didn't get out of the content business at that time.

Though she had contacts in NTT DoCoMo from her dealings with its pocket bell people, Ogawa did not contact those people when the call went out for content providers in the summer of 1998. The reason was a particularly Japanese sensitivity to the political undercurrents that Ogawa saw at DoCoMo at the time. The person in charge of content for the pocket bell and the newly established mobile phone content group (called the Gateway Business Section and presided over by Mari Matsunaga) were somewhat at odds. Wanting to avoid becoming part of any sort of struggle, Ogawa did the Japanese thing, which was to refrain from telling the pocket bell people that she was talking to the mobile phone people, letting their ability to provide compelling content stand on its own. In hindsight, Matsunaga probably would not have cared if she stepped on anyone's toes in getting good content, but Ogawa was being quite prudent. That prudence was a necessary practice for a small company, which still had only 20 employees and was not in a position to step on the toes of any of its customers.

Moving to the DoCoMo Business Model

Up until i-mode, the usual business model was for Index to form a partnership with a carrier, and offer services *to the carrier*. The carrier took the risk of the service proving to be a financial failure. This model presented a safe way of doing business for Index. It was not a very profitable model, however, because regardless of the usefulness of Index's content, its income from the service didn't change from what had originally been agreed on. The relationship that Index maintained with these carriers was a business-to-business relationship, with the carriers purchasing services from Index.

In 1998, when all the mobile phone carriers were planning mobile content strategies, only one company told Ochiai that it didn't want to "pay anything for your content. If that is acceptable, please join our service." That company was NTT DoCoMo. NTT DoCoMo's business model consisted of a business-to-customer relationship, with NTT DoCoMo acting as a broker and making money on the use of its network. This model *has* proven to be successful, and is garnering interest from around the world. At the time, however, no one at Index knew it would be successful, and receiving no payment up front posed a risk for Index.

Compounding the risk was the fact that even if Index wound up snaring a significant percentage of the users of i-mode, that service had no users at all when it started. No one knew that i-mode would be successful, and it was, in fact, rather late to the market, following number three mobile carrier J-Phone's J-Sky mobile Internet service, launched in 1998. Although he was as risk-averse as most businesspeople are, Ochiai saw that the system would actually pay handsomely if Index were successful, in contrast to Index's unprofitable endeavors of the past.

NTT DoCoMo presented Index with a problem, however, by disallowing sites that facilitated dating or person-to-person interaction. This type of interaction was one of the Yoji Baba service's most successful ventures. DoCoMo suggested a fortune-telling site in place of Yoji Baba. This type of service was something that Index had very little experience with, though it had offered horoscopes as a minor feature on Yoji Baba. The bigger problem was that Index's experience in content development had focused on helping young people communicate with one another in virtual peer groups, building networks of a non-wired sort. Leveraging that experience into building a successful fortune-telling site posed a dilemma.

It turns out that this was not as much of a challenge as feared because building a community is a necessary element for most successful sites, including Index's "God of Love" site.

Ochiai, Ogawa, and Watanabe have taken quite a pragmatic approach to what sort of content they want to develop: content that sells. How they go about developing that content is not a big secret, though Ochiai says that they have one big advantage over many of their competitors who have come to the business from doing wired Web sites or game development: "We came to the mobile content business from Niftyserve [an online service much like CompuServe], where everything was text-based. The mobile phone actually enlarged our scope, with its portability and graphics. In the case of someone coming to the business from traditional Web page development or game development, they have to narrow their scope, which has proven impossible for most of them."

When the company started in the content development business, none of its 20 employees were system engineers or had any experience in system operations. The company outsourced server deployment. This lack of system knowledge was not a handicap, says Watanabe, but rather an advantage.

People that Watanabe calls "computerchik" types (those who are into the technology and minutiae of computers) usually don't make great content developers, he says. "Interesting people make interesting content," says Watanabe. He lists four characteristics of interesting and creative people:

- They are able to make and create on their own.
- They believe in themselves.
- They lie.
- They come up with interesting ideas in brainstorming sessions, and can clearly articulate those ideas.

I asked Ochiai about the third characteristic, and he grimaced. "Watanabe," he said "is a liar himself. That's probably where that comes from." The roles of Watanabe, Ochiai, and Ogawa in the corporate culture become slowly clearer.

Ochiai is the businessman looking to make a profit and willing to make the hard choices needed to do that. He is somewhat distant from the day-to-day operations of the company, leaving those to Ogawa. He is a big-picture thinker, and spends much of his time thinking, discussing with affiliated companies, and planning the company's next moves.

Ogawa is an extroverted saleswoman, able to get appointments and put her pride aside to achieve the goal of selling the company's contents. She is also the *nesan*, or older sister, to the mostly female staff at Index. Ochiai attributes part of Index's success in motivating its female employees to the presence of Ogawa. Female leaders and role models are still relatively rare in Japanese corporations, and Ogawa has helped to foster this sort of atmosphere. She has been interviewed by many magazines for her unorthodox management and for bringing her dog to work. Bringing her dog was actually in promotion of a pet site, but it brought the company good publicity and helped to promote a sense that the company was doing something new and special.

Watanabe is the dreamer, the creative sort who contributes to the brainstorming sessions held once a week or so. He was, at first, much more involved in the operations of the company but is now listed as an "advisor." His place in the company seems to be that of resident jester, able to make people laugh and be outrageous.

God of Love – Creating Marketable i-mode Content

God of Love, shown in Figure 16-1, was Index's first service on i-mode. The number of subscribers now stands at about 350,000 people who pay 170 yen (about US$1.50) per month for the service.

Figure 16-1: The opening screen of the God of Love service.

When NTT DoCoMo first suggested a fortune-telling service and Index was considering whether it could do that, one of its considerations was how to sell such a service to its target group of young women, expected to be the biggest users of the new i-mode services.

Index thought that a simple horoscope site would not have a strong enough theme to keep users coming back. Previously, users of its mobile products had clearly showed that people used the services while doing something. It was a more active media because it caught people at a time when they were out and about.

Contents, too, should have an active theme. Additionally, they need to appeal to as broad an audience as possible. Index saw i-mode as a mass-market medium, similar to television in that respect.

Taking into consideration their target audience, the necessity for an active theme, and the mass-market nature of the mobile phone, Ochiai and company settled on their theme: Love.

Love definitely sells in Japan. Many young women saw the movie *Titanic* five to ten times. Love stories, even when not particularly popular in the U.S., regularly top box-office sales in Japan. Index reasoned thus:

All young women have an interest in love. Whether they have a lover, are looking for a lover, or are just getting over a lover, this topic generates a tremendous interest from a broad range of young women.

If the theme for the service were money, work, or relationships, there would be times when people didn't want to think about those issues, and would quit the service. But almost no one in the target market of young women gets tired of thinking about love.

The service consists of various sorts of fortune-telling, using star signs and tarot readings to forecast love's prospects for a certain day, month, group date, or other situation. The service also provides various tests of personality, advice on matters of love, and calendars showing good and bad days.

Following is the service menu, from the beginning:

- Banner (refer to Figure 16-1)
- Notices (what has changed on the service, new contents, and so on)
- Special (seasonal outlooks)
- New on the menu
- God of Love Tonight (sexy fortunes, suitability of various activities tonight)
- Main Menu:
  - Daily Special Love Horoscope
  - Your Fate This Month
  - This Week's Love Outlook
  - Tarot Reading
  - God of Love's Rune Stone Fortune Telling

- About That Special Someone
- Outlook for Group Dates
- The Tao of Love
- Yearly Love Outlook
- Character Tests

◆ i-Appli Character Tests

◆ Free Presents

◆ Free Love Samples

◆ Change Registration Information

◆ Please Read

◆ Tell Your Friends about Us

Though it bills itself as a fortune-telling site, God of Love is also a sort of advice column for those in love or for the lovelorn. It's also pure entertainment. What else could one call a site that mixes all of these elements, after all?

At the beginning of the service and the beginning of i-mode, the site fairly consistently got around 3 to 4 percent of all i-mode customers. Thus, on the first day of i-mode's release—February 22, 1999—i-mode had its 1,000 customers, 40 of whom subscribed to God of Love. As i-mode grew, so did God of Love, though its overall share of i-mode customers fell to a consistent level of about 1.3 percent

Around February 2000, about six months after i-mode really started its furious growth period, the significance of this number became quite significant to Ochiai. If i-mode continued to grow, so too would Index's numbers as a simple function of i-mode's growth.

This expectation assumed that everything would be equal and that the relatively small number of information service providers would stay about the same. That did not happen; the original 65 official sites now amount to more than 1700 on i-mode, with even more on au. Index, too, continued to grow, introducing new sites on i-mode, J-Sky, and EZ Web after August of 1999, when i-mode and the mobile Internet clearly were headed for widespread popularity. The number of sites skyrocketed. Index's overall share of the market decreased from the beginning, when it actually had more subscribers than the entire cellular-based mobile Internet, owing to its presence on PHS and pagers. This share shrank, however, as the mobile Internet overtook it within two months of i-mode's launch. Index's current overall share of subscribers isn't one I have figures for, but should be around 6.5 percent of the market.

Still, Index's growth, after the first spurt, followed basically that of mobile phone growth in general, as shown in Figure 16-2.

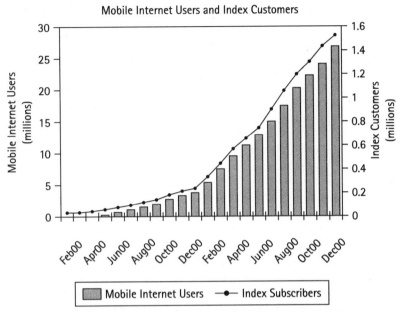

Figure 16-2: The Index's growth very closely mirrored that of the wireless Internet as a whole.

God of Love currently has around 350,000 customers, which is about one-fifth of subscribers to Index sites overall. It is one of 63 sites that Index has on six different services, ranging from pagers to mobile phones. Index would not give subscriber figures for its other services except to say that the smallest had about 10,000 subscribers, and that its total subscriber base was around 2 million (actually, the official figure, as of March 31, is 1.7 million, but an unofficial estimate of the company's current total by Ochiai is 2 million). Index's monthly accounts receivable is currently 300 to 400 million yen (US$2.4 – US$3.2 million dollars), approximately 75 percent of which is comprised of direct income from subscribers and 25 percent of developing sites for other companies, such as Fuji TV and Asahi TV's television sites, TBS's radio sites, and other areas.

At the time of press, Index is in a high-growth period, adding a new site approximately every two weeks. The content of the new sites is worked out during regular brainstorming sessions, in which anyone is welcome to participate. Any new idea has to go through this process of peer review before it is adopted. A three-person team is responsible for developing the site. The leader of the team is called the producer; the other two members are assistant producers. With 63 sites and about 80 employees, of whom approximately 50 work directly on content development, one employee may be a member of as many as ten teams, although the average is four. These teams are completely responsible for the sites they develop, including how much they bring in to the company.

There are, says Ochiai and others at the company, a large number of comments sent to the producers of the sites by subscribers. This feedback is used to gauge what sorts of improvements need to be made and to get ideas for new features on the site. No regular system seems to exist for using this feedback; the responsibility for it, as well as other aspects of the site, is the responsibility of the producer. If the company continues to grow, a solid system of dealing with feedback on more than an ad hoc basis will probably be needed, but right now the process is fairly effective.

The company has, as the preceding suggests, a strong corporate bias in favor of "people, not systems." Although the scalability of Index's approach up until this point is a question, the flexibility and speed the company has shown in developing new sites has shown the strength of a focus on creativity and people rather than on developing an algorithmic process to clearly spell out a methodology for developing and maintaining a site.

When they are developing a new site, members of a team have been known to sleep under their desks for as many as five days straight. I asked Ochiai what he attributed this level of motivation to. "They enjoy it, I think," was his answer. He said, and others agreed, that as a boss he wasn't the sort to say, "Get to work," and, in fact, has told employees to go home and get some rest. They actually refused.

Employees are, says Ochiai "all rich now" because of stock options that all employees had as of the company listing in March of 2001. But, he says, few employees have exercised their options yet, though they are not prevented from doing so. Concerning money as motivator, Ochiai has found that offering people salaries slightly above average and making sure that they increase every year is better for morale. The company has also offered all employees an additional 50,000 yen (around US$400) per month if they move to the same area where the company is located. Many have accepted this offer. Ochiai regards these sorts of motivations as being more effective for people than their stock options are. Japanese companies have generally enjoyed good employee loyalty and a sense of ownership without resorting to stock options, so Ochiai's skepticism about using them as a motivator is understandable.

One unusual feature of the largely female workforce is its loyalty to the company and a willingness to work extraordinarily long hours, which is something that is generally associated more with male employees than female ones. In many workplaces in Japan, there are very good reasons for this, such as limited female promotion, lower pay, blatant discrimination, and now-defunct regulations regulating how much overtime women were allowed to work. That these factors do not exist at Index is evidenced by the results.

Looking Ahead

Index has enjoyed two years of very high growth that has reflected the growth in the number of mobile Internet users. However, more than 75 percent of NTT DoCoMo's mobile users are already i-mode users, so these rates of growth will not continue indefinitely. More prospects for growth exist with au and J-Phone, both of which have a considerably lower rate of adoption among their customers. Still, the majority of revenues and users of Index's services are i-mode users. How Index handles the transition from high subscriber growth to a more profit-oriented growth will determine the future prospects of the company.

One part of this equation is the exclusivity enjoyed up to this point by Index and other official i-mode sites. They have enjoyed the exclusive rights to be included on the menu list, to be saved in the MyMenu area of a user's i-mode menu, to have the charges incurred from their services billed directly to a customer's phone bill, and to automatically receive customer identification information when a customer connects to their site. All these factors have given a very large advantage to Index and other official information providers.

NTT DoCoMo has said, however, that this exclusivity will change. It has promised to open up the i-menu to other portals and to open up its micro-billing system in some way as well. Its timeline for these changes has been vague, although 2002 has been mentioned. What NTT DoCoMo actually intends to do to make the system more open has also been vague.

To information providers such as Index, these changes could have some dramatic consequences, most striking of which is a price pressure that hasn't existed up until now. Index hasn't changed the 170-yen price of God of Love since the site went live in February 1999. If, however, a competing site were offered for free or for a very low price, Index would probably be forced to respond.

One of the attractions of i-mode for content developers was its use of cHTML. But that advantage is like a double-edged sword because it is the same thing that allows any Ken, Shu, or Taro to put up an i-mode site. In the world of cable television, 20 years ago, HBO was able to charge US$20 per month for its service, and willing subscribers were plentiful. Today, however, all kinds of competition exist, and HBO has been forced to respond by developing original programming and by lowering its subscription fee.

So, too, will Index have to respond in some way, though whether 170 yen is a real concern for most users is an open question.

One way that Index has been considering to raise new revenue growth is through advertising. The market for this sort of advertising is still fairly small, but it is growing. Japan's largest advertising firm, Dentsu, and NTT DoCoMo have formed a joint venture, D2, to sell advertising on mobile Internet sites. Hakuhodo, the second-largest advertising firm, has recently launched Keitai Get, a company aimed at making mobile advertising and content an integral part of a company's media mix.

Ochiai says that Index is looking very carefully at what it can do in this area. Index would have to be careful not to inconvenience customers – who are, after all, paying for the privilege of looking at an Index site – while also giving advertisers clear value. Balancing these goals, Ochiai says, is the trick, and Index is currently looking very closely at how to achieve it.

Index has also recently started to focus on mobile commerce, opening online flower and perfume shops in addition to a TV network souvenir shop developed for Fuji TV. Ochiai sees these sorts of shops contributing more to profits in the future, though as of this writing he wasn't able to say how much they contributed to the bottom line.

Index is also cooperating with Takara toys to develop mobile-networked toys that users can control from mobile phones. The company recently demonstrated a robot with a miniature server built-in and with a connection to the wireless network via a built-in wireless telephone. A user can log onto the robot and control it from near or far.

Although he refused to provide details, Ochiai also said that Index is working on a similar sort of service to be able to control VCRs using a mobile phone. This, Ochiai says, has lots of possibilities because users could easily look up a TV schedule on i-mode and then go directly to a page that facilitates taping a program on the users' VCR. Problems exist with this sort of system, though, such as who puts the tape in the VCR. Perhaps the robot could do it?

The theme of Index's approach to the future is leveraging its current position to explore new and possibly profitable forms of business for the future. That future is not so far away, and maintaining the company's current growth could well require many more nights of sleeping under desks.

Index is one of a handful of fast-growing, mobile Internet-related companies that have sprung out of i-mode's huge popularity, and it is one of the more successful of these companies.

Index's workforce has demonstrated creativity in its approach and the capability to do the hard work to make a site successful. Many of the company's employees have become wealthy with stock options, and Index stock has shown to be good value for shareholders. The company has never been in the red. All these factors point to a successful business.

But if the Internet boom and bust have shown anything, it is that past success is no guarantee whatsoever of future success. Ochiai, Ogawa, and Watanabe realize this and are attempting to position their company for continued profitability.

Lessons Learned

Following are some lessons to take from the experiences of Index in the mobile Internet content market:

- ◆ Know your audience. If this takes hiring those people who are your target, do it. If this means a higher level of customer service in terms of answering calls and e-mails, do that.

- ◆ Technology does not make content; people make content. Use the creativity, dedication, and ingenuity of your people to make content that is interesting.

- ◆ Be sure to hire interesting people. Interesting people make interesting content.

- ◆ Going from a wired Web site to a wireless one takes understanding the limits of the medium, and making the most of it. See Chapters 8 and 11 for more on how you can do this.

- ◆ Make sure your content developers interact with users of the content. This will keep them focused on what the users like or don't like.

Chapter 17

Case Study 3: Nikkei

OUR LAST CASE STUDY is of Nihon Keizai Shimbun's i-mode offering. Usually called simply Nikkei, this company plays a role in finance in Japan similar to the role *The Wall Street Journal* plays in the U.S. Its index, the Nikkei Index, is used to gauge the strength or weakness of the markets; its news and real-time stock quotes are used by traders, bankers, businesspeople, and anyone else who needs up-to-the-minute financial information in Japan. According to Mitsutoshi Tanabe, of the Web & Mobile Business Department, which handles Nikkei's Internet presence, Nikkei currently offers these services by both wired Internet and mobile Internet as three different products — including i-mode. Nikkei Net is a free site with revenues coming from advertising. It offers news stories from the current issue of its newspaper, and access to older stories as well. Because nearly all its stories are fully owned by the company rather than by a wire service (unlike many newspapers), Nikkei is free to exploit these assets. Nikkei's other wired service is called Telecom 21, and is a premium service with real-time stock quotes and real-time news feeds. Depending on the service mix, this product costs a customer generally from 10,000 yen to 20,000 yen (US$80 to US$160) per month. It is mostly companies and stock traders that take advantage of this service, and its target market is quite clearly as a business-to-business product.

Presence and Profitability

Mobile Internet services are the third type of delivery medium. Figure 17-1 shows the opening screen for the Nikkei site. Nikkei currently is an official site on DoCoMo's menu, au's EZ Web menu, and J-Phone's J-Sky service menu. At this writing, there are approximately 140,000 customers on these three sites, with about 110,000 on i-mode. Each pays 300 yen, or about US$2.50, per month for news headlines, delayed quotes, and other content from Nikkei newspapers. Tanabe attributes the strength of its presence on i-mode, as compared to the relative lack of customers on EZ Web or J-Sky, to the demographic of the users of these services. The customers of au, J-Phone, and Tuka tend to be younger, whereas DoCoMo tends to have more businesspeople as customers. The nature of its offerings (financial news and data) means that there is a quite specific demographic to aim for, and the demographic of i-mode customers seems to fit well.

Figure 17-1: Nikkei's home page on its free site.

Actually, Nikkei's i-mode service started in August 1999, four months after Nikkei started delivering content to J-Sky and EZ Web. Those services at first used different payment systems to i-mode, though they have copied i-mode's and now are exactly the same, taking 9% of revenue. Tanabe says that the amount of money charged for collecting the money is fine. He compares it to what it cost to collect payment by credit card, which is still quite costly in Japan and cannot be easily used for micropayments.

Currently, Nikkei takes in about 2 billion yen (about US$16 million) per year from its Web and mobile offerings. Only about 30 million yen (about US$2.4 million) is from its mobile sites. On the other hand, says Tanabe, of the 30 to 50 people working on Nikkei Net (the Web site) at any time, none are dedicated solely to mobile content because it is a fairly automatic process to put content on the sites using NT servers with ASP, which is enough like using a PC that anyone on the staff can do the work. They use scripts to format the content for the three formats it is displayed in: cHTML, MML (J-Phone's near-HTML markup language), and WML for EZ Web.

Tanabe says that at a level of 50,000 customers, which is about the number the site had when he started in March 2000, it wasn't much of a business. At 100,000 customers, it was not too bad, and he feels that 200,000 customers is a good level. One limitation on revenues is NTT DoCoMo's requirement that it doesn't use any kind of advertising on premium sites. Part of its mobile site is free, with three daily stories available. This is to attract new users with samples, and they can advertise on this portion of the Web.

When the i-mode service started, there was quite a lot of conflict between Nikkei and Mari Matsunaga, relates Takashi Tanemura (now with Nikkei Interactive USA, but at that time working on the i-mode offering). The conflict, he says, was because of the pricing plans available and the feeling of those at Nikkei that they wanted to provide certain kinds of services that could be offered only at a price higher than the 300 yen (about US$2.50) that is the limit on i-mode. The Tokyo Stock Exchange (TSE) charges news outlets for real-time quotes, and a 300 yen per-month charge would not even come close to covering this fixed cost. Matsunaga's reasoning for the 300-yen limit is well-documented in her book and in other news reports: She thought that people wouldn't be willing to pay for content more expensive than a magazine, which is about 300 yen in Japan. The problem is, in Nikkei's case, "people in general" are not its audience, businesspeople are. Tanabe says that it would need to charge about 2,000 yen (about US$16) per month for real-time stock quotes. He wouldn't make a guess as to how much it would need to charge for real-time news feeds. It is one of the things that Tanemura and Tanabe both see as more in line with Nikkei's strength, and one that they continue to push for.

Advertising, too, is one area in which Tanabe expressed dissatisfaction with the current situation with NTT DoCoMo. This extends, ironically, to NTT East and West, whose L-mode service Nikkei is looking into being listed on. On L-mode, Tanabe says, Nikkei would not even be able to advertise on its free sites. As a newspaper, not being able to leverage the relationships it has with advertisers and advertising companies to create income from its mobile sites is quite unnatural.

As an advertiser, Nikkei has used D2 Communications (NTT DoCoMo and Dentsu's ad agency), which has opt-in e-mail services. Users can select opt-in from the i-mode menu to get sent special offers by advertisers, and do not have to pay the packet-transmission charges normally associated with receiving e-mail. D2 charges approximately 70-80 yen (US$.55 to $.64) per e-mail sent. This is quite high, but the click-through rate is as high as 70%. Tanabe laughs, and says that the 70% might actually come over two or three months' time, but eventually it does seem to come. He also says that this sort of opt-in, targeted e-mail campaign is extremely effective, so that the money is well spent. Nikkei, however, would prefer not to have to pay the money, however well spent.

One aspect of the i-mode system, and that of the EZ Web and J-Sky systems that have come to be modelled on it, is that the content provider, although it has a direct relationship with users of its service for exactly as long as it takes those users to use the service, does not know anything about users aside from unique identifying numbers. It doesn't know their e-mail addresses, age, sex, or professions. This gives NTT DoCoMo sole "possession" of its customers. Building a list of customers for use in the mobile market has been made additionally difficult by the fact that customers were getting lots of spam, and NTT responded by encouraging everyone to change their old e-mail addresses, which were just their phone numbers followed by @docomo.ne.jp. And, not surprisingly, NTT does not offer forwarding services from the old number to the new. Tanabe mentions the experience of another i-mode site, which did mass mailings on its own and had around 40% returned. Because D2C's mailings get sent directly from DoCoMo's database, there is not a problem with inaccurate mailings. This is obviously weighted heavily in D2C's favor and (as half-owner), in DoCoMo's favor.

As the largest financial publication in Japan, and as a fixture in the financial community in the same way *The Wall Street Journal* is in the U.S., Nikkei has clout. It locked horns with DoCoMo over the pricing issue, and did not come out on top. It does not seem pleased with limitations on advertising, but will most likely not come out ahead in that fight, either. DoCoMo claims, and there is nothing thus far that would seem to contradict this claim, that the rules are the same for everyone, take them or leave them. Nikkei actually chose to leave them at first, not joining being listed on the i-menu until six months after the service started, and four months after starting service on rival sites. In Japan, however, a mobile content provider who doesn't serve i-mode customers probably doesn't have many customers.

If anything, in contrast to the other two case studies, Nikkei is closest to being an equal to NTT DoCoMo in terms of size, prestige, and power in its own market. As such, DoCoMo's level playing field and insistence on certain rules has rankled. Nikkei has found a way to work with NTT DoCoMo to make some money in the present situation, while still vigorously pushing NTT DoCoMo to change certain of its policies concerning pricing for content that would allow it to do much more of what it does best, which is delivering current news and financial information as quickly as possible.

Market-Geared Content

Within its premium service, Nikkei provides a range of information to users:

- Stock searches and stock history
- Analyst reports
- News headlines
- Sports news
- i-Appli news agent, which automatically grabs news at intervals
- i-Appli graphs of Nikkei average (from Q4, 2001)

Most of these services, although rather shallow in the depth of content, give a user a good idea of what they want to know more about when they get to their PC. For the sort of users Nikkei is targeting, business users, this is quite acceptable as a goal. The expectation that users will use computers for many of the information-intensive tasks that their jobs require is not unfounded regarding the target group, which very clearly fits into the category of those who use their mobile phones as supplements to their fixed-line Internet access.

One thing that you will notice in Figure 17-1 is the absolute lack of graphics. In fact, Nikkei's only use of graphics is the one graphic on its banner page and miniature banner ads. Because this is the free part of Nikkei's site, ads can be used. Tanabe is the only person I talked to in the mobile content industry who actually was concerned that users had to pay for downloading graphics in their ads. This reflects Nikkei's approach to the service, which is to focus on news content rather than on cute or trendy graphics-heavy design. This focus works well in terms of its market, and in terms of being able to format the content for use on any of three different mobile platforms.

The i-Appli news agent is an interesting product, and bears some explanation. It allows a user to specify how often to update the news it holds, and automatically downloads the news at those intervals. As Tanabe pointed out, by using the i-Appli (especially in Tokyo, where many people ride the subway to work and where mobile phones often do not work), a user can read quite current news without being connected at that moment to the network. If real-time stock quotes become a reality from a pricing standpoint, users could check the prices of all of their stocks every five minutes. This would relieve a Nikkei server from having to check every user's preferences, find out which stocks they wanted tracked, look those up, and send them. The Java midlet could do all of this in one transaction, placing much less a burden on the back end server. Tanabe says that this sort of agent will likely play a bigger role, especially as packet charges come down with the advent of 3G networks.

Tanabe actually sees the 3G networks as the solution to his concern over users having to pay for downloading advertising. Management of Nikkei online assets is centralized in the Electronic Media division, and other Nikkei properties, especially Nikkei-CNBC and TV Tokyo, are pushing to have streaming video on the new 3G service. Nikkei is currently testing the feasibility of this service in-house. Although packet charges are lower than those on the 2G network, Tanabe sees the price of data transmission as the major stumbling block to video on mobile phones. A full 384 Kbps stream would cost about 18 yen per *second*. On the current test setup, streaming video only uses 64 Kbps, and is charged not at data rates, but at voice rates. Even at about 40 yen per minute, however, video does not seem a very attractive proposition. Early reports by test users of the 3G handsets raise basically the same concerns while praising the quality of the video stream.

One of Nikkei's main reasons for both its wired and wireless Internet sites is to leverage the content that it has. A huge amount of information is generated daily by Nikkei bureaus, reporters, and other assets. To not make an effort to utilize these resources to find new business opportunities would be foolhardy, and foolhardy is not something that Nikkei is. If the Japanese Internet advertising market goes the way of the U.S. market, Nikkei and other information providers' revenues from their wired sites ads will likely dry up. That hasn't happened yet, but many people see it as a matter of time. The steady revenue of paying customers will, in that case, take on an added importance.

Tanabe, though not with Nikkei at the time (he was serving a stint at AOL Japan, a Nikkei partner), believes that the pay-for-content model was a big attraction for Nikkei to enter the mobile market. At the time, it had started Nikkei Net, a free service, and felt constrained from charging customers because the common wisdom dictated free content paid for by advertising. Its i-mode and other mobile ventures were in a sense a test case to see whether a for-fee service could fly, what sorts of information people would pay for, and how a payment system for this sort of content could work. Though currently a very small part of Nikkei's overall online revenues, these services are expanding, and look set to contribute a much larger portion of revenues in future. Nikkei estimates that it will have 260,000 paying customers by year-end 2003.

The Lessons

This look at Nikkei's mobile Internet offerings shows a few truisms of the Internet age for content creation companies such as Nikkei:

- Leverage the assets you have built up, whether human, structural, or relationships. Nikkei does not have any dedicated people working solely on wireless content because it doesn't need them. It gives its wireless subscribers content without lots of frills, and this in turn costs little to produce.

- Continue to look at your assets, whether or not they are necessarily marketable at this point. Though DoCoMo has remained negative to higher charges and advertising, persistence on Nikkei's part will probably eventually pay off, and its real sweet spot in the value chain will be achieved. This won't come, however, without a good measure of persistence on Nikkei's part.

- Use presentation and technology where it is of benefit to the customer and the company, and not elsewhere. Fancy graphics neither pleased the customer nor helped the company, and therefore were not used. However, Java midlets helped customers by allowing them to read content in the subway, and helped the company by removing some of the pressure this sort of service might otherwise have created on its servers.

Appendix A

What's On the CD-ROM?

THIS APPENDIX PROVIDES YOU with information on the contents of the CD-ROM that accompanies this book. The CD includes the following software:

- PHP
- MySQL
- Apache
- Adobe GoLive Tryout Version
- Zentek i-Jade Lite Emulator
- Sun Microsystems Forte for Java Community Edition
- Sun Microsystems Java 2 Software Development Kit 1.3 Standard Edition for Windows
- UltraEdit 32 45-day Shareware Version
- NJ Win Tryout Version
- phpMyAdmin
- Music MasterWorks
- Chaku Melo Converter
- Symbol font of the emoji characters

Also included are source code examples from the book and an electronic, searchable version of the book that can be viewed with Adobe Acrobat Reader.

System Requirements

Make sure that your computer meets the minimum system requirements listed in this section. If your computer doesn't match up to most of these requirements, you may have a problem using the contents of the CD.

For Microsoft Windows 9.*x* or Windows 2000, you need the following:

- PC with a Pentium processor running at 120Mhz or faster
- At least 32MB of RAM

- Ethernet network interface card (NIC) or modem with a speed of at least 28,800 bps
- A CD-ROM drive – double-speed (2x) or faster

Using the CD with Microsoft Windows

To install the items from the CD to your hard drive, follow these steps:

1. Insert the CD into your computer's CD-ROM drive.
2. Open the My Computer icon on your desktop.
3. Find the icon for the CD-ROM called i-mode: A Primer, and open it.
4. Open the `index.html` file, which will get you started, by either double clicking it or right-clicking it and selecting the open option.
5. Each of the software packages has installation instructions, usually in a file called `readme`. Please read the instructions thoroughly before attempting to install the software.

What'll You'll Find

The CD-ROM contains source code examples, applications, and an electronic version of the book. Following is a summary of the contents of the CD-ROM, arranged by category.

Source code

Every program in any listing in the book is on the CD in the folder named Sample Code.

Applications

The following applications are on the CD-ROM:

- **PHP:** Installs the PHP scripting language. Requires the Apache server to already be installed.
- **MySQL:** Installs an SQL server.
- **Apache:** Web server that you must install to use PHP.

- **Adobe GoLive (30-day tryout version):** WYSIWYG Web page editor.
- **Zentek i-Jade Lite Emulator:** For i-Appli Java applications.
- **Sun Microsystems Forte for Java Community Edition:** An integrated development environment (IDE) for developing Java applications.
- **UltraEdit 32 (45-day shareware version):** Text editor for creating and editing HTML, PHP, and Java code.
- **NJ Win (tryout version):** Third-party solution for reproducing Chinese, Japanese, or Korean characters on an English or other single-byte character version of Windows.
- **phpMyAdmin:** Free mySQL database adminstration tool that lets you insert records, create tables, download data from the server, and upload data to your mySQL database
- **Music MasterWorks Tryout Version:** A music composition program that can be used to create music in MIDI format.
- **Chaku Melo Converter Freeware:** Application that converts MIDI to MFi (MLD) format or vice versa.

Electronic version of i-mode: A Primer

The complete (and searchable) text of this book is on the CD-ROM in Adobe's Portable Document Format (PDF), readable with the Adobe Acrobat Reader (also included). For more information on Adobe Acrobat Reader, go to www.adobe.com.

Troubleshooting

If you have difficulty installing or using the CD-ROM programs, try the following solutions:

- **Turn off any anti-virus software that you may have running.** Installers sometimes mimic virus activity, and can make your computer incorrectly believe that it is being infected by a virus. (Be sure to turn the anti-virus software back on later.)
- **Close all running programs.** The more programs you're running, the less memory is available to other programs. Installers also typically update files and programs; if you keep other programs running, installation may not work properly.
- **Apache, mySQL, and PHP.** Remember that for Apache, PHP, and mySQL to work correctly together, you need to follow the special instructions in Chapter 12.

If you still have trouble with the CD, please call the Hungry Minds Customer Service phone number: (800) 762-2974. Outside the United States, call (317) 572-3994 or e-mail techsupdum@hungryminds.com. Hungry Minds will provide technical support only for installation and other general quality control items; for technical support on the applications themselves, consult the program's vendor or the (http://book.eimode.com).

Appendix B

A Complete List of Official i-mode Sites

THIS IS A TRANSLATED LIST OF ALL SITES available on the i-mode service as of March 1st, 2001. This is not an official translation, nor are the translated names of services necessarily the translations that the information providers use themselves. Because of the nature of names and cleverly named sites, and the difficulty of translating these from one language to another, I have taken some liberties in this regard in order to give the viewer a better idea of the sorts of things offered on i-mode. That is what this list is, a guide to the sort of services that are available in Japanese on i-mode in Japan. It is not meant to be an exhaustive and official record, nor the basis for any use other than informational ones. Any mistakes, omissions, or mistranslations are completely unintentional, but I cannot be responsible for any damage arising from them.

Menu List

This is the menu in which all content providers are listed, in a system of hierarchical menus. The menu items listed below are those available from this menu:

i-Appli

News/Weather/Information

Mobile Banking

Credit Card/Securities/Insurance

Travel/Transportation/Maps

Shopping/Living

Ringing Tones/Screen Images

Games/Fortune Telling

Entertainment

Town Information/Destinations

Dictionaries/Useful Tools

Mail

Regional

i-Navi Link

i Playstation

Information Service Partners

i-Menu in English

News/weather/information menu

GENERAL NEWS/WEATHER

Asahi Nikkei Sports	100 – 200 yen/mo.
ASCII Magazine	Free
BizTech News	Free
CNN	300 yen/mo.
impressWatch	Free (some parts 300 yen/mo.)
Jiji Wire Service	100 yen/mo/genre
Mainichi Newspaper/Sports Nippon	Free headlines (100 yen/mo./genre for full story)
News Battle News Quiz Game	100 yen/mo.
NHK Japan Public Broadcasting	300 yen/mo.
Nihon Keizai Shimbun	300 yen/mo.
Nihon Keizai Shimbun Clipping Service	200 yen/mo.
Otenki.com (weather)	Free (some parts 100 yen/mo.)
Sankei Newspaper/Sankei Sports	100–300 yen/mo/genre
Teikoku Data Bank	Free (some parts 300 yen/mo.)
The News	Free
TSR Corporate Rating Service	300 yen/mo.
WNI Weather Information	Free (Weather Plus 100 yen/mo.)
Yomiuri Newspaper	200 yen/mo.

LOCAL NEWSPAPERS/REPORTS

Akita Prefecture News	Free
Aomori News	Free
Chugoku Newspaper Company	Free
Chunichi Dragons News	200 yen/mo.
Fukushima News	Free
Hokkaido Newspaper/Hokkaido Sports Newspaper	Free (some parts 100 yen/mo.)
Ishikawa and Toyama News	Free
Iwaki News	Free
Iwate Daily Report	Free
Kanagawa Newspaper News	Free
Kawakita Daily News	Free
Kita Nihon Newspaper News	Free
Kobe Daily Tiger	Free
Kushiro Newspaper Company	Free
Kyoto Newspaper Company	Free
Niigata Daily Report	Free
Nishi Nihon Newspaper Company	Free
Saitama Newspaper	Free
Sanyo Newspaper Company Sanyo	Free
SBS (Shizuoka Broadcasting Service)/Shizuoka Newspaper News	Free
Shikoku Newspaper News	Free
Shimono Newspaper Company	Free
Shinano Mainichi Newspaper News	Free (parts, 50 yen/mo.)
Tokachi Mainichi Newspapers Hokkaido	Free
Tokushima Newspaper	Free
Yamanashi Daily Newspaper	Free

OVERSEAS NEWS

Bridge News Japan	300 yen/mo.
Dow Jones Worldwide Financial Information	300 yen/mo.
Korean Daily News	Free
LP Financial Information Service Company	Free
Peoples Daily News (China)	Free
Pokebras Brazilian Information	Free

Mobile Banking

P=Payment; T=Transfer Funds; I=Balance Inquiry; A=Account Activities; F=Foreign Currency Transaction; D=Fixed Deposit; B=Trust Banking; M=Money Order Information; TB=Telephone Banking

CITY AND REGIONAL BANKS

CITY BANKS

Asahi Bank, Ltd.	P T I A TB
CitiBank Japan, NA	P T I A F TB
Dai-Ichi Kangyo Bank, Ltd.	P T I A TB
Daiwa Bank, Ltd.	I A TB
Fuji Bank, Ltd.	P T I A
Mitsui Sumitomo Bank	P T I A F D B
Sanwa Bank, Ltd.	P T I A D B
Tokai Bank, Ltd.	P T I A TB
Tokyo Mitsubishi Bank, Ltd.	P T I TB

REGIONAL BANKS (FULL SERVICE)

77 Bank, Ltd.	P T I A
Akita Bank, Ltd.	P T I A
Aomori Bank, Ltd.	P T I A
Ashikaga Bank, Ltd.	P T I A
Bank of Fukuoka, Ltd.	P T I A
Bank of Hiroshima, Ltd.	P T I A TB
Bank of Iwate, Ltd.	P T I A
Bank of Nagoya, Ltd.	P T I A
Bank of Yokohama, Ltd.	P T I A
Biwako Bank	P T I A
Chiba Bank, Ltd.	P T I A
Chubu Bank	P T I A
Chugoku Bank	P T I A
Chukyo Bank, Ltd.	P T I A
Daikoh Bank	P T I A
Daishi Bank, Ltd.	P T I A
Daito Bank	P T I A
Fukuho Bank	P T I A
Fukui Bank, Ltd.	P T I A
Fukuoka Chuo Bank	P T I A
Fukuoka City Bank, Ltd.	P T I A
Fukushima Bank	P T I A
Gifu Bank	P T I A
Gunma Bank, Ltd.	P T I A
Hachijuni Bank, Ltd.	P T I A TB
Higashi Nihon Bank	P T I A
Higo Bank, Ltd.	P T I A

Continued

REGIONAL BANKS (FULL SERVICE) *(Continued)*

Hiroshima Sogo Bank, Ltd.	P T I A
Hokkaido Bank, Ltd.	P T I A TB
Hokuen Bank	P T I A
Hokuetsu Bank, Ltd.	P T I A TB
Hokuriku Bank, Ltd.	P T I A
Hokutoh Bank	P T I A
Howa Bank	P T I A
Hyakugo Bank, Ltd.	P T I A
Hyakujushi Bank, Ltd.	P T I A
Ishikawa Bank	P T I A
Iyo Bank	P T I A
Joyo Bank, Ltd.	P T I A TB
Juhachi Bank, Ltd.	P T I A
Juroku Bank, Ltd.	P T I A
Kagoshima Bank, Ltd.	P T I A
Kanagawa Bank	P T I TB
Keiyo Bank	P T I A
Kita Nihon Bank	P T I A
Kiyomizu Bank	P T I A
Kiyou Bank	P T I A
Kochi Bank	P T I
Kumamoto Family Bank	P T I A
Kyoto Bank	P T I A
Michinoku Bank	P T I A
Mie Bank	P T I A
Minami Nippon Bank	P T I A
Minato Bank	P T I A

Miyazaki Taiyo Bank	P T I A
Nagano Bank	P T I A
Nagasaki Bank	P T I A
Nanto Bank, Ltd.	P T I A
Nishikyo Bank	P T I A
Nishi-Nippon Bank, Ltd.	P T I A
Oita Bank, Ltd.	P T I A
Okinawa Bank	P T I A
Okinawa Kaihou Bank	P T I A
Oogaki Kyoritsu Bank Corp.	P T I A
Saga Bank	P T I A
Sakae Bank	P T I A
Sanin Godo Bank	P T I A
Shiga Bank, Ltd.	P T I A
Shikoku Bank, Ltd.	P T I A
Shinwa Bank	P T I A
Shizuoka Bank, Ltd.	P T I A
Shizuoka Chuo Bank	P T I A
Shokugin Shokusan Bank	P T I A
Shonai Bank	P T I A
SurugaBank	P T I A
Tajima Bank	P T I A
Toho Bank, Ltd.	P T I A
Tohoku Bank	P T I A
Tokyo Tomin Bank, Ltd.	P T I A
Tomato Bank, Ltd.	P T I A
Towa Bank	P T I A
Toyama Bank	P T I A

Continued

REGIONAL BANKS (FULL SERVICE) *(Continued)*

Tsukuba Bank	P T I A
Yachiyo Bank	P T I A TB
Yamagata Bank, Ltd.	P T I A
Yamagata Shiawase Bank	P T I A
Yamaguchi Bank	P T I A
Yamanashi Chuo Bank, Ltd.	P T I A

REGIONAL/LOCAL BANKS (BALANCE INQUIRY/ACCOUNT DETAILS ONLY)

Aichi Bank

Awa Bank, Ltd.

Bank of Ikeda, Ltd.

Chiba Kogyo Bank, Ltd.

Ehime Bank

Ibaraki Bank

Kagawa Bank

Musashino Bank, Ltd.

Niigata Chuo Bank, Ltd.

Sapporo Bank

Senshu

Tochigi Bank

Tokushima Bank

Tokyo Sowa Bank, Ltd.

Tottori Bank

Wakayama Bank

NATIONAL CREDIT UNIONS

HOKKAIDO

Asahikawa Credit Bank	P T I A TB
Daichi Credit Bank	P T I A
Date Credit Bank	P T I A
Esashi Credit Bank	P T I A
Furano Credit Bank	P T I A
Hidaka Credit Bank	P T I A
Hokkai Credit Bank	P T I A
Hokumon Credit Bank	P T I A
Kitami Credit Bank	P T I A
Kushiro Credit Bank	P T I A
Monbetsu Credit Bank	P T I A
Muroran Credit Bank	P T I A
Nayoro Credit Bank	P T I A
Obihiro Credit Bank	P T I A
Oshima Credit Bank	P T I A
Sapporo Credit Bank	P T I A TB
Wakkanai Credit Bank	P T I A

TOHOKU

Abukuma Credit Bank	P T I A
Aizu Credit Bank	P T I A
Akita Fureai Credit Bank	P T I A
Aomori Credit Bank	P T I A

Continued

TOHOKU *(Continued)*

Fukushima Credit Bank	P T I A
Himawari Credit Bank	P T I A
Hokujo Credit Bank	P T I A
Morioka Credit Bank	P T I A
Nihonmatsu Credit Bank	P T I A
Sennan Credit Bank	P T I A
Shinjo Credit Bank	P T I A
Shiogama Credit Bank	P T I A
Shirakawa Credit Bank	P T I A
Shirakawa Credit Bank	P T I A
Sukagawa Credit Bank	P T I A
Tsuruoka Credit Bank	P T I A
Yonezawa Credit Bank	P T I A

KANTO DISTRICT, KOSHINETSU

Adachi Credit Bank	P T I A TB
Akaho Credit Bank	P T I A
Aoki Credit Bank	P T I A
Arai Credit Bank	P T I A
Arakawa Credit Bank	P T I A TB
Asahi Credit Bank	P T I A
Ashikaga Credit Bank	P T I A
Credit Bank	P T I A
Dai Sakae Credit Bank	P T I A
Funabashi Credit Bank	P T I A
Heisei Credit Bank	I A

Higaxhi Chofu Credit Bank	P T I A
Hiratsuka Credit Bank	P T I A
Ina Credit Bank	P T I A
Isesaki Credit Bank	P T I A
Ishioka Credit Bank	P T I A
Jonan Credit Bank	P T I A
Kameari Credit Bank	P T I A
Kanuma Mutual Credit Bank	P T I A
Karasuyama Credit Bank	P T I A
Kawaguchi Credit Bank	P T I A
Kawasaki Credit Bank	P T I A
Kiryu Credit Bank	P T I A
Kofu Credit Bank	P T I A
Kofu Shoko Credit Bank	P T I A
Matsumoto Credit Bank	P T I A TB
Meguro Credit Bank	P T I A
Minato Credit Bank	P T I A
Miura Fujisawa Credit Bank	P T I A
Nagano Credit Bank	P T I A
Narita Credit Bank	P T I A TB
Nikko Credit Bank	I A
Ohji Credit Bank	P T I A
Ootsuki Credit Bank	P T I A
Oyama Credit Bank	P T I A
Sagami Credit Bank	P T I A
Saitama Ken Credit Bank	P T I A
Sanjo Credit Bank	P T I A
Sano Credit Bank	P T I A
Setagaya Credit Bank	P T I A

Continued

KANTO DISTRICT, KOSHINETSU *(Continued)*

Shiba Credit Bank	P T I A
Shonan Credit Bank	P T I A
Sugamo Credit Bank	P T I A
Suwa Credit Bank	P T I A
Taiyo Credit Bank	P T I A
Takasaki Credit Bank	P T I A
Taki Nogawa Credit Bank	P T I A TB
Tochigi Credit Bank	P T I A
Tokyo industry Credit Bank	P T I A
Tone-gun Credit Bank	P T I A
Ueda Credit Bank	P T I A
Utsunomiya Credit Bank	P T I A
Yokohama Credit Bank	P T I A

TOKAI

Aihoku Credit Bank	P T I A
Bisai Credit Bank	P T I A
Chita Credit Bank	P T I A
Chunichi Credit Bank	P T I A
Enshu Credit Bank	P T I A
Fuji Credit Bank	P T I A
Fujinomiya Credit Bank	P T I A
Gamagoori Credit Bank	P T I A
Gifu Credit Bank	P T I A
Hamamatsu Credit Bank	P T I A
Iwata Credit Bank	P T I A

Izu Credit Bank	P T I A
Kakegawa Credit Bank	P T I A
Kita Ise Credit Bank	P T I A
Kuwana Credit Bank	P T I A
Mishima Credit Bank	P T I A
Nishio Credit Bank	P T I A
Numazu Credit Bank	P T I A
Okazaki Credit Bank	P T I A
Oogaki Credit Bank	P T I A
Sanjyu Credit Bank	P T I A
Seino Credit Bank	P T I A
Seki Credit Bank	P T I A
Seto Credit Bank	P T I A
Shimada Credit Bank	P T I A
Shimoda Credit Bank	P T I A
Shizukyo Credit Bank	P T I A
Shizuoka Credit Bank	P T I A
Suruga Credit Bank	P T I A
Takayama Credit Bank	P T I A
Tono Credit Bank	P T I A
Toyohashi Credit Bank	P T I A
Toyokawa Credit Bank	P T I A
Toyota Credit Bank	P T I A
Tsu Credit Bank	P T I A
Ueno Credit Bank	P T I A
Yahata Credit Bank	P T I A
Yaizu Credit Bank	P T I A

HOKURIKU

Echizen Credit Bank	P T I A
Fukui Credit Bank	P T I A
Hokuriku Credit Bank	P T I A
Ishido Credit Bank	P T I A
Kanazawa Credit Bank	P T I A
Kohama Credit Bank	P T I A
Kono Credit Bank	P T I A
Noto Credit Bank	P T I A
Sabae Credit Bank	I A
Shinminato Credit Bank	P T I A
Toyama Credit Bank	P T I A
Tsuruga Credit Bank	P T I A

KANSAI DISTRICT

Amagasaki Credit Bank	P T I A
Awaji Credit Bank	P T I A
Ayabe Credit Bank	P T I A
Banshu Credit Bank	P T I A
Fukuchiyama Credit Bank	P T I A
Hachiko Credit Bank	P T I A
Hanna Credit Bank	P T I A
Hikone Credit Bank	P T I A
Himeji Credit Bank	P T I A
Hirakata Credit Bank	P T I A

Hyogo Credit Bank	P T I A
Jyusan Credit Bank	P T I A
Kansai Nishinomiya Credit Bank	P T I A
Kinoku Credit Bank	P T I A
Kobe Credit Bank	P T I A
Koto Credit Bank	P T I A
Kyoto Chuo Credit Bank	P T I A
Kyoto Credit Bank	P T I A
Kyoto Miyako Credit Bank	P T I A
Kyoto north city Credit Bank	P T I A
Maizuru Credit Bank	P T I A
Nagahama Credit Bank	P T I A
Naka Hyogo Credit Bank	P T I A
Nara central Credit Bank	P T I A
Nara Credit Bank	P T I A
Nishi Hyogo Credit Bank	P T I A
Nisshin Credit Bank	P T I A
Osaka City Credit Bank	P T I A
Osaka Credit Bank	P T I A
Oumihachiman Credit Bank	P T I A
Senshu Credit Bank	P T I A
Senyo Credit Bank	P T I A
Settsu Credit Bank	P T I A
Sougo Credit Bank	P T I A
Tajima Credit Bank	P T I A
Tanyo Credit Bank	P T I A
Yamato Credit Bank	P T I A

CHUGOKU

Hagi Credit Bank	P T I A
Hiroshima Credit Bank	P T I A
Hofu Credit Bank	P T I A
Kamome Credit Bank	P T I A
Kurashiki Credit Bank	P T I A
Kurayoshi Credit Bank	P T I A
Kure Credit Bank	P T I A
Mizushima Credit Bank	P T I A
Okayama Credit Bank	P T I A
Shimane Credit Bank	P T I A
Shimonoseki Credit Bank	P T I A
Tamashima Credit Bank	P T I A
Tottori Credit Bank	P T I A
Toyoura Credit Bank	P T I A
Yamaguchi Credit Bank	P T I A
Yonago Credit Bank	P T I A
Yoshinan Credit Bank	P T I A

SHIKOKU

Anan Credit Bank	I A
Ehime Credit Bank	P T I A
Hatada Credit Bank	P T I A
Kanonji Credit Bank	P T I A
Kawanoe Credit Bank	P T I A
Takamatsu Credit Bank	P T I A TB

KYUSHU

Amakusa Credit Bank	P T I A TB
Iizuka Credit Bank	P T I A
Kagoshima Credit Bank	P T I A
Kagoshima Mutual Credit Bank	P T I A
Karashima Credit Bank	P T I A TB
Kita Kyushu Yahata Credit Bank	P T I A
Konabe Credit Bank	P T I A
Kumamoto First Credit Bank	P T I A
Moji Credit Bank	I TB
Oita Mirai Credit Bank	P T I A
Ookawa Credit Bank	P T I A
Tachibana Credit Bank	P T I A
Tohga Credit Bank	P T I A
Wakamatsu Credit Bank	P T I A TB

TRUST ASSOCIATIONS AND OTHERS

HOKKAIDO

Abashiri Trust Association	P T I A

TOHOKU

Kita Gun Trust Association	P T I A
Tsubasa Trust Association	P T I A

KANTO DISTRICT, KOSHINETSU

Akagi Trust Association	P T I A
Kamitsuke Trust Association	P T I A
Kimizu Trust Association	P T I A
Nagano Prefecture Trust Association	P T I A
Sakae Trust Association	P T I A
Sanjyo Trust Association	P T I A
Tanimura Trust Association	P T I A
Tsuru Trust Association	P T I A

TOKAI

Hida Trust Association	P T I A

HOKURIKU

Toyama Prefecture Trust Association	P T I A

KANSAI DISTRICT

Hyogo Prefecture Trust Association	P T I A
Kansai Industrial Bank of Japan, Ltd.	A TB
Tanyo Trust Association	P T I A

CHUGOKU

Hiroshima City Trust Association	P T I A
Okayama Prefecture Trust Association	P T I A

KYUSHU

Kagoshima Kogyo Trust Association	P T I A
Ooita Prefecture Trust Association	P T I A

POSTAL SAVINGS

Postal savings	TB

JA (AGRICULTURAL COOPERATIVE CREDIT UNIONS)

JA AICHI PREFECTURE

JA Aichi	P T I A
JA Aichi Atsumi-cho	P T I A
JA Aichi Central	P T I A
JA Aichi East	P T I A
JA Aichi Mikawa	P T I A
Ja Aichi Moto	P T I A
JA Aichi North	P T I A

Continued

JA AICHI PREFECTURE *(Continued)*

JA Aichi Trust Association	P T I A
JA Bisai City	P T I A
JA Gamagoori City	P T I A
JA Green Trust	P T I A
JA Himawari	P T I A
JA Inazawa	P T I A
JA Inazawa City	P T I A
JA Kaifu	P T I A
JA Kaifu East	P T I A
JA Kaifu South	P T I A
JA Kasugai City	P T I A
JA Komaki City	P T I A
JA Miyoshi-cho	P T I A
JA Nagoya	P T I A
JA Nishimikawa	P T I A
JA Owari	P T I A
JA Shimoyama-mura	P T I A
Ja Takakuraji	P T I A
JA Tenpaku Trust	P T I A
JA Toyo	P T I A
JA Toyohashi	P T I A
JA Toyota City	P T I A
JA West Kasugai	P T I A
JA Yamabiko	P T I A
JA Yotsuba	P T I A

JA SHIZUOKA PREFECTURE

Aira Izu Agricultural Cooperative	P T I A
JA Enshu Central Agricultural Cooperative	P T I A
JA Enshu Yumesaki Agricultural Cooperative	P T I A
JA Fuji City Agricultural Cooperative	P T I A
JA Fuji Reclaimed Fuji Agricultural Cooperative	P T I A
JA Fujinomiya Agricultural Cooperative	P T I A
JA Gotenba Agricultural Cooperative	P T I A
JA Hainan Agricultural Cooperative	P T I A
JA Hakonan Tobu Agricultural Cooperative	P T I A
JA Hamamatsu Agricultural Cooperative	P T I A
JA Izu Taiyo Agricultural Cooperative	P T I A
JA Izunokuni Agricultural Cooperative	P T I A
JA Kakegawa City Agricultural Cooperative	P T I A
JA Mikkabi Agricultural Cooperative	P T I A
JA Mishimakonan Agricultural Cooperative	P T I A
JA Nansun South Agricultural Cooperative	P T I A
JA Oikawa Agricultural Cooperative	P T I A
JA Sampoharakaitaku Agricultural Cooperative	P T I A
JA Shimizu City Agricultural Cooperative	P T I A
JA Shizuoka City Agricultural Cooperative	P T I A
JA Surugaji Agricultural Cooperative	P T I
JA Susono City Agricultural Cooperative	P T I A
JA Topia Hamamatsu Agricultural Cooperative	P T I A
Shizuoka Prefecture Agricultural Cooperative Credit Union	P T I A

JA TOKYO METROPOLIS

JA Akigawa	P T I A
JA Aoba	P T I A
JA Hachioji	P T I A
JA Machida City	P T I A
JA Mainzu	P T I A
JA Nishitama	P T I A
JA Setagaya Meguro	P T I A
JA Tokyo Aguri	P T I A
JA Tokyo Central	P T I A
JA Tokyo Credit Association	P T I A
JA Tokyo Midori	P T I A
JA Tokyo Minami	P T I A
JA Tokyo Mirai	P T I A
JA Tokyo Musashi	P T I A
JA Tokyo Shimasho	P T I A
JA Tokyo Smile	P T I A
JA west Tokyo	P T I A

JA HIROSHIMA PREFECTURE

JA Aki	P T I A
JA Edajima	P T I A
JA Fuchu City	P T I A
JA Fukuyama City	P T I A
JA Fukuyama North	P T I A
JA Hiroshima Central	P T I A

JA Hiroshima Chiyoda	P T I A
JA Hiroshima City	P T I A
JA Hiroshima Credit Association	P T I A
JA Hiroshima Yutaka	P T I A
JA Hiroshimaasa	P T I A
JA Innoshima	P T I A
JA Kannobe	P T I A
JA Kei	P T I A
JA Kirikushi	P T I A
JA Kure	P T I A
JA Matsunaga	P T I A
JA Mihara	P T I A
JA Mukohigashi-cho	P T I A
JA Mukojima-cho	P T I A
JA Nomijima	P T I A
JA Numakuma	P T I A
JA Onomichi City	P T I A
JA Ootake City	P T I A
JA Saeki Central	P T I A
JA Sanji	P T I A
JA Sera-gun	P T I A
JA Setoda	P T I A
JA Shinsekikogen	P T I A
JA Shinshi	P T I A
JA Shobara	P T I A
JA Takata	P T I A
JA Takehara	P T I A
JA Yachiyo-cho	P T I A

JA WAKAYAMA PREFECTURE

JA Arida	P T I A
JA Green Hidaka	P T I A
JA Hashimoto City	P T I A
JA Hiokigawa	P T I A
JA Ito	P T I A
JA Kami Tomita	P T I A
JA Katsuragi-cho	P T I A
JA Kinan	P T I A
JA Kinosato P T I A	P T I A
JA Kishu Central	P T I A
JA Kushimoto-cho	P T I A
JA Minabe	P T I A
JA Nagamine	P T I A
JA Naka Hetchi	P T I A
JA Shingu	P T I A
JA Shirahama Credit Association	P T I A
JA Susami-cho	P T I A
JA Tanabe City	P T I A
JA Tonda	P T I A
JA Wakayama	P T I A
JA Wakayama Credit Association	P T I A
JA Wakayama Inami	P T I A
JA Yukaide	P T I A

NATIONAL LABOR UNION CREDIT UNION

Hokkaido Labor Union Credit Union	P T I A
Okinawa Prefecture Labor Union Credit Union	P T I A
Shizuoka Prefecture Labor Union Credit Union	P T I A
Tokai Labor Union Credit Union	P T I A
Yamaguchi Prefecture Labor Union Credit Union	P T I A

NET BANK

Japan Net Bank	P T I A

Credit card/securities/insurance

STOCKS/SECURITIES
All providers have the following services available:

- Stock Quotes
- Market Condition News
- Purchase and Sale of Stocks
- Balance Inquiry

All provide these services free to registered customers. An application to the company is necessary before using the services

DLJ Direct

kabu.com Securities

Morningstar Fund Ratings

Nikko Beans Securities

Nikko Securities

Nomura Securities

Tokyo Mitsubishi TD Waterhouse Securities

Yamato Securities

CREDIT CARDS
All companies listed provide account activity inquiries for no fee.
   It is necessary to have a credit card from one of the companies in order to be able to use this service. Some companies may also require a separate application.
   The following designate additional services available: P=Point Club Inquiries; R=Revolving Credit Changes; T=Card Loss or Theft Reporting; A=ATM Location Guide; C=Sales Campaign Information; L=Loan Application.

Aeon Credit Service	P C L
DC Card	P R T A
JCB Card	P
KC Card	P T C
Million Card	P C
Mitsui Sumitomo Card	P R L
NICOS CARD	P T C
OMC (Daiei) Card.	P L
Orico Card	P C
Saison Card	P C
TS3 Card Toyota Finance	None given
UC Card	A C

INSURANCE
An application to each insurance company is necessary before using the services.
A= Insurance ATM; C= Contract Inquiries; P= Product Information; I= Consultation Information Offered

AIU Insurance Company	C P
Asahi Life	A

Dai-Ichi Seimei	A C P I
Meiji Life	A C
Mitsui Life	A I
Nihon Seimei	A C
Sumitomo life	A C I
Yasuda Life	A I

Travel transport maps

TRAIN GUIDANCE

Jordan AD Train Guidance	Free
JR.East Travel Navigator	Free (parts 100 yen/mo.)
Toshiba Corp. i-station	Free (parts 100 yen/mo.)

AIRLINE INFORMATION

S= Check Seat Availability; R= Domestic Reservations; T= Ticketless Reservations; M= Mileage Club Information

ANA All Nippon Airways	S R T M
JAL Japan Air Lines	S R T M
JAS Japan Air System	S R T M
NWA Northwest Airlines	S R M

HOTEL/LODGING/TRAVEL

R=Hotel Room Reservations; A=Check Room Availability; D=Discount Travel Offered

Business Hotel Reservation Center	R
Countdown PB Community	R D (150 yen/mo.)
HIS Overseas Travel Navigator	R D
JTB Mobile Crew	R D
MyTrip.Net Package Tour Reservations	R
Nippon Travel Agency Co., Ltd.	A
Prince Hotel/Seibu Golf Course Reservations	R
Tokyu Kanko	R

RENT-A-CAR/TRAFFIC INFORMATION

R=Reservation; A=Check Availability/Rates; M=Members/Mileage Club Information; I=Route Information; C=Traffic Conditions E=E-mail Notifications of Important Traffic Information

Access C&E Traffic Information	I C I (180 yen/mo.)
ATIS Traffic Information	I C (200 yen/mo.)
JH Expressway Navigation	C
Nippon Rent-a-Car	R A M
Orix Rent-a-Car	R A M

MAPS

iMapFan	Free (parts 300 yen/mo.)
Zenrin Mobile Map	300 yen/mo.

Shopping/living

TICKETS

R=Reservations; S=Special Discounts; E=E-mail Notifications

e+ Entertainment Plus Simplicity	Event tickets	R S E Free
i mode Ticket Pia	Concert tickets	R S Free (charges apply to parts)
Lawson Tickets	Sports/music tickets	R S (200 yen/mo.)
Playguide	Event/travel tickets	R S (150 yen/mo.)

BOOKS/CDS/GAMES

E=E-mail notifications; P=Online Purchasing; O=Order Tracking; S=Uses Store for Payment and/or Delivery

7dream.com	CDs, DVDs, software	P S Free
Book Service	Books	P
Famima-I	CDs, DVDs, games	P S
HMVJapan	CDs, DVDs, games	E P S
Honyasan (Bookstore)	Books	P
IndiesMusic	Independently produced CDs	P Free (one part 300 yen/mo.)
JBOOK Bunkyodo	Books, CDs, DVDs, magazines, games	P
Kinokuniya Bookstore	Books	P O S
Lawson I-Convenience Store	Books, CDs, hotel reservations	P S
MUSICNAVI	CDs	P

Continued

BOOKS/CDS/GAMES *(Continued)*

PlayStation.com	Playstation games/hardware	E P
Tsutaya Online	CDs, games, videos, etc.	E P S Free (some parts are charged)
TV Panic Game Store	Game software	P S

PART-TIME JOBS/EMPLOYMENT

Dispatch Net	Short-term Jobs	Free
From A Recruit	Job listings	Free
Job Site Dispatch	Short-term jobs	Free
Mobile an (arbeit news)	Part-time job LISTINGS	E-mail notifications of jobs Free (parts 100 yen/mo.)
Recruit Navi	Job search for students	Free (parts 300 yen/mo.)

CARS

Auto ASCII	News, maintenance information	Free
autoBank Car Information	Used car information	Free
Car Information Goo	Used car information	Free
Car Mode I	Reference	Free
Car Sensor	Used car information, news, shops	Free

Gulliver Car Information	Used car information	Free
i-GAZOO.com	New/used/auctioned car information	Free

REAL ESTATE RENTAL INFORMATION

Able Rental Information	National apartment listings	Free
CHINTAI Rental Search	National rental search	Free
E Size Residential Information	Home rentals and sales	Free
Home Mate	National apartment listings	Free
JPM Rental Market	National home rental listings	Free

STUDY/QUALIFICATION

Always Snoopy	English conversation, translation	100 yen/mo.
Cinema English Conversation	Vocabulary study, short lessons, quizzes	100 yen/mo.
COOL ENGLISH	Study using games	200 yen/mo.
Kanji World	Study kanji for work or school	100 yen/mo.
Menkyo King/Juken King	Study for qualifications, entrance exams	300 yen/mo.
OkiDoki English Conversation	English conversation	Free (parts 200 yen/mo.)
TOEIC Friend	Study for TOEIC English test	300 yen/mo.

FASHION

f-mode	Fashion news, columns	300 yen/mo.
Kosuiyasan (Parfumery)	Indexes and sells perfumes	Free
Maga Sheik	Buy C. Itoh or Oggi fashion products	Free

COSMETICS/HEALTH

@Cosme	Product rankings, professional information	Free
Aru Aru Encyclopedia	Information on beauty from TV show	Free (one part is 100 yen/mo.)
Calorie Diet	Diet and health and menu planners	200 yen/mo.
Color Karina	Information exchange on cosmetics	Free
Heart Healing Mode	Refreshment for the heart	300 yen/mo.
Shiseido.mode	Beauty, fashion women's health guide	Free

PETS

Cat Time	Information on caring for cats	200 yen/mo.
Pet Pet	Reference to pet shops, vets, others	Free

LIFE INFORMATION

Flower Shop Index	Find a local flower shop	Free
Hibiya Kadan	Aromatherapy, flower orders, etc.	Free
I Flower Shop	Flower delivery orders	Free
Pocket Size	Discount restaurant and rental residence information	Free (some parts charged)

GOURMET INFORMATION

Enoteka Wine Shop	Wine information/education	300 yen/mo.
Gourmet Navi	Restaurant guide by train station	Free
Isaizu Discount Gourmet	Restaurant guide, discount coupons	Free (parts are 300 yen/mo.)
Sapporo Beer Hall Guide	Guide, reservation, discounts	Free
Suntory Bar Navigator	National bar guide	Free
Taru Liquor Department Store	Liquor portal site, also sells liquor	Free
Tazaki's Wine	Wine information	300 yen/mo.
Tokyo ZAGAT Survey	Restaurant guide	Free

RECIPES

A-Dish from Ajinomoto Co.	10-minute recipes	Free
Bob & Angie's Osaka Gas Recipe	Daily recipe, including calorie information	Free (one part is 100 yen/mo.)

Continued

Appendixes

RECIPES (Continued)		
Kikkoman Recipe	10-, 20-, or 30-minute recipes	Free
Nestle Japan I recipe	Dessert and drink recipes	Free

Ringing tones/screen pictures

RINGING TONES/KARAOKE		
R=Ringing Tones; K=Karaoke		
5 Melodies for 100 Yen	Store your favorites songs here	100 yen/mo.
Bimani Hits!	Ringing tones from Konami	300 yen/mo.
Chaku On	Screens and ringing tones from Capcom games	200 yen/mo.
Chaku Paradise	Ringing tones from cartoons	150 yen
Cinema Melo Chu	Movie and J-POP ringing tones	100 – 300 yen/mo.
CM Commercial Life Melo	Ringing tones from TV commercials	100 yen/mo.
CoolSound	Hip-hop, Euro, R&B, Techno ringing tones	100 – 300 yen/mo.
Disney-i	Ringing tones and screen images from Disney movies	100 yen/mo.
Enix e melody	Original composition ringing tones	100 yen/mo.
Fuji Melo	Ringing tones from Fuji TV shows	80 yen/mo.
Full of Music	New melodies and screens e-mailed daily	100 yen/mo.
Get de I-melody	8,000 ringing tones, w/4,000 16-chord tones	100 – 300 yen/mo.

GIGA Networks	New melodies available on day of release	R 100 – 200 yen/mo.
Half Note Jazz	Jazz ringing tones w/ 16 chords	150 yen/mo.
IkaraX	Karaoke, with music information	200 yen/mo.
Melo Chara Namco	Melodies from Namco's popular games	R 300 yen/mo.
Melo DAM/ Karaoke DAM	20,000+ ringing tones	R K 100 – 300 yen/mo.
Melo Goat	1,000 songs freely downloaded, sent to friends	90 yen/mo.
Melody Pa!	Ringing tones	R K 100 – 200 yen/mo.
MelodyClip Interbrain	Ringing tones, screen images	300 yen/mo.
NHK Melody File	Japanese pop ringing tones	100 – 300 yen/mo.
Petit Melo Republic	Newly released ringing tones	100 – 300 yen/mo.
Poke Melo Joysound	4,500 ringing tones	R 90 – 300 yen/mo.
Roland Mobile Juke	Karaoke using I-Appli	200 yen/mo.
Roland Ringing Tones	J-Pop ringing tones	R 50 – 200 yen/mo.
Sega Music Networks	2,400 ringing tones, 16,000 karaoke	R K 10 – 300 yen/mo.
Shinsei Commercial Songs	New and old songs from commercials	20 – 300 yen/mo.
Special Screen/ Melody	A variety of screens and ringing tones	300 yen/mo.
TBS Original	Ringing tones from T BS shows, announcers	150 yen/mo.
Tele Melo Music	Ringing tones from TV shows	200 yen/mo.

Continued

RINGING TONES/KARAOKE *(Continued)*

VIBEbeep!!	Hit songs from this Japanese MTV	100 yen
Victor Sound Chop	Karaoke, animated cartoon screens, 4,000 16-chord ringing tones	50 – 300 yen/mo.
Word Melo Chaosu	Inputted names turned into ringing tones	140 yen/mo.
Yamaha Melocha	Original ringing tones	R K 90 – 300 yen/mo.

CHARACTERS

@Gachapin & Mook	Postcards, fortunes, etc. from TV characters	150 yen/mo.
@Sakura Momoko	Greeting card, IQ tests	200 yen/mo.
Ai-yai-ya!	Fortunes, etc. from Nokia Japan	200 yen/mo.
Always Charapa!	Popular characters sent every day	100 yen/mo./character
Always, Snoopy NEC.	Peanuts comic characters	100 yen/mo.
Anime Pa	Daily comic strips	300 yen/mo.
Appli Screen Clock	Screen clock that automatically updates weather information	200 yen/mo.
Chakuchara!	Foreign character startup screens	100 yen/mo.
Chara de Hoo!	Animated characters, fortunes	300 yen/mo.
Chara Mail!	Cute clocks and timers	300 yen/mo.
Character Mail Service	Mail done with cute characters	300 yen/mo.

Appendix B: A Complete List of Official i-mode Sites

Character Secretary	Manages schedule data	300 yen/mo.
CharaStar	Bakabon by mail	100 yen/mo.
Colorful Charapa!	Color characters for startup screen	100 yen/mo./character
Comic@mail Panasonic	Character by mail	180 yen/mo.
Cyborg 009	Screens and games from this *manga* comic	Free
Disney-i	Disney character startup screens	100 yen/mo.
Hasegawa-kun	Daily screens of baseball anime star	100 yen/mo.
I Love Pandas!	Panda photo, illustrated screens	100 yen/mo.
i Rika-chan (Barbie)	Character screens, tools, fortunes	200 yen/mo.
Inbox Charapa!	Characters for memo pad, clock	100 yen/mo.
ISANRIO	Information	Free
i-world Masterpiece Theater	My friend Rascal characters	100 yen/mo.
i-world Masterpiece Theater	Flanders Dog characters, tools	250 yen/mo.
i-world Masterpiece Theater.	Famous character startup screens	200 yen/mo.
Kitaro Mania Index	Ghostly card games, images	200 yen/mo.
NHK Character	Daily screens by e-mail	100 yen/mo.
Pro Baseball Charapa!	Daily pictures from your favorite team	100 yen/mo./team
TV Tokyo Anime	News, ringing tones, startup screens	300 yen/mo.
Universal Studios Japan	Information	Free
Yoshimoto Channel	Mucchiko character screens	200 yen/mo.

VISUAL

3DCG Digital Beauties	Digital imaging brings us to this!?	200 yen/mo.
Aqua Mode	Fish image screens	300 yen/mo.
Aquarium Information Bureau	Aquarium information, screens	100 yen/mo.
Artist C.The End	Artist site	300 yen/mo.
Cat Comics/Dog Comics	Weekly comic strip	200 yen/mo.
Cool Screen Cybird	Pop art, club graffiti, typography	100 yen/mo.
CS Plaza Hide	Swimsuit idols, cards can be traded	200 yen/mo.
CS Plaza Ningendamono	Brush calligraphy screens	200 yen/mo.
Jungle Jam Jam Origianl Screens	Make your own screens!	300 yen/mo.
Natsu Melo Bromide	Bromide pictures of stars	300 yen/mo.
PhotoNet	Photo posting and exchange	Free
Picture Club Cybird	Take a picture at an arcade, see it on your I-mode	300 yen/mo.
Puppy Kitty Baby Hamster	Cute screens of baby animals	100 yen/mo.
Screen Art Designer	Design your own color, words, etc.	200 yen/mo.
Screen Art Paradise	Lassen, Tenno, Ikeda, etc. famous artists	300 yen/mo.
Screen Gallery the Palace	Adult photo screens	300 yen/mo.
Start Screen Collection	Variety of celebrity photos, pop art	300 yen/mo.
Train Club	Over 500 images from JR East railways	300 yen/mo.

Ugoku-ID Art Publishing	Design a character with your name	200 yen/mo.
Yamagishi Photo Gallery	Idol photographer's site	300 yen/mo.

HANDSET MAKER SITES

@F Mobile Supporter	Images, melodies	Free
ERICSSON CAFÉ	Ringing tones, games	Free
i-NOKIA	Bilingual characters, news, etc.	Free
MyD-style Mitsubishi Electric Corp.	Tools/information	Free
N Land for Everyone NEC	High-quality ringing tones, screens	Free
P-SQUARE Panasonic user	I-Appli Games	Free
Sony i-Effect	Support site	Free

Games/fortune telling

GAMES

BANDAI CO., LTD. CHANNEL

Chara Game	I-Appli character game	150 yen/mo.
Game! Mahjongg Game	Play against other players	300 yen/mo.
Go Life Japan	Solve weekly go problems	300 yen/mo.

Continued

BANDAI CO., LTD. CHANNEL *(Continued)*

i Horse	Multiplayer horseracing simulation	300 yen/mo.
Love Mail	Virtual lover sends e-mail	300 yen/mo.
Mystique Grapple	Multirole player game	300 yen/mo.
Primary School Friends	Virtual primary school	300 yen/mo.
Puzland	Various kinds of puzzles, rankings	200 yen/mo.
Saku to Golf	Golf game with virtual caddy	200 yen/mo.
Vegas Everywhere	Virtual gambling games	300 yen/mo.

TOMY-WEB

Chu Chu Hamster	Raise a virtual hamster	200 yen/mo.
Moba Game Pon	Virtual character game	200 yen/mo.
Star Wars Picture	All 110 Star Wars characters in this game	200 yen/mo.

NAMCO STATION

Appli Carrot	Like Pac-man	300 yen/mo.
Compa Time	Have group dates with stars!	300 yen/mo.
Drama Drama No. 4	Your choices affect your partner	200 yen/mo.
LVLovers!	Realistic gambling games	100 yen/mo.
LVLovers! HILIMIT	More games, better graphics	300 yen/mo.

Namco i-land_	Four family games	300 yen/mo.
Uki Uki Niki	Baby monkey talks and keeps diary	180 yen/mo.

COULD BE DWANGO

Doki Paku Mushroom	Mushroom hunting game	200 yen/mo.
Fishing Fool	Try your luck at 900 fishing spots	300 yen/mo.
Heart A Flutter	Group dating and travel game	300 yen/mo.
Kome Poko	Become friends with rice	300 yen/mo.
Sea Trade Generation	Remember Trader? Multi-player	300 yen/mo.

KONAMI NET

Band Age	Experience rock 'n roll life	300 yen/mo.
Dance Dance Revolution	I-Appli dancing/music game	300 yen/mo.
e-Game Center	Konami's arcade games in miniature	200 yen/mo.
Fascinating Memorial	Online version of popular game	100/300 yen/mo.
i-Appli Power Pro	Four baseball games	300 yen/mo.
Konami Deluxe Pack	Choose a new game every month	300 yen/mo.
Loving Marriage	You are the hero of this lovely RPG	300 yen/mo.
Mahjong Yaroze!	Play against the computer	250 yen/mo.
Money Game	Fantasy stock trading game	300 yen/mo.

Continued

KONAMI NET *(Continued)*

Portable President	Claw your way into daddy's good graces	300 yen/mo.
Quiz School	Compete with others while learning	200 yen/mo.
Sports Series	Train baseball or soccer players	300 yen/mo.
Sushi Seal	Raise this cute character and play	300 yen/mo.

SAKUMA'S SUGUROKU

53 Stations of Tokkaido	Ancient Japan is the setting for this game	150 yen/mo.
Haiku Road	Train yourself in typing with haiku	150 yen/mo.
Westward Ho!	Another RPG, this time with 81 obstacles	150 yen/mo.

WEBBY HUDSON

Face Friend Net	Make friends using a face you create	100 yen/mo.
MailDrama Cosmos	This time go to Kanazawa, where drama awaits	300 yen/mo.
Miracle GP	Contend with others by time and rank	300 yen/mo.
Miracle Quest	An RPG with multiple quests	300 yen/mo.
Morita Shogi I mode	Play against other players online	300 yen/mo.

Appendix B: A Complete List of Official i-mode Sites

Mugwort School	Make friends; have adventures at school	300 yen/mo.
Northward Mail Drama	Eight girls living in Hokkaido	300 yen/mo.
Petit Poochi Collection	Checklists and to-do lists done cutely	150 yen/mo.
Ringing Appli	Enjoy various games online or off	300 yen/mo.
SUPER i-soccer	Build your fantasy team	300 yen/mo.
Taisen Sojo	Play chess, backgammon, others against computer	300 yen/mo.
Unlikely Stories	UFOs, ghost stories, and other adventures	300 yen/mo.

SEGA MODE

I'm an Alien!	Make lots of alien friends, and try to become a space master!	300 yen/mo.
Lets Make a Rally Team!	You become a rally team manager	300 yen/mo.
Little Black Book	Cute scheduler works for all!	300 yen/mo.
Sakura Daisen	Sakura Daisen and other popular games	300 yen/mo.
Sakura Daisen Lover Letter	A new adventure. Exchange mail with members	300 yen/mo.
Sonic Café	Play Nights, Che Che Puzzle, or Samba de Amigo	300 yen/mo.
UFO Catcher	Catch UFOs and score free screen images	300 yen/mo./center
Venus Paradise	Network RPG where the winner gets love	300 yen/mo.

ANOTHER YOU ADVENTURES BY RIVER HILL SOFTWARE

Bus Adventure	Begin another life in a country in the sky!	300 yen/mo.
Dear My Restaurant	You are a restaurant cook	300 yen/mo.
Excavation!	Explore yourself while looking for treasure	300 yen/mo.

GAME TAITO

Battle Gear	Use a real car in this racing game	300 yen/mo.
Game Pack	These I-Appli games are just like arcade games	300 yen/mo.
Let's Go by Train	A train game where you are conductor	200 yen/mo.
Taito I-box	Novel-style train game	200 yen/mo.

OTHER GAMES

8ing Net Magazine	Play RPG Baron House Summer Vacation	300 yen/mo.
Atlas Web-i	Summons the devil on your mobile and then negotiates	300 yen/mo.
Battle Tokyo 23	A futuristic sim game in post-apocalypse Tokyo	300 yen/mo.
Capcom Pack	Dinogenesis and an ancient simulation game	300 yen/mo.
Chinese Restaurant	Join the owner in making Chinese food	300 yen/mo.
Choro Q	Racing game where you get a rating	Free
Disney-i Academy	Join Mickey and Donald answering quiz questions	300 yen/mo.
Disney-i Game World	Look for characters in a Disneyland game!	300 yen/mo.
Dragonquest Net	Hear melodies or play games based on Dragonquest	100 yen/game

Appendix B: A Complete List of Official i-mode Sites

Final Championship Panasonic	A trivia game with questions on everything from science to Ultraman	300 yen/mo.
Fire Pro Wrestling	Raise your wrestler in this sim game	300 yen/mo.
Flash	A mysterious forest experience adventure	300 yen/mo.
GameGuild (NEC)	Game information	300 yen/mo.
Genki Game Launch	Launch a new video game in this sim	300 yen/mo.
Get!! Petit Appli	Get 3 i-Appli games every month	100 yen/mo.
Guitar Man Ai	Play popular Guitar Man character Puma	Free
Hamsters World Tour	Get letters from a globe-trotting hamster	300 yen/mo.
Ito's Bad Dreams	A horror comic book	300 yen/mo.
K-1 Grand Prix	Official public K-1 site with ringing tones, information, etc.	300 yen/mo.
Let's Kiss!	Tired of your real-world partner? Join the virtual kissing game	100 yen/mo.
Let's Play Horseracing	Use real races and forms to play this game	200 yen/mo
Life	Like the board game of the same name	200 yen/mo.
Life Raising BT	Raise your own character, and communicate with others	300 yen/mo.
Love English Lesson	A sim game of international love	300 yen/mo.
Love Parlour	Play pachinko, and pit yourself against other players	300 yen/mo.
Magi Mame	Raise a magician; trade characters	300 yen/mo.
Make a Square Head Round	A sim game based on a real-life advertising miss	300 yen/mo.
My Prince	A girl's game for exchanging e-mails with a prince	300 yen/mo.
Naomi's Detective File	A detective story/game played by e-mail	300 yen/mo.
Nepos Napos	A fantasy world above whale clouds	200 yen/mo.
Pencil Puzzle Dayo!	Pencil puzzle word games	300 yen/mo.
Pet's Outing	Your pet travels Japan making friends	300 yen/mo.
Petit Love	Fall in love by matching up in this game	200 yen/mo.

Continued

OTHER GAMES *(Continued)*

Post Pet	Incredibly popular mail character comes to i-mode	200 yen/mo.
RoboRobo	Teach the robot new words and make friends	300 yen/mo.
Rosenqueen Land	Popular PS game on i-mode	Free
Samurai Romanesque	Samurai RPG	300 yen/mo.
Sankyo Pachi Mode	Manage a pachinko parlor in this sim game	300 yen/mo.
Sanrio i-Park	Hello Kitty in all her glory	300 yen/mo.
Sasurai Hunter Z	RPG, where you are hunting for over 300 things	300 yen/mo.
Space Battle Role	Keep galactic peace as an agent	300 yen/mo.
Street Quest	A Shibuya RPG game	300 yen/mo.
Tenya no Oyaji	Rescue your family by running a shop	300 yen/mo.
The Silent Fleet	A submarine sim game	200 yen/mo.
Where Does it End Kakkeko	A sim game of raising a pet	300 yen/mo.
Yarusho	Four player mahjong net game	280 yen/mo.

FORTUNE-TELLING

Animal Fortune-Telling	From popular TV show on NTV	300 yen/mo.
Artology	Imai tells your fortune by colors and forms	100 yen/mo.
Campus Life Helper	Advice on classes, peer groups, student matters	300 yen/mo.
Disney-i Magical Fortune Disney	Includes fortunes, stories, and an IQ test	200 yen/mo.
Face Reading FAQ	An entertainment site about reading faces	300 yen/mo.
Good Fortune in the Stars	Western fortune teller Urana gives the best advice	300 yen/mo.
Hachiro's Psychology	Take psychology tests and get advice	150 yen/mo.

Appendix B: A Complete List of Official i-mode Sites

Hello Kitty Fortune Calendar	Fortunes, biorhythms, etc.	300 yen/mo.
House of Tarot	Leave it to the cards	200 yen/mo.
Life Consulting	Advice on work, dating, and life	300 yen/mo.
Love Fortune-telling	Checks the compatibility of a prospective date (for sex, too!)	300 yen/mo.
Love Fountain	Love advice from the fountain	200 yen/mo.
Love in the Zoo	Diet, character, love, sex, etc.	200 yen/mo.
Love Navi Robot	Love advice from the stars and robots	150 yen/mo.
Love Seer Mr. Moon	Based on the 48-month lunar calendar	100 yen/mo.
Love Watcher	Using the 24 stars, Mme. Love helps you navigate	200 yen/mo
Loving Egg	Like Tamagotchi telling fortunes	300 yen/mo.
Mail Address Fortune Telling	Uses a 2600-year history as its base(!?)	200 yen/mo.
Mailpoli	Tells you the character of friends or lovers by their e-mail address	100 yen/mo.
Mari Origin Marias	Fortunes based on eight-star astrology	200 yen/mo.
Name Checker	Checks the luck of names your are considering for your child	100 yen/mo.
Oz Dream Interpretation	Keep a daily dream diary and find out about yourself	300 yen/mo.
Palace of Bad Girls	Amusing fortunes for bad girls	100 yen/mo.
Ryuj Reads the Stars	Psychology advice from astrology	200 yen/mo.
Secret Way	A 1,000-year-old secret fortune-telling method is back!	300 yen/mo.
Skill at Love	Advice on dating, to make you a better lover	300 yen/mo.
Stella Fortune Diary	A scheduler with fortunes	200 yen/mo.
Takashima Fortuneteller	Daily fortunes based on your birth year	200 yen/mo.
The Love God	Popular Izumi produces tarot readings and fortune telling	150 yen/mo.
What's in a Name	Tells your fortune by the characters in your name	300 yen/mo.
What's Your Price?	Based on your weekly problems, it gives you a price	100 yen/mo.

Sports and entertainment

BASEBALL/SOCCER

Baystars News	300 yen/mo.
Chiba Lotte Marines Mail Info	300 yen/mo.
Chunichi Dragons News	200 yen/mo.
Daiei Hawks Town News	300 yen/mo.
Hanshin Tigers Entertainment	300 yen/mo.
J's Goal J-League Soccer Info	200 yen/mo.
Kintetsu Buffaloes Mail Info	300 yen/mo.
Kobe Shimbun's Hanshin Tigers News	200 yen/mo.
NPB Baseball News Flashes	150 yen/mo.
Orix Blue Wave News	300 yen/mo.
RCC Broadcasting's Hiroshima Carp News	Free (parts 180 yen/mo.)
Seibu Lions Information	200 yen/mo.
Show Up! Pro Baseball Info	200 yen/mo.
Show Up! Pro Soccer Info	100 yen/mo.
Sports i Kyodo News Agency	300 yen/mo.
Sports Nippon Baseball News	100 yen/mo.
Yomiuri Giants Info	200 yen/mo.

OUTDOOR

Fishing Channel Mail	Fishing tour information	300 yen/mo.
Fishing King	Fishing information	300 yen/mo.
Legend of the Wave	Surfing information	300 yen/mo.
POP Snow Navi	Ski information	Free

MARTIAL ARTS

K-1 Grand Prix Official Site	Kickboxing on steroids	300 yen/mo.
Martial Arts Special Selection	Wrestling news	200 yen/mo.
Pro Wrestling/Martial	Information, melodies, etc	300 yen/mo.

MOTORSPORT

Formula Nippon Official Site	News, interviews	Free (parts 300 yen/mo.)
Motor Sports Channel	News, race queen photos, etc.	Free (GT channel 300 yen/mo.)

GOLF

Golf Par 72	Weather, news, information	Free (parts 200 yen/mo.)
The Golf	Discounts, real-time information	300 yen/mo.
Web Golf	Reserve tee times at over 2300 courses	Free

TENNIS

Tennis Net	News and information on the pro tennis tour	Free

MUSIC INFORMATION

Avex Net Mobile	Latest Avex artist news, melodies	200 yen/mo.
CD Songlet	News about all genres of music	200 yen/mo.
HMV Japan	70,000 CDs, DVDs, and games	Free
ISony	Information on Sony music acts	Free
J Music Pocket	Japan recording association site	Free
Mobile MC	CD reviews, music news, concert information	Free
Mobile Studio	Music, movie, entertainment information	100-200 yen/mo.
MobileViewsic	J-pop concert reports, news	Free
Mostly Classic	Online concert information, ringing tones	Free (members 200 yen/mo.)
National Club Information	Information on 210 clubs, events, etc.	Free (300 yen/mo. for event information)
Oricon Hit Next	CD review, ringing tones, etc.	100 yen/mo.
Pocket Vibe Music Channel	Latest music information, mail magazine, too	150 yen/mo.
Poke Shock Wave	Visual shock music site	Free (parts 300 yen/mo.)
Remix-I	Online version of Remix magazine	300 yen/mo.
Tsutaya Online	Music information, samples, ringing tones	Free (parts charged 100 or 200 yen/mo.)
Victor Music	Register for album/artist news	Free

MOVIES

Cinema Magazine i-Flix	Online version of movie magazines	Free (parts 300 yen/mo.)
Geo Movie Database	Information on old and new movies	Free
Universal	Movie, music, screens, melodies	200 yen/mo.

PRIZE/LOTTERY/HORSE RACING

Aqua Navi Kyotei	Motorboat race information	Free
Daily Horse Saburo	E-mails of favorite horse/race results	200 yen/mo.
Free! Prize ID	Register for free contests	Free
General Horseracing Channel	Information, photos, games	250 yen/mo.
JRA VAN	Odds, information from horse racing	300 yen/mo.
Local Horse Racing Channel	News on regional races	300 yen/mo.
Lotto Corner	Get lotto results and stats	Free
Online Booky	Ranking, odds, local race information	300 yen/mo.
Pachi Slot & Pachinko information	Manage your gambling habit,	150 yen/mo./menu
Pachinko Club	Melodies, characters, information from	300 yen/mo. favorite games
Pachinko Now!	Data on how to win	300 yen/mo.
Prize Puzzler	Solve puzzles for prizes	300 yen/mo.
Public Contest Guide	Guide to getting free stuff	300 yen/mo.
Sankei Keiba Net Jr.	Daily horse racing news, results	Free (members 200 yen/mo.)
The Prize	Guide to free contests/offers	300 yen/mo.

TELEVISION

Asahi TV	Program information, fortunes, screens, melodies	280 yen/mo.
Fuji TV	Program tie-in, fortunes, baseball news	200 – 300 yen/mo.
Fuji TV Entertainment	Official sites for singers, boxers, actors	300 yen/mo.
i-NTV	News updates seven times daily	Free
My Nippon TV (NTV)	Announcer's diary, games, cooking	300 yen/mo.
Sky Perfect TV	Italian soccer information, program information	Free
TBS	Program information, news	Free (parts 150 yen/mo.)

LOCAL BROADCASTERS

AKT Akita	Weather, traffic, program information	Free
Biwa Lake Broadcasting	Fishing, sightseeing, events	Free
BSN Niigata	News, weather, announcer	Free
CBC Central Japan Broadcasting	Program, news, weather, ticket information	Free
Ciao! Hiroshima Tvi	News, ski information	Free
EAT Ehime Asahi TV	News, emergency hospital information	Free
FBS Fukuoka Broadcasting	Announcer diary, soccer information, recipes	Free
Fuji/Yamanashi Broadcasting	Real-time picture of Fuji, event, news, movie information	Free
Fukui TV	News, cooking, shopping information	Free
HAB Hokuriku Asahi Broadcasting	Local information, news	Free
HBC Hokkaido	News, weather, hot spring information	Free
Hot 6 GoGo TV Shizuoka	Hot spring, shopping, movie information	Free
i Kochi Broadcasting	News, weather, program information	Free
i Sanin Broadcasting	Gourmet, program information	Free
IBC Iwate	News, weather, announcer information	Free
KBC Kyushu Asahi Broadcasting	Hawks information, etc.	Free
KFB Fukushima	News, weather, fortunes	Free
KHB Higashi Nippon	News, program, event information	Free
KTK Kanazawa TV	Movie, program news	Free
KTN TV Nagasaki	News, weather, event, announcer information	Free
MBC Minami Nihon Broadcasting	News, traffic, weather, program	Free
MBS Mainichi Broadcasting	Idol news, fortune telling, daily news	Free
MRT Miyazaki Broadcasting	News, weather, program, announcer information	Free
Nagoya TV	Outdoor, gourmet, travel information	Free
NBC Nagasaki TV	Program information, notices	Free

OBC Oita Broadcasting	News, announcer information	Free
OHK Okayama	News, weather, program, recipe, event, ski information	Free
OX Sendai	Weather, announcer essays	Free
RBC Ryuku Broadcasting	Okinawan culture, information	Free
RCC Chugoku Broadcasting	Program, fortune, CD, shopping information	Free
RKB Mainichi Broadcasting	News, Daiei Hawks information, character downloads	Free
RKK Kumamoto	Fortunes, announcer, program, local information	Free
RNB Nankai Broadcasting	Gourmet, news, weather, announcer	Free
RNC Nishi Nihon Broadcasting	News, weather, program, gourmet information	Free
RSK Sanyo Broadcasting	News, weather, traffic, fortune, announcer information	Free
SBC Shinetsu	News, weather, program information	Free
Sendai Broadcasting	Program information	Free
STV Sapporo TV	News, weather, announcer information	Free
TBC Tohoku Broadcasting	News, weather, program, event information	Free
TKU Kumamoto TV	News, weather, program, announcer, photo information	Free
TNC Nishi Nippon TV	Program, announcer information	Free
Tokai TV	Soap opera information, announcer, events	Free
TOS TV Oita	Program, weather, event, announcer information	Free
TUF Fukushima TV	News, weather, program, logos	Free
TV Ehime	Fishing, news, cooking, bulletin board	Free
TV Osaka	Program information	Free
TYS Yamaguchi TV	News, event, travel information	Free
UHB Hokkaido	Announcers, events	Free
YTS Yamagata TV	Traffic, program, gourmet guide	Free

TELEVISION PROGRAM INFORMATION

Aru Aru	Quizzes, information	Free (parts 100 yen/mo.)
i TV Eye	TV guide for all of Japan	Free
TV Guide	Check by time, actor, or genre	Free
TV Program News	Synopses of popular programs	100 yen/mo.

FM

Air-G FM Hokkaido	Requests, real-time song information, program information	Free
FM 802	Program, event information	Free
FM North Wave	Requests, music information	Free
FM Osaka 851	Program information, requests, top-100 charts	Free
Hiroshima FM	Soccer results, program information	Free
J-Wave	Song requests	Free
NACK 5 FM Saitama	Artist information, program guide	Free
TFM 80.0 Tokyo FM	Real-time program information	Free

MAGAZINES

Fami Tsu Con	Video game information	200 yen/mo.
i Bomb Atelier	Idol information	300 yen/mo.
i Weekly Diamond	Popular manga magazine	200 yen/mo.
iHappie	Fashion magazine	200 yen/mo.
iShogakukan	Comic entertainment	100 yen/mo./menu

MagaSeek	Catalog shopping	Free
National Pachinko/Slot Machine Info	Information on parlors, machines	Free (parts 300 yen/mo.)
Newtype-i	Cartoons, melodies, information	100 – 200 yen/mo.
OL Board	Working women's magazine	200 yen/mo.
PopTeen-net	Online version of popular teen magazine	190 yen/mo.
Walkers I	Entertainment information	Free (parts 250 yen/mo.)
Weekly Young Jump	Popular manga magazine	Free (parts 100/300 yen/mo.)

VARIETY

Famous and Kings Shogi News	Latest shogi news, commentary	200 yen/mo.
i Animate	Comic, game, information	300 yen/mo.
i Love Magic	Your mobile turns into a magic wand	150 yen/mo.
Naito, God of Good Fortune	Shogi questions, advice	100 yen/mo.
So-net	Tennis, cosmetic, and post pet information	Free

PUBLIC ENTERTAINMENT

Amuse	Southern Allstars, Fukuyama, others	200 yen/mo.
Artist Channel Bandai	Pop idols have diaries, fan club information, and appearance dates here	300 yen/mo./site
Asian Star	Vivian Sue and other stars information	300 yen/mo.
Battle Auction	Idols run these auctions	300 yen/mo.

Continued

PUBLIC ENTERTAINMENT (Continued)

Cartoon Voice Gran Prix	Voice actors' site	300 yen/mo.
Click-station	Young stars share their diaries	200 yen/mo.
CS Plaza	Holipro idols messages, etc.	300 yen/mo.
E. Yazawa	Yazawa's site answers Qs, etc.	300 yen/mo.
Enka Paradise	Messages, screen images	300 yen/mo.
Fan Club Net	Information from many popular stars' fan clubs	300 yen/mo.
Geinottodetai	Entertainment news, books, etc.	100 yen/mo.
Gendai Net	Site of Nikkei Gendai magazine	300 yen/mo.
i Takarazuka Opera	Information, screens, ringing tones from this all-female troupe	300 yen/mo.
Idol Paradise	Messages, screen images, etc.	300 yen/mo.
Idorama	Idols play characters in sim dramas	300 yen/mo.
Innatural	General music information site	300 yen/mo.
i-Rising	Information on Max, Da Pump, and other idols	300 yen/mo.
My Idol	Real idols play games and send e-mails	300 yen/mo.
Nakata Channel	Soccer player's official mobile site	300 yen/mo.
Pocket Morning Musume	Group's members offer their hellos	300 yen/mo.
Pop I-message	Messages from pop stars	300 yen/mo.
Pyramid Channel	Ten stars, original images, music	300 yen/mo.
SpoNichi Entertainment	Entertainment news	100 yen/mo.
Stardust i-Web	Images, mail from real pop stars	300 yen/mo.
Talent Entertainment Kingdom	Shociku's stars information	300 yen/mo.
Tommy's Diary	Messages, screen images	300 yen/mo.
Yes! Yoshimoto	YES fm's Yoshimoto's information site	Free

Town information/administration

ADMINISTRATION

Post Office General Site	Information on postal rates, postal savings, and postal insurance	Free
Telemo I	Site for local weather and traffic information	Free (parts 100 yen/mo.)

TOWN INFORMATION

Art Navi	National art gallery guide, artsy screens	Free (parts 300 yen/mo.)
i-Apartment Mansion	Find an apartment or condo here	Free
i-nagano.com	Events, gourmet, shopping, hot springs, etc.	Free
Niigata Komachi Week	Weekly news, restaurant reviews, used car ads	Free
Town Navi	National information on towns with coupons, images, e-mail	Free
Yamanashi Navi	A database of 5,000 tourist spots, news, etc.	Free

RAILROAD SERVICE INFORMATION TIMETABLE

JR East Travel Notices	When a train falls behind or if there is an accident, check here	Free (one part 200 yen/mo.)

PART-TIME JOBS/EMPLOYMENT

Cyber QJ&J	From casual jobs to full-time employment	Free
i-Working	Jobs in Niigata, Nagano, and Fukushima	Free

SPORT

Golf Reservation I	Reserve golf courses and get information	Free

Dictionaries/tools

DICTIONARIES/HANDY INFORMATION

DoCoMo Information Dial	Telephone guide to handy information such as weather	Free
Handy Guide to Social Grace	Reference on manners	Free
i-Imidas	Popular reference tool in your pocket	100 yen/mo.
Ima Hima2!	Find friends who are free for lunch	100 yen/mo.
i-mode Handy Dial	Report lost items or credit cards	Free
i-mode Handy Memo	Save memos here to you have when you need them	Free
i-Townpage	NTT directory assistance	Free
Iwanami Japanese Language Dictionary	Definitive reference	Free
Kiss Kiss Kiss	Your weather, fortune, and recipes automatically in your date book	200 yen/mo.
Learning I Channel	Dictionaries, drug encyclopedia, etc.	Free (one part 150 yen/mo.)
Petit Poochi Collection	Manage your books, movies, or music with this tool	150 yen/mo.
Sanseido Dictionary	English<->Japanese, Japanese	Free (one part 50 yen/mo.)

PACKAGE TRACKING

FedEx Tracking	Free
Footwork Tracking	Free
Fukuyama Tracking	Free
Meitetsu Tracking	Free
Nittsu Pelican Tracking	Free
Post Office Tracking	Free
Sagawa Kyubin Tracking	Free
Seibu Transportation Tracking	Free
Seino Kangaroo Tracking	Free
Yamato Tracking	Free

E-MAIL

Chara Mail	More than 600 characters	300 yen/mo.
Character Mail Service	Characters send you e-mail, including themselves!	300 yen/mo./character
EnterMail	A mail game for 2–4 people	200 yen/mo.
Gravure Idol Mail	A photo of the idol with a personal note to you!	300 yen/mo.
i-mode Mail Plus	Send to mailing lists, at certain times, other tools	Free
Imunote	Cute animals exchange diary entries	100 yen/mo.
Inbox for 90 yen	Cute characters in your mail	90 yen/mo.
Mail Anime	Sound and animation	100 yen/mo.
Mail de Girl	Your own girl mail supporter. She becomes your girlfriend, but let's see how long that lasts	230 yen/mo.
Mail Doll	All sorts of new features such as mail balloon mail	160 yen/mo.
Mail Fortune	Your fortune told according to your e-mail address	200 yen/mo.

Continued

E-MAIL *(Continued)*

Melo Goat	Freely download and send to friends 1,000 songs	90 yen/mo.
Melody by Mail	All you can hear for a pittance!	100 yen/mo.
Melopoly	Mail templates for friends or lovers	100 yen/mo.
Photo Mail	Use a digital camera, upload the image, and always have it with you	200 yen/mo.
Remote Mail	Check your home and work mail accounts	200 yen/mo.
Sugoneta	Send your screens to everyone	100 yen/mo.

Regional

HOKKAIDO

All-Japan Weather	Weather information for the whole country	Free (parts 100 yen/mo.)
Fishland Online	Hokkaido's premier fishing place	Free
Hako Navi	Town information for Hakkodate	Free
Hokkaido Police	Guidance on procedures, accidents, etc.	Free
I Fishy	Information on more than 300 fishing spots in Hokkaido	140 yen/mo.
i-nfo@movie	Movie information for all of Hokkaido	Free
JR Hokkaido	Time and schedule information	Free
Kushiro Information	Events, sightseeing, facilities, trash-collection schedule	Free
My Town Asahikawa	Sightseeing, event, information	Free
North Weather	Weather for Hokkaido	Free (premium 100 yen/mo.)
Nostalgic Otaru	Restaurant, shopping, leisure information	Free
Papa Mama Yellow Pages Sapporo	1800 day care, hospitals, etc.	Free

Part-Time Job Hokkaido	2,500–3,500 jobs listed	Free
Sapporo City General Information	Tourist information, events, hospitals	Free
Sapporo Room Search	Rental information	Free
Yellow Pages Sapporo	More than 5,000 shops listed	Free

TOHOKU

Akita Town Information	Dining, drinking, movies, shopping	Free
Aomori TJ	Dining, events, concerts, movies, information from 67 municipalities	Free
Fukushima Mobile Prefecture Office	Tourist information, administrative procedures, etc.	Free
Fukushima Navi	Dining, leisure, movies for Fukushima, Koriyama, Aizu, Iwaki	Free
Iwate Information	Fresh information	Free
Sendai TJ	Dining, movies, events, tickets	Free
Tohoku Road Information	Road conditions in real-time	Free
Yamagata TJ	Noodles, hot springs, souvenirs, music, movies	Free

TOKAI

Apartment News	More than 10,000 listings with daily updates	Free
Domo Job Part Time	Shizuoka employment information	Free
EL Town Shizuoka	Shop, town information	Free
FishOn	Fishing information for central Japan	Free (parts 300 yen/mo.)

Continued

TOKAI (Continued)

FMA 80.7 FM Aichi	Event, program, and guest information	Free
i-Navitar Tokai	Maps of Tokai/Hokuriku	Free
Job Can	Employment information for Shizuoka	Free
K-MIX Shizuoka	Radio top-30 hits, requests, etc.	Free
Marutoku Town Tokai	A bargain guide to Tokai	100 yen/mo.
Meitetsu Navi Nagoya	Train, bus, tickets, etc.	Free
Mobile Aichi Police	Accident reporting, license renewal information, etc.	Free
Mobile Shizuoka	Information, tourist information, Fuji images	Free
Shizuoka Real Estate Info	Sales and rental, more than 4,000 listings	Free
Surf Information Tokai	Izu, Hamamatsu, Shizuoka surf conditions, shops	Free (parts 100 yen/mo.)
Tokai Going Out Guide	Camping, skiing, hot springs, leisure	Free
Town Check	Guide to more than 9,000 shops in Tokai	Free
ZIP-FM FM Nagoya	Hit charts and event information	Free

HOKURIKU

@Fukui	Dining, drinking spots database with more than 1500 entries	Free
Fishing Genius	Fishing information	200 yen/mo.
Fukui TV	TV news, weather, cooking, program information	Free
i-Toyama.com	Life, dining, ticket, information	Free
K805 FM Ishikawa	Program, music information	Free
Lodging Selection	i-mode-only special rates, hints	Free
Sumairu	Real estate information for Ishikawa, Toyama, Fukui	Free

KANSAI AND CHUGOKU DISTRICTS

American Village Cyber Mall	Shop information, fashion information	Free
a-Station 894 AM	Music charts, timetables, town information	Free
Bike Info Mj Net	Information on more than 7,000 used motorbikes	Free
Check & Check I	Unusual dining information	Free (parts 200 yen/mo.)
Cinema Navi	Movie times, information, events, presents	Free
City Home	Home-rental information	Free
DaiMaru Department Store	Bargain information	Free
Digiba Nara	Dining, shopping, events, coupons	Free
From Nishinomiya	Emergency information, events, baseball	Free
Hard Town	Restaurant, store information, coupons	Free (parts 200 yen/mo.)
HEP Hankyu Entertainment Park	Event information, shop information	Free
i-agasu Wakayama	Dining, hot springs, ramen information	Free
JR West Japan	Train times and information	Free
Kintetsu Train	Reservation, timetable, fares	Free
Kobe City Walk	100,000 listings of restaurants, movies, music	Free
Kyoto City Information	Hospital, statistics, subway/bus times	Free
Kyoto Miya Call	Enjoy traditional Kyoto dining	Free (parts 100 yen/mo.)
Mini mini	Apartment living information	Free
Okayama Bus	Timetables, bus locations in real time	Free
Osaka Prefectural Police	Traffic information, event information	Free
Osaka ToyonoNet	Toyono district municipal information	Free
Pachinko Info	Event information, newly opened, presents, P/L table, hall guide	Free

Continued

KANSAI AND CHUGOKU DISTRICTS *(Continued)*

Public Road Information	Road construction, etc. information for Chugoku	Free
Purple Nara	Dining, shopping, leisure, event, coupon	Free
Room Navi	Apartment listings with pictures	Free
Town Info @Chugoku	Movie, dining, ticket information	Free
Yellow Book Information Service	Yellow pages	Free

SHIKOKU

Car Info Mj Net	Information on more than 13,000 used cars	Free
Corporate Careerasaurus	Careers in Ehime and Kochi	Free
Duke Live Info	Concert, event tickets	Free
Ehime Town Information SPC	Dining, shopping, movie, leisure information	Free
Family Info Shikoku	Living, dining, fashion, leisure	Free
Hand Navi Shikoku	Navigation in your hand	Free
iKochi Broadcasting	News, weather, programs	Free
Kagawa Info Navi	Child-rearing, hospitals, life	Free
Kagawa Town Info	Life, dining, leisure	Free
Kochi Prefecture Menu	Tourist, road information	Free
Kochi Town Info	Life, leisure information	Free
Music Café Apex	CD ordering, reference	Free
Real Estate Info Shkoku SPC	Rental, details, locations	Free
RNC Nishi Nippon Broadcasting	News, weather, program information	Free
Shikoku Road Conditions	Information from Ministry. of Construction	Free
Tokushima Mobile Prefetural Office	Police, hospital, event, tourist information	Free
Tokushima Town Info	Dining, information, fortunes	Free

KYUSHU

Corporate Get Together	Corporate information; more than 6,000 entries	Free
JR Kyushu	Timetables, around the stations	Free
Kagoshima Navi TJ	Leisure, hot springs, movies	Free
Kita Kyushu Navi	Dining, traffic, hotel information	Free
Kumamoto Navi	Tourist, hot springs, hotel	Free
Kurume Navi	Event, dining, tourist	Free
Miyazaki Navi	Dining, movies, event, surfing	Free
Nagasaki Navi	Event, movies, hot springs, dining	Free
Oita Navi	Dining, movies, hot springs, dialect	Free
Okinawa Navi	Dining, movies, flight information, hotel, tourist information	Free
Onsen Play Guide	Hot spring guide	Free
Pocket First Aid	First-aid information, hospital telephone numbers	Free
Private Railway in Your Pocket	Nishi Nippon Railway train and bus information	Free
Real Estate Agent Info Kyushu	Find an agent in any of the nine prefectures	Free
Saga Navi	Hot springs, dining	Free

I-NAVI LINK

AutoBank Car Info	Information on more than 50,000 used cars	Free
i Townpage NTT	An address can be displayed by inputting a number	Free
iMap Fan	Nationwide map coverage	Free
Zagat Tokyo Survey	Restaurant guide	Free

I-PLAYSTATION

Anywhere Together	Playing online makes this game even better	300 yen/mo.
Chekittipi	The first PS 2 i-mode game	300 yen/mo.
Playstation News	Latest information on games that connect	Free

English menu

NEWS/INFORMATION

Asahi Shimbun	Large daily's online version	100 yen/mo.
Bridge	Foreign exchange, market news	300 yen/mo.
Chosun Ilbo	South Korea's largest daily newspaper	Free
CNN	Headlines are free, full story costs	300 yen/mo.
Dow Jones	Financial news	300 yen/mo.
Nikkei News	Translated stories from Nihon Keizai Shimbun	Free
Peoples Daily	News with a Chinese twist	Free
Pokebras	Brazilian news (Portuguese only)	Free
Stock Smart	Stock quotes from many markets	300 yen/mo.
Weathernews	Weather for all of Japan	Free

ENTERTAINMENT

Disney-i	Ringing tones, information	100 yen/mo.
Fortune i	Horoscope and Tarot reading	170 yen/mo.
Ima hima 2	Find your friends and make plans	100 yen/mo.
Miracle GP	Car-racing game	300 yen/mo.
Universal J	Information on Universal Studios Japan theme park	Free

Appendix B: A Complete List of Official i-mode Sites

RINGING TONES

Bemani Hits	Ringing tones from Beat Mania, Dance Dance Revolution games	300 yen/mo.
Pokemelo JOY	Harmonic ringing tone downloads	300 yen/mo.
Vibe Beep	Lots of foreign hit songs	100 yen/mo.

DATABASE

Cooking Japan	Recipes and tips	Free
i-Townpage	English version of yellow pages has 70,000 listings	Free
Kobe City Walk	Plan your own thematic tour	Free
Kyoto Info	Restaurants, hotels, stores	Free
Tokyo Food Page	Where to eat	Free
Tokyo Q	What's going on, dining information, etc.	Free
Tokyo Wine News	Wine information, news, reviews	Free

OTHERS

Citibank	Account information	Free
Ericcson Café	Games, ringing tones, Swedish cooking	Free
FedEx	Package tracking	Free
Northwest Airlines	Frequent flyer information, flight status, telephone. numbers	Free
TMTDW	Tokyo Mistubishi TD Waterhouse Securities site	Free

Appendix C

i-mode Java API

AT THE TIME OF WRITING, the official DoCoMo DoJa API documentation is not available in English. I used several Japanese versions, as well as several other Japanese sources with information about the API as guides, and came up with this very rough version of documentation of the API in English. Because this book's aim is not specifically to teach readers how to program Java midlets on i-mode phones, and because doing so would be a book in itself, I have not given detailed explanations of each method, field, or constructor. This API is more a guide to what is in the DoJa specification, and is probably more useful to experienced Java programmers than those just starting.

AudioPresenter

Library: com.nttdocomo.ui
Type: Class
Inheritance:

```
java.lang.Object
   |
   +--com.nttdocomo.ui.AudioPresenter
```

```
public class AudioPresenter
extends java.lang.Object
implements MediaPresenter
```

`AudioPresenter` is an audio control class. Sound data is received and the `playback` event is sent to the listener. When model-specific sound data is received on a certain handset that cannot play it, a `UIException` may be thrown. Although data can be handled by two methods (the `setSound` and `setData` methods), they are mutually exclusive, and cannot both be present in one object.
Exceptions:
   P503i: Although this model is set up so that the key-input sound plays by default, the key input sound stops playing during playback of sound data. However, when a `ListBox` in the high-level API (`Panel`) is in the selection state, a key input sound plays, and a quite jarring noise occurs.

Minimum specified functions:

- No `setAttribute` methods are carried out for this class. Therefore, the attribute for passing this method is not defined, either.
- The listener is only guaranteed to be notified of the `AUDIO_COMPLETE` event.
- The `setData` method does not work. (An `UnsupportedOperationException` is thrown.)
- The `getAudioPresenter` method can return a minimum of one instance.

Exceptions:

J2ME Wireless SDK for DoJa API Release2.2: The usual MLD extension, used with i-melodies, cannot be used; instead, the .MID extension, which is for standard MIDI files, must be used.

Related items: `MediaSound`, `MediaData`, and `UnsupportedOperationException`

FIELDS

Public static final int AUDIO_COMPLETE	Event number (3) showing playback is completed
Public static final int AUDIO_PLAYING	Event number (1) showing playback started
Public static final int AUDIO_STOPPED	Event number (2) showing playback was interrupted
Protected static final int MAX_VENDOR_ATTR	Maximum (127) of a vendor's unique attributes
Protected static final int MAX_VENDOR_AUDIO_EVENT	A vendor-unique audio event maximum (127)
Protected static final int MIN_VENDOR_ATTR	The minimum value of a vendor unique attribute (64)
Protected static final int MIN_VENDOR_AUDIO_EVENT	A vendor-unique audio event minimum value (64)

CONSTRUCTOR

Protected AudioPresenter()	An application must not generate an object using this constructor

METHODS

Public static AudioPresenter getAudioPresenter()	Acquires an effective instance of `AudioPresenter`
Public MediaResource getMediaResource()	Returns the `MediaSound` or `MediaData` currently saved as a `MediaResource`
Public void play()	Plays the `MediaSound` or `MediaData` currently saved
Public void setAttribute(int *attr*, int *value*)	Sets an attribute value for playback
Public void setData(MediaData *data*)	Sets `MediaData` for playback
Public void setMediaListener(MediaListener *listener*)	Sets a listener to receive the playback status of media system data
Public void setSound(MediaSound *sound*)	Sets a `MediaSound` for playback
Public void stop()	Stops playback

Methods inherited from java.lang.Object class: `equals`, `getClass`, `hashCode`, `notify`, `notifyAll`, `toString`

Button

Library: com.nttdocomo.ui
Type: Class
Inheritance:

```
java.lang.Object
  |
  +--com.nttdocomo.ui.Component
        |
        +--com.nttdocomo.ui.Button
              |
              +--Interactable
```

```
public final class Button
extends Component
implements Interactable
```

The `Button` class is a component for the high-level API and defines various styles of buttons.

About the display method of a label — although it is model-specific, typically a layout in the center is advisable. Moreover, the size of a button changes depending on how a label is displayed on the screen. What a handset does with a label that doesn't fit on the screen depends on the model.

Related items: `Panel` and `Panel.add` (`com.nttdocomo.ui.Component`)

CONSTRUCTORS

Public Button()	A label generates an empty ("") instance
Public Button(java.lang.String *label*)	An instance with a specific label is generated

METHODS

Public void requestFocus()	Sets focus to a button
Public void setEnabled(boolean *b*)	Sets a button to an active or an inactive state. `true` is active; `false` inactive
Public void setLabel(java.lang.String *label*)	Sets a button's label text

Methods inherited from `com.nttdocomo.ui.Component` class: `getHeight`, `getWidth`, `getX`, `getY`, `setBackground`, `setForeground`, `setLocation`, `setSize`, `setVisible`

Methods inherited from `java.lang.Object` class: `equals`, `getClass`, `hashCode`, `notify`, `notifyAll`, `toString`,

Canvas

Library: com.nttdocomo.ui
Type: Class
Inheritance:

```
java.lang.Object
   |
   +--com.nttdocomo.ui.Frame
          |
          +--com.nttdocomo.ui.Canvas
```

public abstract class **Canvas**
extends Frame

Canvas is an abstract class that processes a low-level display. It also prepares the processEvent method for processing low-level events other than display, such as key events or timer events. By passing the object that inherited this class to the setCurrent method of Display class, the Canvas is displayed on the screen.

  Related item: Display.setCurrent (com.nttdocomo.ui.Frame)

  Fields inherited from the com.nttdocomo.ui.Frame class: SOFT_KEY_1, SOFT_KEY_2

CONSTRUCTOR

Public Canvas()	Constructor

METHOD

Public Graphics getGraphics()	Returns a graphics instance for drawing to Canvas
Public int getKeypadState()	Returns the state of a keypad
Public abstract void paint(Graphics g)	Draws to Canvas
Public void processEvent(int type, int param)	Is called when a low-level event is received
Public void repaint()	Requests redrawing of the whole Canvas
Public void repaint(int x, int y, int width, int height)	Requests r-drawing of a part of Canvas

  Methods inherited from com.nttdocomo.ui.Frame class: getHeight, getWidth, setBackground, setSoftLabel

  Methods inherited from java.lang.Object class: equals, getClass, hashCode, notify, notifyAll, toString

Component

Library: com.nttdocomo.ui
Type: Class
Inheritance:

```
java.lang.Object
  |
  +--com.nttdocomo.ui.Component
```

```
public abstract class Component
extends java.lang.Object
```

Known subclasses: `Button, ImageLabel, Label, ListBox, TextBox, Ticker, VisualPresenter`

`Component` is an abstract class showing the components used by the high-level API.

CONSTRUCTOR

Component()	Sets a component to a visible state by this constructor

METHODS

Public final int getHeight()	Returns the height of a component
Public final int getWidth()	Returns the width of a component
Public int getX()	Returns X-coordinates of a component
Public int getY()	Returns Y-coordinates of a component
Public void setBackground(int c)	Sets the background color of a componen
Public void setForeground(int c)	Sets the foreground color of a component
Public void setLocation(int x, int y)	Sets the position of a component
Public void setSize(int width, int height)	Sets the size of a component
Public void setVisible(boolean b)	Sets a component to visible (`true`) or invisible (`false`)

Methods inherited from java.lang.Object class: `equals, getClass, hashCode, notify, notifyAll, toString`

ComponentListener

Library: com.nttdocomo.ui
Type: Interface

`Super interface: EventListener public interface ComponentListener extends EventListener`

`ComponentListener` is the interface that defines the listener for a component event.
   Related items: `Panel.setComponentListener (com.nttdocomo.ui.ComponentListener)`

FIELDS

public static final int BUTTON_PRESSED	Returns event code (1) when a button is pushed
public static final int SELECTION_CHANGED	Returns event code (2) when changing the item in which a component is contained
public static final int TEXT_CHANGED	Returns event code (3) when the text of a component in which a text input is possible is decided

---

METHOD

public void componentAction (Component *source*, int *type*, int *param*)	Receives an event of a component

---

ConnectionException

Library: com.nttdocomo.io
Type: Class
Inheritance:

```
java.lang.Object
   |
   +--java.lang.Throwable
         |
         +--java.lang.Exception
               |
               +--java.io.IOException
                     |
                     +
                     |
com.nttdocomo.io.ConnectionException
```

public class **ConnectionException**
extends java.io.IOException

ConnectionException is used as an exception in the com.nttdocomo.io package defined by DoCoMo Profile-1.0 and in the input/output processing of the com.nttdocomo.net package.

FIELDS

public static final int HTTP_ERROR	HTTP server error. (10)
public static final int ILLEGAL_STATE	The state of an object is inaccurate. (1)
public static final int IMODE_LOCKED	i-mode is in a locked state. (6)
public static final int NO_RESOURCE	Failure obtaining a resource. (2)
public static final int NO_USE	Outside the i-mode service area. (4)
public static final int OUT_OF_SERVICE	Outside the network service area. (5)
public static final int RESOURCE_BUSY	A resource is being used. (3)
public static final int SCRATCHPAD_OVERSIZE	The maximum capacity of a scratch pad was exceeded. (11)
public static final int STATUS_FIRST	Head status code of a system definition. (0)
public static final int STATUS_LAST	The last status code of a system definition. (32)
public static final int SYSTEM_ABORT	Processing was stopped by the system. (9)
public static final int TIMEOUT	Processing was not completed by the timeout. (7)
public static final int UNDEFINED	Exception with no definition. (0)
public static final int USER_ABORT	Processing was stopped by the user. (8)

METHOD	
ConnectionException(int stat)	An exception instance with defined status is built
public synchronized int getStatus()	Acquires an exception status

Methods inherited from `java.lang.Throwable` class: `fillInStackTrace, getMessage, printStackTrace, toString`

Methods inherited from java.lang.Object class: `equals, getClass, hashCode, notify, notifyAll`

Dialog

Library: com.nttdocomo.ui
Type: Class
Inheritance:

```
java.lang.Object
   |
   +--
com.nttdocomo.ui.Frame
        |
        +--com.nttdocomo.ui.Dialog
```

```
public final class Dialog
extends Frame
```

`Dialog` is a class that offers a dialog function to check user response.
Exceptions:

- ◆ F503i: A window similar in appearance to a screen using `Panel` is displayed. A message is displayed into a component the same in appearance as the `TextBox` component. A `Button` component is displayed and operated similarly.

- ◆ P503i and P503iS: Usually, a window with a character display or item selection button of the same appearance as that used for checking the menu is displayed. An item selection button is chosen by the right and left buttons, and is selected with the (center) select button. Other items are assigned to a soft key.

- ◆ N503i: A window with a title bar is displayed. The icon corresponding to the kind of dialog is displayed in the window's central upper area, and the message is displayed below it. The button looks the same as the `Button` component in how it is displayed and operated. Use caution because it is cut off on the right and left when a title exceeds the maximum length.
- ◆ SO503i: A window is displayed. At this time, the drawing area becomes the bottom of a window and is darkened. A button is displayed and operated in the same way as the `Button` component.
- ◆ D503i: The whole screen becomes a dialog box, and a title is displayed on the top line. A message can be scrolled and can be displayed altogether. A button attached to a key is arranged at the lower part of the screen.

Other information about models is as follows:

MAXIMUM TITLE LENGTH

Kind of Dialog	F503i		P503i		N503i		SO503i		D503i	
1- or 2-bit char.	1	2	1	2	1	2	1	2	1	2
Information	18	9	15	7	16	8	16	8	16	8
Warning	16	8	15	7	16	8	14	7	16	8
Error	16	8	15	7	16	8	14	7	16	8
Yes/No	18	9	15	7	16	8	14	7	16	8
Yes/No/Cancel	18	9	15	7	16	8	16	8	16	8

MAXIMUM MESSAGE LENGTH

Kind of Dialog	F503i		P503i		N503i		SO503i		D503i	
1- or 2-bit char.	1	2	1	2	1	2	1	2	1	2
Information	*	*	15*4	7*4	16*3	8	16*5	8*5	16*5	8*5
Warning	*	*	15*4	7*4	16*3	8	16*5	8*5	16*5	8*5
Error	*	*	15*4	7*4	16*3	8	16*5	8*5	16*5	8*5
Yes/No	*	*	15*4	7*4	16*3	8	16*5	8*5	16*5	8*5
Yes/No/Cancel	*	*	15*4	7*4	16*3	8	16*5	8*5	16*5	8*5

SELECTION DISPLAY

Kind of Dialog	F503i	P503i	N503i	SO503i	D503i
Information	OK	OK	OK	OK	OK
Warning	OK	OK	OK	OK	OK
Error	OK	OK	OK	OK	OK
Yes/No	YES NO	YES NO	(hai) (iie)	YES NO	YES NO
Yes/No/Cancel	YES NO Cancel	YES NO (kyanseru)	(hai) (iie) (kyanseru)	YES NO (kyanseru)	(modoru) (hai) (iie)

ICONS DISPLAYED

Kind of Dialog	F503i	P503i	N503i	SO503i	D503i
Information	--	Bird	(i)	--	--
Warning	ISO Warning	Bird	ISO Warning	ISO Warning	--
Error	ISO Warning	Bird	X	ISO Warning	--
Yes/No	--	--	(?)	--	--
Yes/No/Cancel	--	--	(?)	--	--

FIELDS

public static final int BUTTON_CANCEL	Constant that shows that the dialog display result was a CANCEL button
public static final int BUTTON_NO	Constant that shows that the dialog display result was the NO button
public static final int BUTTON_OK	Constant that shows that the dialog display result was the OK button
public static final int BUTTON_YES	Constant that shows that the dialog display result was a YES button

Continued

FIELDS *(Continued)*

public static final int DIALOG_ERROR	Code that shows an error dialog
public static final int DIALOG_INFO	Code that shows an information dialog
public static final int DIALOG_WARNING	Code that shows a warning dialog
public static final int DIALOG_YESNO	Code that shows the selection dialog of YES or NO
public static final int DIALOG_YESNOCANCEL	Code that shows the dialog of YES, NO, or CANCEL (cancellation)

Fields inherited from com.nttdocomo.ui.Frame class: SOFT_KEY_1, SOFT_KEY_2

CONSTRUCTOR

public Dialog(int *type*, java.lang.String *title*)	The kind and title of a dialog box are received, and a dialog box is generated

METHODS

public void setBackground(int *c*)	Sets the background color of a dialog box
public void setSoftLabel(int *key*, java.lang.String *label*)	Sets the label of a soft key
public void setText(java.lang.String *msg*)	Sets the characters to display
public int show()	Displays a dialog box on the screen

Methods inherited from com.nttdocomo.ui.Frame class: getHeight, getWidth
Methods inherited from java.lang.Object class: equals, getClass, hashCode, notify, notifyAll, toString

Display

Library: com.nttdocomo.ui
Type: Class
Inheritance:

```
java.lang.Object
   |
   +--com.nttdocomo.ui.Display
```

public class **Display**
extends java.lang.Object

Display is a class that performs the processing and management of display and keys. Because this class makes static declarations of all the methods and variables that are contained in the class, generating an instance is not required.

FIELDS

public static final int KEY_0	Constant for the number key 0(=0x00)
public static final int KEY_1	Constant for the number key 1(=0x01)
public static final int KEY_2	Constant for the number key 2(=0x02)
public static final int KEY_3	Constant for the number key 3(=0x03)
public static final int KEY_4	Constant for the number key 4(=0x04)
public static final int KEY_5	Constant for the number key 5(=0x05)
public static final int KEY_6	Constant for the number key 6(=0x06)
public static final int KEY_7	Constant for the number key 7(=0x07)
public static final int KEY_8	Constant for the number key 8(=0x08)
public static final int KEY_9	Constant for the number key 9(=0x09)
public static final int KEY_ASTERISK	Constant for the asterisk key (=0x0a)
public static final int KEY_DOWN	Constant for the down key (=0x13)
public static final int KEY_LEFT	Constant for the left key (=0x10)

Continued

FIELDS *(Continued)*

public static final int KEY_POUND	Constant for the pound key (=0x0b)
public static final int KEY_PRESSED_EVENT	Constant for a key down event (=0)
public static final int KEY_RELEASED_EVENT	Constant for a key up event (=1)
public static final int KEY_RIGHT	Constant for the right key (=0x12)
public static final int KEY_SELECT	Constant for the select key (=0x14)
public static final int KEY_SOFT1	Constant for the soft 1 key (=0x15)
public static final int KEY_SOFT2	Constant for the soft 2 key (=0x16)
public static final int KEY_UP	Constant for the up key (=0x11)
protected static final int MAX_VENDOR_EVENT	The maximum event value of a vendor-unique event
protected static final int MAX_VENDOR_KEY	The maximum number of a vendor-unique key
public static final int MEDIA_EVENT	A comparison constant for distinction of media events at event processing
protected static final int MIN_VENDOR_EVENT	The minimum event value of a vendor-unique event
protected static final int MIN_VENDOR_KEY	The minimum number of a vendor-unique key
public static final int RESET_VM_EVENT	A comparison constant for distinction of reset events at event processing
public static final int RESUME_VM_EVENT	A comparison constant for distinction of stop events at event processing
public static final int TIMER_EXPIRED_EVENT	A comparison constant for distinction of timer events at event processing
public static final int UPDATE_VM_EVENT	A comparison constant for distinction of update events at event processing

Constructor: `Display()`

METHODS

public static final Frame getCurrent()	Returns an instance of the class derived from the current Frame class
public static final int getHeight()	Acquires the height of the screen
public static final int getWidth()	Acquires the width of the screen
public static final boolean isColor()	Returns whether the screen supports color
public static final int numColors()	Returns the number of colors that can be used on a screen
public static final void setCurrent(Frame frame)	Sets an instance of the class derived from the current Frame class

Methods inherited from java.lang.Object class: `equals`, `getClass`, `hashCode`, `notify`, `notifyAll`, `toString`

EventListener

Library: com.nttdocomo.util
Type: Interface
List of known sub interfaces: `ComponentListener`, `KeyListener`, `MediaListener`, `SoftKeyListener`, `TimerListener`

```
public interface EventListener
```

This interface defines all of the listeners that use it in DoCoMoProfile-1.0. All interfaces that receive notices of events are derived from this interface.

FocusManager

Library: com.nttdocomo.ui
Type: Interface

```
public interface FocusManager
```

FocusManager is an interface for managing the component focus on `Panel`. Use of this interface depends on model-specific APIs, so it is not accessible except from these APIs.

Font

Library: com.nttdocomo.ui
Type: Class
Inheritance:

```
java.lang.Object
  |
  +--com.nttdocomo.ui.Font

public class Font
extends java.lang.Object
```

This class manages the text font, its size, and its style. `Font` is model-dependent as to how the specified font displays, including these:

- ◆ **F503i and N503I:** Even if the size of a font is changed in the midlet, the font size stays the same.
- ◆ **P503i, SO503i, D503i, and P503is:** Using `large` for the font size returns a font the same size as `medium`; setting the font to `small` returns a small font.

FIELD

public static final int FACE_MONOSPACE	Constant for monospace face specification
public static final int FACE_PROPORTIONAL	Constant for proportional face specification
public static final int FACE_SYSTEM	Constant for system face specification
public static final int SIZE_LARGE	Constant for large size specification
public static final int SIZE_MEDIUM	Constant for medium size specification
public static final int SIZE_SMALL	Constant for small size specification
public static final int STYLE_BOLD	Constant for bold style specification
public static final int STYLE_BOLDITALIC	Constant for bold and italic style specification
public static final int STYLE_ITALIC .	Constant for italic style specification
public static final int STYLE_PLAIN	Constant for plain style specification
public static final int TYPE_DEFAULT	Constant for default font specification
public static final int TYPE_HEADING	Constant for title font specification

Constructor: Font()

METHODS

public int getAscent()	The ascent value (length from the baseline to the top) of a font is returned
public int getBBoxHeight(java.lang.String str)	The height of the bounding box of a specified string is returned
public int getBBoxWidth(java.lang.String str)	The width of the bounding box of a specified string is returned
public static Font getDefaultFont()	An instance of the Font class used as a default by a specific model is returned
public int getDescent()	The descent value (length from the baseline to the bottom) of a font is returned
public static Font getFont(int type)	An instance of Font class with a setup specified by the argument is returned
public int getHeight()	The height of a font is returned
public int getLineBreak(java.lang.String str, int off, int len, int width)	The number of characters before a specified string reaches a specified width is returned
public int stringWidth(java.lang.String str)	The width of a specified string is returned

Methods inherited from java.lang.Object class: equals, getClass, hashCode, notify, notifyAll, toString

Frame

Library: com.nttdocomo.ui
Type: Class
Inheritance:

```
java.lang.Object
   |
   +--com.nttdocomo.ui.Frame
```

```
public abstract class Frame
extends java.lang.Object
```

Known subclasses of com.nttdocomo.ui.Frame: `Canvas`, `Dialog`, and `Panel`
`Frame` is the class that manages the display on the screen. An event is sent to a subclass from this class. However, `Frame()` is not called other than the instance called by the `setCurrent` method of the `Display` class.
Related item: `Display.setCurrent (com.nttdocomo.ui.Frame)`

FIELDS

Public static int SOFT_KEY_1	Constant showing soft key 1
Public static int SOFT_KEY_2	Constant showing soft key 2

CONSTRUCTOR

Frame()	Because an abstract declaration of this class is made, an instance cannot be generated by this constructor

METHODS

Public int getHeight()	Returns the height of a frame
Public int getWidth()	Returns the width of a frame
Public void setBackground(int *c*)	Changes a background color to the specified color
Public void setSoftLabel(int *key*, java.lang.String *label*)	Sets a label string as a soft key

Methods inherited from java.lang.Object class: `equals, getClass, hashCode, notify, notifyAll, toString`

Graphics

Library: com.nttdocomo.ui
Type: Class
Inheritance:

```
java.lang.Object
  |
  +--com.nttdocomo.ui.Graphics
```

public class **Graphics**
extends java.lang.Object

Graphics is the class that performs drawing control in the Canvas class. (It is the so-called graphical context.)

FIELDS

public static final int AQUA	Constant for light-blue. (=3, or 0x00,0xff,0xff)
public static final int BLACK	Constant for black. (=0, or 0x00,0x00,0x00)
public static final int BLUE	Constant for blue. (=1, or 0x00,0x00,0xff)
public static final int FUCHSIA	Constant for purple. (=5, or 0xff,0x00,0xff)
public static final int GRAY	Constant for gray. (=8, or 0x80,0x80,0x80)
public static final int GREEN	Constant for dark green. (=10, or 0x00,0x80,0x00)
public static final int LIME	Constant for lime green. (=2, or 0x00,0xff,0x00)
public static final int MAROON	Constant for maroon. (=12, or 0x80,0x00,0x00)
public static final int NAVY	Constant for navy blue. (=9, or 0x00,0x00,0x80)
public static final int OLIVE	Constant for dark yellow. (=14, or 0x80,0x80,0x00)
public static final int PURPLE	Contant for purple. (=13, or 0x80,0x00,0x80)
public static final int RED	Constant for red. (=4, or 0xff,0x00,0x00)
public static final int SILVER	Constant for silver. (=15, or 0xC0,0xC0,0xC0)
public static final int TEAL	Constant for teal. (=11, or 0x00,0x80,0x80)
public static final int WHITE	Constant for white. (=7, or 0xff,0xff,0xff)
public static final int YELLOW	Constant for color yellow. (=6, or 0xff, 0xff,0x00)

CONSTRUCTOR

Protected Graphics()	An application must not make an instance using this constructor

METHODS

public void clearRect(int *x*, int *y*, int *width*, int *height*)	The rectangle's specified range reverts to the background color
public Graphics copy()	Returns another instance of its own instance
public void dispose()	After this method has been called, a Graphics object cannot be accessed
public void drawChars(char[] *data*, int *x*, int *y*, int *off*, int *len*)	Draws part of a string in a specified position
public void drawImage (Image img, int *x*, int *y*)	Draws an image
public void drawLine(int *x1*, int *y1*, int *x2*, int *y2*)	Draws a straight line from the specified starting point (x1,y1), to the specified terminal point (x2,y2)
public void drawPolyline(int[] xPoints, int[] yPoints, int nPoints)	Draws a line that connects the specified points
public void drawRect(int x, int y, int width, int height)	Draws a rectangle of a width and height specified from the starting point downward and right
public void drawString(java.lang.String str, int x, int y)	Draws a string in a specified position
public void fillPolygon(int[] *xPoints*, int[] *yPoints*, int *nPoints*)	Draws a polygon using the points in the x and y arrays
public void fillRect(int *x*, int *y*, int *width*, int *height*)	Draws a rectangle of specified dimensions
public static int getColorOfName(int *name*)	Returns a color code from the name of the color constant
public static int getColorOfRGB (int *r*, int *g*, int *b*)	Generates a color code from an RGB value

public void lock()	Starts a lock for double buffering
public void setColor(int c)	Sets the color code used for drawing
public void setFont(Font f)	Sets the font used for drawing by an instance of Font class
public void setOrigin(int x, int y)	Sets the starting point of drawing coordinates
public void unlock(boolean forced)	Ends the lock for double buffering

Methods inherited from java.lang.Object class: `equals, getClass, hashCode, notify, notifyAll, toString`

HttpConnection

Library: com.nttdocomo.io
Type: Interface
Super interfaces:

```
javax.microedition.io.Connection,
javax.microedition.io.ContentConnection,
javax.microedition.io.InputConnection,
javax.microedition.io.OutputConnection,
javax.microedition.io.StreamConnection
```

```
public interface HttpConnection
extends javax.microedition.io.ContentConnection
```

HttpConnection is the interface that defines communication by HTTP. When an instance of the class is passed a URL starting with `http:` or `https:`, the instance opens a method of the `Connector` class of the `javax.microedition.io` package. A `SecurityException` is thrown when the user does not receive network approval.

FIELDS

Public static final java.lang.String GET	Constant that shows GET method
Public static final java.lang.String POST	Constant that shows POST method
Public static final java.lang.String HEAD	Constant that shows HEAD method

Continued

Appendixes

FIELDS *(Continued)*

Public static final int HTTP_ACCEPTED	Constant for HTTP status code (202)
Public static final int HTTP_BAD_GATEWAY	Constant HTTP status code (502)
Public static final int HTTP_BAD_METHOD	Constant for HTTP status code (405)
Public static final int HTTP_BAD_REQUEST	Constant for HTTP status code (400)
Public static final int HTTP_CLIENT_TIMEOUT	Constant for HTTP status code (408)
Public static final int HTTP_CONFLICT	Constant for HTTP status code (409)
Public static final int HTTP_CREATED	Constant for HTTP status code (201)
Public static final int HTTP_ENTITY_TOO_LARGE	Constant for HTTP status code (413)
Public static final int HTTP_EXPECT_FAILED	Constant for HTTP status code (417). *Not currently part of DoCoMo API, but expected to be included in future
Public static final int HTTP_FORBIDDEN	Constant for HTTP status code (403)
Public static final int HTTP_GATEWAY_TIMEOUT	Constant for HTTP status code (504)
Public static final int HTTP_GONE	Constant for HTTP status code (410)
Public static final int HTTP_INTERNAL_ERROR	Constant for HTTP status code (501), *but* this status code should actually be for HTTP_NOT_IMPLEMENTED. This field is an apparent mistake in this API, and SHOULD NOT BE USED!
Public static final int HTTP_LENGTH_REQUIRED	Constant for HTTP status code (411)
Public static final int HTTP_MOVED_PERM	Constant for HTTP status code (301)
Public static final int HTTP_MOVED_TEMP	Constant for HTTP status code (302)
Public static final int HTTP_MULT_CHOICE	Constant for HTTP status code (300)
Public static final int HTTP_NO_CONTENT	Constant for HTTP status code (204)
Public static final int HTTP_NOT_ACCEPTABLE	Constant for HTTP status code (406)
Public static final int HTTP_NOT_AUTHORITATIVE	Constant for HTTP status code (203)
Public static final int HTTP_NOT_FOUND	Constant for HTTP status code (404)
Public static final int HTTP_NOT_IMPLEMENTED	Constant for HTTP status code (501)
Public static final int HTTP_NOT_MODIFIED	Constant for HTTP status code (304)
Public static final int HTTP_OK	Constant for HTTP status code (200)

Public static final int HTTP_PARTIAL	Constant for HTTP status code (206)
Public static final int HTTP_PAYMENT_REQUIRED	Constant for HTTP status code (402)
Public static final int HTTP_PRECON_FAILED	Constant for HTTP status code (412)
Public static final int HTTP_PROXY_AUTH	Constant for HTTP status code (407)
Public static final int HTTP_REQ_TOO_LONG	Constant for HTTP status code (414)
Public static final int HTTP_RESET	Constant for HTTP status code (205)
Public static final int HTTP_SEE_OTHER	Constant for HTTP status code (303)
Public static final int HTTP_SERVER_ERROR	Constant for HTTP status code (500). *This should not be used, as it is an apparent mistake, where HTTP_INTERNAL_ERROR should be using this status code
Public static final int HTTP_TEMP_REDIRECT	Constant for HTTP status code (307). *Though this is not defined in DoCoMo's API, it is thought that it will be defined in future
Public static final int HTTP_UNAUTHORIZED	Constant for HTTP status code (401)
Public static final int HTTP_UNAVAILABLE	Constant for HTTP status code (503)
Public static final int HTTP_UNSUPPORTED_RANGE	Constant for HTTP status code (416). *Though this is not defined in DoCoMo's API, it is thought that it will be defined in future
Public static final int HTTP_UNSUPPORTED_TYPE	Constant for HTTP status code (415)
Public static final int HTTP_USE_PROXY	Constant for HTTP status code (305)
Public static final int HTTP_VERSION	Constant for HTTP status code (505)

METHODS

Public void close()	Closes connection and frees the resource
Public void connect()	Connects by HTTP, sends the request, and waits for a response

Continued

METHODS *(Continued)*

Public long getDate()	Returns message date returned by HTTP response header
Public java.lang.String getEncoding()	Returns an encoding when encoding is included in the response header
Public long getExpiration()	Returns the term of validity of an HTTP response header
Public java.lang.String getHeaderField(java.lang.String name)	Returns a value corresponding to an appointed key from HTTP response header
Public long getLastModified()	Returns a message modification date
Public long getLength()	Returns the content length of a response header
Public int getResponseCode()	Returns the HTTP response code
Public java.lang.String getResponseMessage()	Returns the HTTP response message
Public java.lang.String getType()	The content type of a response header
Public java.lang.String getURL()	Returns a URL string for a connection
Public java.io.InputStream openInputStream()	Returns the `InputStream` corresponding to the content body (minus the header) of a response
Public java.io.OutputStream openOutputStream()	Returns the `OutputStream` corresponding to the content body (minus the header) of a request
Public void setIfModifiedSince(long ifmodifiedsince)	Sets the value of an "If-Modified-Since" header
Public void setRequestMethod(java.lang.String method)	Sets the kind of request method
Public void setRequestProperty(java.lang.String key, java.lang.String value)	Sets the property name and property value that are included in a request header

Method inherited from javax.microedition.io.InputConnection interface: `openDataInputStream`

Method inherited from javax.microedition.io.OutputConnection interface: `openDataOutputStream`

IApplication

Library: com.nttdocomo.ui
Type: Class
Inheritance:

```
java.lang.Object
  |
  +--com.nttdocomo.ui.IApplication
```

```
public abstract class IApplication
extends java.lang.Object
```

IApplication is a class that starts and manages application instances. All applications must be public classes that are inherited from this class. After an application suspends for reasons such as a telephone call, where an instance of `setCurrent` method of the `Display` class exists, a `Display.RESUME_VM_EVENT` event is notified of the instance and the `resume` method of this class is called.

CONSTRUCTOR

Iapplication()	An application must not generate an instance using this constructor

METHODS

Public final java.lang.String[] getArgs()	Returns the starting parameters of an application, stored in the Parameter field of the jar file
Public static final Iapplication getCurrentApp()	Returns the application presently being executed
Public final java.lang.String getSourceURL()	Returns the URL of the application download source
Public void resume()	Is called after a momentary pause of an application

Continued

METHODS *(Continued)*	
Public abstract void start()	Is called at the start of the execution of an application
Public void terminate()	Terminates the currently executing application

Methods inherited from java.lang.Object class: `equals, getClass, hashCode, notify, notifyAll, toString`

Image

Library: com.nttdocomo.ui
Type: Class
Inheritance:

```
java.lang.Object
   |
   +--com.nttdocomo.ui.Image
```

```
public abstract class Image
extends java.lang.Object
```

`Image` is a class showing an image (still picture). The `getImage` method of the `MediaImage` class generates an instance of this class.

Related items: `MediaImage.getImage(), Graphics.drawImage (com.nttdocomo.ui.Image, int, and int), ImageLabel,` **and** `ImageLabel.setImage (com.nttdocomo.ui.Image)`

CONSTRUCTOR

Protected Image()	An application cannot generate an instance using this constructor

METHODS

public abstract void dispose()	Prepares the cancellation of an instance
public int getHeight()	Returns the height of the image currently held
public int getWidth()	Returns the width of the image currently held

Methods inherited from java.lang.Object class: `equals`, `getClass`, `hashCode`, `notify`, `notifyAll`, `toString`

ImageLabel

Library: com.nttdocomo.ui
Type: Class
Inheritance:

```
java.lang.Object
  |
  +--com.nttdocomo.ui.Component
         |
         +--com.nttdocomo.ui.ImageLabel
```

`public final class` **ImageLabel**
`extends Component`

`ImageLabel` is a component class that displays an image (still picture) in the high-level API (`Panel`). When an image is set to an invalid instance of the `Image`, `ImageLabel` is set to the background color the same as if it had not been set.

It is not possible to dispose of an instance of the `MediaImage` class that generated an instance of the `Image` class made into the display object, nor of the instance of the `Image` class during the time the `ImageLabel` class is being used.

Certain `display` methods are model-specific: when an image is larger than the display; when the image for a display is smaller than a component; and component size.

Typically, when an image is smaller than a component, it is displayed in the center. The following show some model-specific issues:

- ◆ **SO503i:** Transparency does not work even if it is a transparent GIF.

- ◆ **F503i, P503i, N503i, and P503iS:** If a transparent GIF is set, transparency works.

CONSTRUCTORS

Public ImageLabel()	Builds an instance without specifying the instance of Image class to display
Public ImageLabel(Image image)	Image class is specified, and an instance is built

METHODS

Public void setImage(Image image)	Sets an instance of Image class to display

  Methods inherited from com.nttdocomo.ui.Component class: `getHeight, getWidth, getX, getY, setBackground, setForeground, setLocation, setSize, setVisible`
  Methods inherited from java.lang.Object class: `equals, getClass, hashCode, notify, notifyAll, toString`

Interactable

Library: com.nttdocomo.ui
Type: Interface

```
public interface Interactable
```

  List of related classes: `ListBox, TextBox, and Button`
  `Interactable` is the interface that defines the relationship between a user and a component that has a dialog.

METHODS

Public void requestFocus()	Sets focus to a component
Public void setEnabled(boolean b)	Sets a component to a true or false state

KeyListener

Library: com.nttdocomo.ui
Type: Interface
Super interface: `EventListener`

```
public interface KeyListener
extends EventListener
```

`KeyListener` is the interface that defines the high-level API listener that receives key input.

METHODS

public void keyPressed(Panel *panel*, int *key*)	It is called when a key is pushed
public void keyReleased(Panel *panel*, int *key*)	It is called when a key is released

---

Label

Library: com.nttdocomo.ui
Type: Class
Inheritance:

```
java.lang.Object
   |
   +--com.nttdocomo.ui.Component
          |
          +--com.nttdocomo.ui.Label
```

```
public final class Label
extends Component
```

`Label` is the component class that displays a string in the high-level API (`Panel`). Only one line of text display is supported. If you want to use more, use a `TextBox`. The `Label` component is model-specific as to how a specified alignment actually displays on the screen, but you can generally expect that left, center, and right alignment will be correct. As for size, the size of the string serves as a minimum, whereas the maximum size is the width that can be displayed on a screen. However, the behavior in a case where a label is too large to fit on a screen is model-specific.

FIELDS

public static final int CENTER	Constant showing central alignment of the label string
public static final int LEFT	Constant showing left alignment of the label string
public static final int RIGHT	Constant showing right alignment of the label string

CONSTRUCTORS

Label()	Builds an instance in which the string is not set
Label(java.lang.String *text*)	Specifies a label string and builds an instance
Label(java.lang.String *text*, int *alignment*)	Specifies a label string and alignment and builds an instance

METHODS

public void setAlignment(int *alignment*)	Sets alignment
public void setText(java.lang.String *text*)	Sets a label string

Methods inherited from com.nttdocomo.ui.Component class: `getHeight, getWidth, getX, getY, setBackground, setForeground, setLocation, setSize, setVisible`

Methods inherited from java.lang.Object class: `equals, getClass, hashCode, notify, notifyAll, toString`

LayoutManager

Library: com.nttdocomo.ui
Type: Interface

```
public interface LayoutManager
```

LayoutManager is the interface used when the Panel class lays out a component. The actual use of this class to layout components is model-specific.

ListBox

Library: com.nttdocomo.ui
Type: Class
Inheritance:

```
java.lang.Object
  |
  +--com.nttdocomo.ui.Component
        |
        +--com.nttdocomo.ui.ListBox
```

```
public final class ListBox
extends Component
implements Interactable
```

Related Interface: Interactable

ListBox is the list box component class used by the high-level API (Panel). It is a component class expressing a list box, a check box, or a radio button, determined by the argument passed to the constructor. The following types of list can be specified:

- ◆ **Option menu:** This option creates a single-selection menu. Although a default selection is displayed, if this component is in focus, all the items that can be chosen will be displayed. Though the number of items that can be displayed is restricted to the height of a screen, an item can be chosen by scrolling. The width of the item display is model-specific.

- ◆ **Single selection list:** This creates a selection list of two or more items, from which only one item can be chosen.

- ◆ **Radio button list:** This creates a list for choosing only one choice, using radio buttons.

- ◆ **Check box list:** This creates a check box list from which any number of selection item can be chosen.

- ◆ **A numbered list:** This creates a numbered list with quick access by the number keys.

FIELDS

public static final int CHECK_BOX	Constant that shows a check box
public static final int CHOICE	Constant that shows a popup menu from which one item can be chosen
public static final int MULTIPLE_SELECT	Constant that shows a selection list from which two or more items can be chosen
public static final int NUMBERED_LIST	Constant that shows a number selection list from which one item can be chosen with a number
public static final int RADIO_BUTTON	Constant that shows a radio button that can choose one item
public static final int SINGLE_SELECT	Constant that shows a selection list from which one item can be chosen

CONSTRUCTORS

public ListBox(int *type*)	A list box without a selection item is built
ListBox(int *type*, int *rows*)	A list box without a selection item is built

METHODS

public void append(java.lang.String *item*)	Adds a new item to a list item
public void deselect(int *index*)	Puts the specified list item into a deselected state
public java.lang.String getItem(int *index*)	Returns the list item of a specified position
public int getItemCount()	Returns the number of list items
public int getSelectedIndex()	Returns the smallest value among the list items chosen

public boolean isIndexSelected(int *index*)	Returns whether the specified list item is chosen
public void removeAll()	Deletes all list items
public void requestFocus()	Sets focus on a list box
public void select(int *index*)	Changes a specified list item into a selected state
public void setEnabled(boolean *b*)	User selection of a list box sets this method to an active or an inactive state
public void setItems(java.lang.String *items*)	Sets up a list item in String arrangement

Methods inherited from com.nttdocomo.ui.Component class: `getHeight, getWidth, getX, getY, setBackground, setForeground, setLocation, setSize, setVisible`

Methods inherited from java.lang.Object class: `equals, getClass, hashCode, notify, notifyAll, toString`

MediaData

Library: com.nttdocomo.ui
Type: Interface
Super interface: `MediaResource`

```
public interface MediaData
extends MediaResource
```

`MediaData` is a media system data management and expression interface. Use of this interface is not defined and it becomes model-specific as to what data can be expressed. Moreover, the class that actually uses this interface is also model-specific.

Methods inherited from com.nttdocomo.ui.MediaResource interface: `dispose, unuse, use`

MediaImage

Library: com.nttdocomo.ui
Type: Interface
Super interface: `MediaResource`

```
public interface MediaImage
extends MediaResource
```

MediaImage is a media system image management and expression interface. When treating a media resource as a picture, a required definition and required implementation are defined. An instance is acquired from the `getImage` method of the `MediaManager` class. The class that actually implements this interface is model-specific.

An image to hold is taken out as an instance of the `VisualPresenter` class or `Image` class, and is used by the `drawImage` method of the `Graphics` class.

It is model-specific as to what image data this interface can express.

Related items: `MediaManager.getImage(java.lang.String)`, `VisualPresenter.setImage(com.nttdocomo.ui.MediaImage)`, `Graphics.drawImage(com.nttdocomo.ui.Image, int, int)`, `Image`

METHODS

Public int getHeight()	Acquires the height of the image currently held
Public Image getImage()	Acquires an image (still picture) as an instance of the Image class
Public int getWidth()	Acquires the width of the image currently held

Methods inherited from com.nttdocomo.ui.MediaResource interface: `dispose`, `unuse`, `use`

MediaListener

Library: com.nttdocomo.ui
Type: Interface
Super interface: `EventListener`

```
public interface MediaListener
extends EventListener
```

`MediaListener` is the interface that defines the listener that can register with the `MediaPresenter` interface.

METHODS

Public void mediaAction(MediaPresenter source, int type, int param)	Receives notice of an event from `MediaPresenter` interface

MediaManager

Library: com.nttdocomo.ui
Type: Class
Inheritance:

```
java.lang.Object
  |
  +--com.nttdocomo.ui.MediaManager
```

```
public final class MediaManager
extends java.lang.Object
```

`MediaManager` is a media management processing class. Shared processing of acquired media system data is performed. An instance of a class that implements the `MediaResource` interface by either the `getData`, `getImage`, or `getSound` method is acquired. Minimum checks for each method (validity and accessibility checks) are passed in a URL argument to the instance of the implementing class.

In each instance in which an implementing class performs a `use` method inherited from the `MediaResource` interface, data will actually be taken from the URL that the implementing class holds, and it will be changed to an internal format that can be used. In this case, an exception by the `ConnectionException` class or the `UIException` class may be thrown if needed. After processing results from a `use` method inherited from `MediaResource` interface, the internal format is effective and can be used. Execution of an `unuse` method inherited from the `MediaResource` interface cancels data in the internal format. In this state, internal data can be again built by the `use` method. If the `dispose` method inherited from `MediaResource` interface is called, all the data that an instance of an implementing class holds will be canceled. Thereafter, no calling of methods can be performed.

In the minimum specification, the `getData` method returns `null`.

Related items: `MediaResource.use()`, `MediaResource.unuse()`, `MediaResource.dispose()`, `MediaData`, `MediaImage`, `MediaSound`, `MediaPresenter.setData(com.nttdocomo.ui.MediaData)`, `AudioPresenter.setData(com.nttdocomo.ui.MediaData)`, `AudioPresenter.setSound(com.nttdocomo.ui.MediaSound)`, `VisualPresenter.setData(com.nttdocomo.ui.MediaData)`, `VisualPresenter.setImage(com.nttdocomo.ui.MediaImage)`

Constructor: `MediaManager()`

METHOD

Public static final MediaData getData(java.lang.String *location*)	Acquires an instance of the class that implemented the `MediaData` interface from the URL specified by location
Public static final MediaImage getImage(java.lang.String *location*)	Acquires an instance of the class that implemented the `MediaImage` interface from the URL specified by location
Public static final MediaSound getSound(java.lang.String *location*)	Acquires an instance of the class that implemented `MediaSound` interface from the URL specified by location

Methods inherited from java.lang.Object class: `equals, getClass, hashCode, notify, notifyAll, toString`

MediaPresenter

Library: com.nttdocomo.ui
Type: Interface
List of known implementing classes: `AudioPresenter` and `VisualPresenter`

`public interface` **`MediaPresenter`**

`MediaPresenter` is the interface that performs the shared definition of media system data playback. If data is playable, an implementing class evaluates it and plays it back. It is model-specific regarding what kind of data is playable, and the action in a case where data is not playable is also model-specific (a `UIException` class may be thrown).

Related items: `MediaManager` and `MediaResource`

METHODS

Public MediaResource getMediaResource()	Returns media data currently held as an instance of the `MediaResource` interface
Public void play()	Starts playback of media data currently held
Public void setAttribute(int *attr*, int *value*)	Sets the attribute value for playback
Public void setData(MediaData *data*)	Sets an instance of the class that implemented the `MediaData` interface as a reproduced object
Public void setMediaListener (MediaListener *listener*)	Determines a listener
Public void stop()	Stops playback

MediaResource

Library: com.nttdocomo.ui
Type: Interface
List of known sub interfaces: `MediaData`, `MediaImage`, and `MediaSound`

```
public interface MediaResource
```

`MediaResource` is a media system resource management and expression interface. This interface defines the contents that the class handling media data implements. Media data signals the start of the `use` method defined by this interface, and signals the end by the `unuse` method, similarly defined by this interface, after the end of a cycle. When an instance of the implementing class itself becomes unnecessary, a `dispose` method defined by this interface must be called, and you have to declare cancellation beforehand. After the `dispose` method is called a `UIException` class is thrown if this interface is used.

METHODS

Public void dispose()	Cancels a media resource
Public void unuse()	Declares an end to the use of a media resource
Public void use()	Declares the initializing of a media resource

MediaSound

Library: com.nttdocomo.ui
Type: Interface
Super interface: `MediaResource`

```
public interface MediaSound
extends MediaResource
```

`MediaSound` is the media system's sound data management and expression interface. This interface defines the data handled as sound from media system data. An instance of the class that implemented the interface is acquired by the `getSound` method of the `MediaManager` class. The sound plays by passing the acquired instance to the `setData` method of the `AudioPresenter` class. The kind of sound data this interface can express is model-specific.

    Related items: `MediaManager.getSound(java.lang.String)`, `AudioPresenter.setSound(com.nttdocomo.ui.MediaSound)`

    Methods inherited from com.nttdocomo.ui.MediaResource interface: `dispose`, `unuse`, `use`

Panel

Library: com.nttdocomo.ui
Type: Class
Inheritance:

```
java.lang.Object
   |
   +--com.nttdocomo.ui.Frame
         |
         +--com.nttdocomo.ui.Panel
```

```
public class Panel
extends Frame
```

`Panel` is a class used as the frame within which high-level API components are displayed. Instances of classes derived from various `Component` classes are put in this frame class, and displayed. `Panel` class can set a title by the `setTitle` method, a title clearly different from other components by its location in the upper part of the display. Moreover, when a user scrolls, the title remains on the display without scrolling. Specific scrolling behavior is model-specific.

Related items: Canvas, Dialog, FocusManager, LayoutManager, ComponentListener, SoftKeyListener, KeyListener, EventListener

Fields inherited from com.nttdocomo.ui.Frame class: SOFT_KEY_1, SOFT_KEY_2

CONSTRUCTOR

Public Panel()	Builds an empty instance

METHODS

Public void add(Component *c*)	Adds component
Public FocusManager getFocusManager()	Returns an instance of the class that implemented the FocusManager interface set up
Public void setBackground(int *c*)	Determines a background color
Public void setComponentListener (ComponentListener *listener*)	Sets an instance of the listener that receives events generated by the component
Public void setFocusManager(FocusManager *fm*)	Sets an instance of the class that implements the FocusManager interface
Public void setKeyListener(KeyListener *listener*)	Sets the listener object that receives the key push event and key release event
Public void setLayoutManager(LayoutManager *lm*)	Sets an instance of the class that implemented LayoutManager interface
Public void setSoftKeyListener(SoftKeyListener *listener*)	A listener object that receives a soft key pushed or soft key released event
Public void setTitle(java.lang.String *title*)	Sets a title string

Methods inherited from com.nttdocomo.ui.Frame class: getHeight, getWidth, setSoftLabel

Method inherited from java.lang.Object class: `equals, getClass, hashCode, notify, notifyAll, toString`

PhoneSystem

Library: com.nttdocomo.ui
Type: Class
Inheritance:

```
java.lang.Object
  |
  +--com.nttdocomo.ui.PhoneSystem
```

```
public class PhoneSystem
extends java.lang.Object
```

`PhoneSystem` is a class to control the settings of a mobile phone. It offers a means to control native resources of a mobile phone.

FIELDS

Public static final int ATTR_BACKLIGHT_OFF	Attribute constant that shows back light off
Public static final int ATTR_BACKLIGHT_ON	Attribute constant that shows back light on
Public static final int DEV_BACKLIGHT	Device constant that shows a back light
Public static final int MAX_VENDOR_ATTR	A vendor-unique attribute maximum value (=127)
Public static final int MIN_VENDOR_ATTR	A vendor-unique attribute minimum value (=64)

Constructor: `PhoneSystem()`

METHOD

Public static void setAttribute(int attr, int value)	Performs native resource control

---

Methods inherited from java.lang.Object class: `equals`, `getClass`, `hashCode`, `notify`, `notifyAll`, `toString`

ShortTimer

Library: com.nttdocomo.ui
Type: Class
Inheritance:

```
java.lang.Object
   |
   +--com.nttdocomo.ui.ShortTimer

public final class ShortTimer
extends java.lang.Object
implements TimeKeeper
```

Implementing interface: `TimeKeeper`

`ShortTimer` is a class that offers a short-time, one shot and a short-time, interval timer function. It is used by designating the object derived from the `Canvas` class to receive the timer conditions met event. It works only when an object derived from the `Canvas` class is designated by the `setCurrent` method of the `Display` class and is active. After becoming inactive, even if it becomes active again, `ShortTimer` does not resume automatically. Setting it in a state where it is not active is invalid. Although a dialog is displayed, a notice of an event is not performed.

A timer event calls the `processEvent` method of the `Canvas` class. An instance of this class is reserved to native resources at the time of object generation, and the resource is not opened until the `dispose` method performs.

Constructor: `ShortTimer()`

METHODS

Public void dispose()	Cancels a timer
Public int getResolution()	Returns the minimum time resolution that ShortTimer class can use. Model-specific
Public static ShortTimer getShortTimer (Canvas *canvas*, int *id*, int time, boolean *repeat*)	An instance of the ShortTimer class in which information based on the contents of a setting is built and returned

Continued

METHODS *(Continued)*	
Public void start()	Starts a timer.
Public void stop()	Suspends a timer

Methods inherited from java.lang.Object class: `equals, getClass, hashCode, notify, notifyAll, toString`

SoftKeyListener

Library: com.nttdocomo.ui
Type: Interface
Super interface: `EventListener`

```
public interface SoftKeyListener
extends EventListener
```

In the high-level API (Panel), the `SoftKeyListener` interface defines the input listener of a soft key.

METHODS	
Public void softKeyPressed(int *softKey*)	This method is called when a soft key is pushed
Public void softKeyReleased(int *softKey*)	This method is called when a soft key is released

TextBox

Library: com.nttdocomo.ui
Type: Class
Inheritance:

```
java.lang.Object
   |
   +--com.nttdocomo.ui.Component
         |
         +--com.nttdocomo.ui.TextBox
```

Appendix C: i-mode Java API

```
public final class TextBox
extends Component
implements Interactable
```

Implementing interface: Interactable

TextBox a text input component class used by the high-level API. A component used for one line or more of text input. In two or more lines mode, when a string cannot be displayed on a screen a vertical scroll bar is used. In the one-line mode, it is model-specific as to behavior when a string cannot display on the screen.

FIELDS

Public static final int ALPHA	Constant that sets the initial input mode to alphabet input. (=1)
Public static final int DISPLAY_ANY	Constant that shows the mode that a displayed string is in. (=0)
Public static final int DISPLAY_PASSWORD	Constant that shows the mode that a hidden string is in. (=1)
Public static final int KANA	Constant that sets initial input mode to kana Chinese character input. (=2)
Public static final int NUMBER	Constant that sets the initial input mode to number input.

CONSTRUCTOR

Public TextBox(java.lang.String *text*, int *columns*, int *rows*, int *mode*)	The initial string, the number of columns, the number of rows, and the display mode are specified, and an instance is built.

METHODS

Public java.lang.String getText()	Returns a text string
Public void requestFocus()	Focuses as a text box

Continued

Appendixes

METHODS *(Continued)*

Public void setEditable(boolean *b*)	Sets whether it is possible to edit a text string
Public void setEnabled(boolean *b*)	Sets a text box to an enabled or disabled state
Public void setInputMode(int *mode*)	Sets the initial input mode of text input
Public void setText(java.lang.String *text*)	Sets a text string in the box

Methods inherited from com.nttdocomo.ui.Component class: `getHeight, getWidth, getX, getY, setBackground, setForeground, setLocation, setSize, setVisible`

Methods inherited from java.lang.Object class: `equals, getClass, hashCode, notify, notifyAll, toString`

Ticker

Library: com.nttdocomo.ui
Type: Class
Inheritance:

```
java.lang.Object
  |
  +--com.nttdocomo.ui.Component
     |
     +--com.nttdocomo.ui.Ticker
```

```
public final class Ticker
extends Component
```

The `Ticker` class behaves as a marquee attribute in the high-level API. Initial size becomes the height and size at which the specified string is displayed.

CONSTRUCTORS

Public Ticker()	A display string builds an empty ("") object
Public Ticker(java.lang.String *text*)	A display string is specified and a `Ticker` object is built

METHOD

Public void setText(java.lang.String *text*)	Sets a text string

Methods inherited from com.nttdocomo.ui.Component class: `getHeight, getWidth, getX, getY, setBackground, setForeground, setLocation, setSize, setVisible`

Methods inherited from java.lang.Object class: `equals, getClass, hashCode, notify, notifyAll, toString`

TimeKeeper

Library: com.nttdocomo.util
Type: Interface
List of known implementing classes: `ShortTimer` and `Timer`

```
public interface TimeKeeper
```

`TimeKeeper` is the interface that defines a timer. This interface specifies the definition that the class implementing the timer processing should have.

METHODS

Public void dispose()	Cancels a timer
Public int getResolution()	Returns the minimum time resolution that `TimeKeeper` can handle. Model-specific
Public void start()	Starts the operation of a timer
Public void stop()	Suspends a timer

Timer

Library: com.nttdocomo.util
Type: Class
Inheritance:

```
java.lang.Object
   |
   +--com.nttdocomo.util.Timer
```

```
public final class Timer
extends java.lang.Object
implements TimeKeeper
```

Implementing interface: TimeKeeper

The `Timer` class supports a one-time or an interval timer. When a timer meets the conditions, a listener's `timerExpired` method registered as a `TimerListener` interface object is called. A native resource is reserved when a timer is set, and opened up when it is finished.

CONSTRUCTOR

Public Timer()	An instance of a one-time timer, with an interval of 0 milliseconds, and no listener is built

METHOD

Public void dispose()	Cancels a timer
Public int getResolution()	Returns the minimum time resolution that `Timer` can handle. Model-specific
Public void setListener (TimerListener *listener*)	Sets the listener (an instance of the class implementing the `TimerListener` interface) that receives a timer event
Public void setRepeat(boolean *b*)	Specifies an interval timer or a one-time timer
Public void setTime(int *time*)	Specifies the interval of a timer
Public void start()	Starts a timer
Public void stop()	Suspends a timer

Methods inherited from `java.lang.Object` **class:** `equals, getClass, hashCode, notify, notifyAll, toString`

TimerListener

Library: com.nttdocomo.util
Type: Interface
Super interface: EventListener

```
public interface TimerListener
extends EventListener
```

This component is the interface that defines the listeners of a timer event.

METHOD

Public void timerExpired(Timer *source*)   A method called when a timer event occurs

UIException

Library: com.nttdocomo.ui
Type: Class
Inheritance:

```
java.lang.Object
   |
   +--java.lang.Throwable
         |
         +--java.lang.Exception
               |
               +--java.lang.RuntimeException
                     |
                     +--com.nttdocomo.ui.UIException
```

```
public class UIException
extends java.lang.RuntimeException
```

`UIException` is the common exception class published in DoCoMo Profile-1.0.

FIELDS

Public static final int BUSY_RESOURCE	Constant showing an exception caused by a resource being used. (3)
Public static final int ILLEGAL_STATE	Constant showing an exception caused by an illegal state. (1)
Public static final int NO_RESOURCES	Constant showing an exception caused by an unavailable resource. (2)
Public static final int STATUS_FIRST	The first status code this exception can have. (0)
Public static final int STATUS_LAST	The last status code this exception can have. (63)
Public static final int UNDEFINED	Constant showing an undefined exception status. (0)
Public static final int UNSUPPORTED_FORMAT	Constant showing that the exception is caused by being passed a format that is not supported.

CONSTRUCTORS

Public UIException()	Builds an exception without specifying a state code and a message
Public UIException(int *status*)	Specifies only a status code, and builds an exception
Public UIException(int *status*, java.lang.String *msg*)	Specifies a status code and a detailed message, and builds an exception

METHOD

public int getStatus()	Acquires a status code

Methods inherited from `java.lang.Throwable` class: `fillInStackTrace, getMessage, printStackTrace, toString`

Method inherited from `java.lang.Object` class: `equals, getClass, hashCode, notify, notifyAll`

UnsupportedOperationException

Library: com.nttdocomo.lang
Type: Class
Inheritance:

```
java.lang.Object
  |
  +--java.lang.Throwable
       |
       +--java.lang.Exception
            |
            +--java.lang.RuntimeException
                 |
                 +
                 |
com.nttdocomo.lang.UnsupportedOperationException
```

`public class` **UnsupportedOperationException**
`extends java.lang.RuntimeException`

An `UnsupportedOperationException` is thrown when an unsupported operation or method is called.

CONSTRUCTORS

Public UnsupportedOperationException()	Builds an instance without a detailed message
public UnsupportedOperationException (java.lang.String *msg*)	Builds an instance with a detailed message

---

Methods inherited from `java.lang.Throwable` class: `fillInStackTrace, getMessage, printStackTrace, toString`

Methods inherited from `java.lang.Object` class: `equals, getClass, hashCode, notify, notifyAll`

URLDecoder

Library: com.nttdocomo.net
Type: Class
Inheritance:

```
java.lang.Object
   |
   +--com.nttdocomo.net.URLDecoder
```

public class **URLDecoder**
extends java.lang.Object

URLDecoder is a class with the function of changing URL encoding (x-www-form-urlencoded form) into a string.
  Constructor: public URLDecoder()

METHOD

---

public static java.lang.String decode (java.lang.String *str*)	Performs and returns URL decoding of the specified string

---

Methods inherited from java.lang.Object class: equals, getClass, hashCode, notify, notifyAll, toString

URLEncoder

Library: com.nttdocomo.net
Type: Class
Inheritance:

```
java.lang.Object
   |
   +--com.nttdocomo.net.URLEncoder
```

public class **URLEncoder**
extends java.lang.Object

The `URLEncoder` class changes a string into URL encoding (`x-www-form-urlencoded` form).

Constructor: `public URLEncoder()`

METHOD

public static java.lang.String encode (java.lang.String *str*)	Performs and returns URL encoding of the specified string

Methods inherited from `java.lang.Object` class: `equals`, `getClass`, `hashCode`, `notify`, `notifyAll`, `toString`

VisualPresenter

Library: com.nttdocomo.ui
Type: Class
Inheritance:

```
java.lang.Object
   |
   +--com.nttdocomo.ui.Component
          |
          +--com.nttdocomo.ui.VisualPresenter
```

```
public class VisualPresenter
extends Component
implements MediaPresenter
```

Implemented interface: `MediaPresenter`

`VisualPresenter` is the component class that plays back the media data that can be displayed in the high-level API. Visual media data is received and played back, or sent as an event to a playback listener. The display position of an image can be controlled by setting the position as an attribute using constant `IMAGE_XPOS` and `IMAGE_YPOS` of this class, which also enable control of scrolling effects and so on. Although data is treated by two methods, the `setImage` method and `setData` method, they are mutually exclusive, and cannot both be present in one object. Visual data that cannot be reproduced on a handset is treated as model-specific, and a `UIException` may be thrown.

Minimum requirements: A listener is not set.

An `UnsupportedOperationException` is thrown if the `setData` method is called.

Related item: `AudioPresenter`

FIELDS

public static final int IMAGE_XPOS	Attribute value that shows X-coordinate position to display an image.
public static final int IMAGE_YPOS	Attribute value that shows Y-coordinate position to display an image.
protected static final int MAX_VENDOR_ATTR	The maximum vendor unique attribute number. (=127)
protected static final int MAX_VENDOR_VISUAL_EVENT	The maximum vendor unique event number. (=127)
protected static final int MIN_VENDOR_ATTR	The minimum vendor unique attribute number. (=64)
protected static final int MIN_VENDOR_VISUAL_EVENT	The minimum vendor unique event number. (=64)
public static final int VISUAL_COMPLETE	Event number that shows playback has stopped (completion). (3)
public static final int VISUAL_PLAYING	Event number that shows playback start. (1)
public static final int VISUAL_STOPPED	Event number that shows reproduction discontinuation. (2)

CONSTRUCTOR

VisualPresenter()	Builds an empty instance

METHODS

public MediaResource getMediaResource()	Returns the instance of the class that implemented the `MediaImage` interface or `MediaData` interface currently held as an instance of the `MediaResource` interface
public void play()	Starts playback of data
public void setAttribute(int attr, int value)	Sets the attribute value for playback
public voidsetData(MediaData data)	Sets the data to playback by the object that implemented the `MediaData` interface
public void setImage(MediaImage image)	Sets the data to play back by the object that implemented the `MediaImage` interface
public void setMediaListener(MediaListener listener)	Sets a listener to receive the playback situation of media system data
public void stop()	Stops playback of data

Methods inherited from com.nttdocomo.ui.Component class: `getHeight, getWidth, getX, getY, setBackground, setForeground, setLocation, setSize, setVisible`

Methods inherited from java.lang.Object class: `equals, getClass, hashCode, notify, notifyAll, toString`

Appendix D

cHTML and XHTML Basic Tags

THIS CHART DEMONSTRATES WHICH TAGS ARE AVAILABLE in HTML and in which version, cHTML, and XHTML Basic. You should note that XHTML Basic does not contain tags that are specifically used for formatting the style or appearance of text. An Extensible Stylesheet Language (XSL) module can be expected when i-mode begins to use X-HTML Basic, but that is not available as of this writing.

Elements	Attributes	HTML	cHTML	XHTML Basic
!-	—	2	Y	Y
!doctype	—	2	Y	Y
&xxx;	—	2	Y	Y
a	name=	2	Y	—
	href="URL"		Y	Y
	rel=		—	Y
	rev=		—	Y
	title=		—	Y
	urn=(deleted from HTML3.2)		—	—
	accesskey=		Y	Y
	hreflang=		—	Y
	tabindex=		—	Y
	type=		—	Y
	utn	3	?	
	methods=(deleted from HTML3.2)		—	—

Continued

Elements	Attributes	HTML	cHTML	XHTML Basic
abbr	—	4	—	Y
acronym	—	4	—	Y
address	—	2	—	Y
applet	—	3.2	—	—
area	shape=	3.2	—	—
	coords=		—	—
	href="URL"		—	—
	alt=		—	—
	nohref		—	—
b	—	2	—	—
base	href="URL"	2	Y	Y
basefont	size=	3.2	—	—
bdo	—	4	—	—
big	—	3.2	—	—
blockquote	—	3.2	Y	Y
body	—	2	Y	Y
	bgcolor=	3.2	—	—
	background=	3.2	—	—
	text=	3.2	—	—
	link=	3.2	—	—
	vlink=	3.2	—	—
	alink=	3.2	—	—
br	—	2	Y	Y
	clear=all/left/right	3.2	Y	—
button	—	4	—	—
caption	—	3.2	—	Y
center	—	3.2	Y	—

Elements	Attributes	HTML	cHTML	XHTML Basic
cite	—	2	—	Y
code	—	2	—	Y
col	—	4	—	—
colgroup	—	4	—	—
dd	—	2	Y	Y
del	—	4	—	—
dfn	—	3.2	—	Y
dir	—	2	Y	—
	compact		—	—
div	—	3.2	Y	Y
	align=left/center/right		Y	—
dl	—	2	Y	Y
	compact		—	—
dt	—	2	Y	Y
em	—	2	—	Y
fieldset	—	4	—	—
font	size=n	3.2	—	—
	size=+n/-n		—	—
	color=		2 (color)	—
form	action=	2	Y	Y
	method=get/post		Y	Y
	enctype=		Y	Y
frame	—	4	—	—
frameset	—	4	—	—
head	—	2	Y	Y
hn	—	2	Y	Y
	align=left/center/right	3.2	Y	—

Continued

Elements	Attributes	HTML	cHTML	XHTML Basic
hr	–	2	Y	–
	align=left/center/right	3.2	Y	–
	size=	3.2	Y	–
	width=	3.2	Y	–
	noshade	3.2	Y	–
html	–	2	Y	Y
	version=	3.2	Y	Y
i	–	2	–	–
iframe	–	4	–	–
img	Src=	2	Y	Y
	align=top/middle/bottom	2	Y	–
	align=left/right	3.2	Y	–
	width=	3.2	Y	Y
	height=	3.2	Y	Y
	hspace=	3.2	Y	–
	vspace=	3.2	Y	–
	alt=	2	Y	Y
	border=	3.2	Y	–
	usemap=	3.2	–	–
	longdesc=	–	–	Y
	ismap=	2	–	–
input	type=text	2	Y	Y
	name=		Y	Y
	size=		Y	Y
	maxlength=		Y	Y
	tabindex=		–	Y
	accesskey=		Y	Y
	action=		Y	Y

Elements	Attributes	HTML	cHTML	XHTML Basic
	type=password	2	Y	Y
	name=		Y	Y
	size=		Y	Y
	maxlength=		Y	Y
	tabindex=		–	Y
	accesskey=		Y	Y
	value=		Y	Y
	type=checkbox	2	Y	Y
	name=		Y	Y
	tabindex=		–	Y
	accesskey=		Y	Y
	value=		Y	Y
	checked		Y	Y
	type=radio	2	Y	Y
	name=		Y	Y
	tabindex=		–	Y
	accesskey=		Y	Y
	value=		Y	Y
	checked		Y	Y
	type=hidden	2	Y	Y
	name=		Y	Y
	value=		Y	Y
	type=image	2	–	–
	name=	2	–	–
	src=	2	–	–
	align=top/middle/bottom/left/right	3.2	–	–

Continued

Elements	Attributes	HTML	cHTML	XHTML Basic
	type=submit	2	Y	—
	name=		Y	—
	tabindex=		—	Y
	accesskey=		Y	Y
	value=		Y	—
	type=reset	2	Y	—
	name=		Y	—
	tabindex=		—	Y
	accesskey=		Y	Y
	value=		Y	—
	type=file	3.2	—	—
	name=		—	—
	value=		—	—
ins	—	4	—	—
isindex	—	2	—	—
	prompt=	3.2	—	—
kbd	—	2	—	Y
label	—	4	—	—
legend	—	4	—	—
li	—	2	Y	Y
	type=1/A/a/I/i	3.2	—	—
	type=circle/disc/square	3.2	—	—
	value=	3.2	—	—
link	href="URL"	2	—	Y
	rel=		—	Y
	rev=		—	Y
	urn=		—	—
	methods=		—	—

Elements	Attributes	HTML	cHTML	XHTML Basic
	title=		—	—
	charset=		—	Y
	hreflang=		—	Y
	media=		—	Y
	type=		—	Y
	id=		—	—
listing	—	2	—	—
map	name=	3.2	—	—
menu	—	2	Y	—
	compact		—	—
meta	name=	2	Y	Y
	http-equiv=		—	Y
	content=		—	Y
nextid	n=	2	—	—
noframes	—	4	—	—
noscript	—	4	—	—
object	—	4	—	Y
ol	—	2	Y	Y
	type=1/A/a/I/i	3.2	—	—
	start=	3.2	—	—
	compact	2	—	—
optgroup	—	4	—	—
option	—	2	Y	Y
	selected		Y	Y
	value=		—	Y
p	—	2	Y	Y
	align=left/center/right	3.2	Y	—

Continued

Elements	Attributes	HTML	cHTML	XHTML Basic
param	—	4	—	Y
plaintext	—	2	Y	Y
pre	—	2	Y	Y
	width=	3.2	—	—
q	—	4	—	Y
	cite=		—	Y
s	—	2	—	—
samp	—	2	—	Y
script	—	3.2	—	—
select	name=	2	Y	Y
	size=		Y	Y
	tabindex=		—	Y
	multiple		Y	Y
small	—	3.2	—	—
span	—	4	—	Y
strike	—	2	—	—
strong	—	2	—	Y
style	—	2	—	—
sub	—	3.2	—	—
sup	—	3.2	—	—
table	—	3.2	—	Y
	align=left/center/right etc.		—	—
	border=		—	Y
	width=		—	—
	cellspacing=		—	—
	summary=		—	Y
	cellpadding=		—	—
tbody	—	4	—	—

Elements	Attributes	HTML	cHTML	XHTML Basic
td	—	3.2	—	Y
	align=left/center/right		—	Y
	valign=top/middle/bottom/baseline	—	Y	
	rowspan=		—	Y
	colspan=		—	Y
	width=		—	—
	height=		—	—
	abbr=		—	Y
	nowrap		—	—
textarea	name=	2	Y	Y
	rows=		Y	Y
	accesskey=		Y	Y
	tabindex=		—	Y
	cols=		Y	Y
tfoot	—	4	—	
th	—	3.2	—	Y
	align=left/center/right		—	Y
	valign=top/middle/bottom/baseline	—	Y	
	rowspan=		—	Y
	colspan=		—	Y
	width=		—	—
	height=		—	—
	abbr=		—	Y
	nowrap		—	—
thead	—	4	—	—

Continued

Elements	Attributes	HTML	cHTML	XHTML Basic
title	—	2	Y	Y
tr	—	3.2	—	Y
	align=left/center/right		—	Y
	valign=top/middle/bottom/baseline		—	—
tt	—	2	—	—
u	—	3.2	—	—
ul	—	2	Y	Y
	type=disk/circle/square	3.2	—	—
	compact	2	—	—
var	—	2	—	Y
xmp	—	2	—	—

Appendix E

Emoji Symbol Codes

THESE CODES ARE USED ON I-MODE PHONES to bring up the emoji symbol characters. They are also supported by competitor J-Phone on its J-Sky mobile Internet service. A symbol font with these characters is included on the enclosed CD-ROM for preview purposes. The encodings, however, do not match up, so if you use it to preview these characters, be sure to replace these characters with the encodings below before using them for i-mode.

The decimal and hex codes are used thus:



This would be the sun character below.

Symbol	Name	S-JIS Hex Code	S-JIS Decimal Code
☀	Sun	F89F	63647
☁	Cloud	F8A0	63648
☂	Rain	F8A1	63649
☃	Snow	F8A2	63650
⚡	Lightning	F8A3	63651
🌀	Typhoon	F8A4	63652
⋮⋮	Fog	F8A5	63653
☂	Light rain	F8A6	63654
♈	Aries	F8A7	63655
♉	Taurus	F8A8	63656
♊	Gemini	F8A9	63657
♋	Cancer	F8AA	63658
♌	Leo	F8AB	63659
♍	Virgo	F8AC	63660

Continued

Symbol	Name	S-JIS Hex Code	S-JIS Decimal Code
♎	Libra	F8AD	63661
♏	Scorpio	F8AE	63662
♐	Sagitarius	F8AF	63663
♑	Capricorn	F8B0	63664
♒	Aquarius	F8B1	63665
♓	Pisces	F8B2	63666
	Sports	F8B3	63667
	Baseball	F8B4	63668
	Golf	F8B5	63669
	Tennis	F8B6	63670
	Soccer	F8B7	63671
	Ski	F8B8	63672
	Basketball	F8B9	63673
	Motor sports	F8BA	63674
	Pager	F8BB	63675
	Train	F8BC	63676
M	Metro (subway)	F8BD	63677
	Bullet train	F8BE	63678
	Car (sedan)	F8BF	63679
	Car (SUV)	F8C0	63680
	Bus	F8C1	63681
	Ship	F8C2	63682
	Airplane	F8C3	63683
	House	F8C4	63684
	Building	F8C5	63685
	Post office	F8C6	63686
	Hospital	F8C7	63687
BK	Bank	F8C8	63688

Appendix E: Emoji Symbol Codes

Symbol	Name	S-JIS Hex Code	S-JIS Decimal Code
	ATM	F8C9	63689
	Hotel	F8CA	63690
	Convenience store	F8CB	63691
	Gas station	F8CC	63692
	Parking lot	F8CD	63693
	Traffic light	F8CE	63694
	Restroom	F8CF	63695
	Restaurant	F8D0	63696
	Coffee shop	F8D1	63697
	Bar	F8D2	63698
	Beer	F8D3	63699
	Hamburger (fast food)	F8D4	63700
	Boutique	F8D5	63701
	Beauty salon	F8D6	63702
	Karaoke	F8D7	63703
	Movie	F8D8	63704
	Upward right arrow	F8D9	63705
	Amusement park	F8DA	63706
	Music	F8DB	63707
	Art	F8DC	63708
	Stage play	F8DD	63709
	Event	F8DE	63710
	Ticket	F8DF	63711
	Smoking allowed	F8E0	63712
	Smoking not allowed	F8E1	63713
	Camera	F8E2	63714

Continued

Appendixes

Symbol	Name	S-JIS Hex Code	S-JIS Decimal Code
	Bag	F8E3	63715
	Book	F8E4	63716
	Ribbon	F8E5	63717
	Present	F8E6	63718
	Birthday	F8E7	63719
	Telephone	F8E8	63720
	Mobile phone	F8E9	63721
	Memo	F8EA	63722
	Television	F8EB	63723
	Video game	F8EC	63724
	CD	F8ED	63725
	Heart	F8EE	63726
	Spade	F8EF	63727
	Diamond	F8F0	63728
	Club	F8F1	63729
	Eyes	F8F2	63730
	Ear	F8F3	63731
	Hand (rock)	F8F4	63732
	Hand (scissors)	F8F5	63733
	Hand (paper)	F8F6	63734
	Downward right arrow	F8F7	63735
	Upward left arrow	F8F8	63736
	Foot	F8F9	63737
	Shoe	F8FA	63738
	Glasses	F8FB	63739
	Wheelchair	F8FC	63740
	No moon	F940	63808
	Almost full moon	F941	63809

Appendix E: Emoji Symbol Codes

Symbol	Name	S-JIS Hex Code	S-JIS Decimal Code
	Half moon	F942	63810
	Crescent moon	F943	63811
	Full moon	F944	63812
	Dog	F945	63813
	Cat	F946	63814
	Resort	F947	63815
	Christmas	F948	63816
	Downward left arrow	F949	63817
	Phone to	F972	63858
	Mail to	F973	63859
	Fax to	F974	63859
	i-mode	F975	63860
	i-mode with box	F976	63861
	Mail	F977	63863
	DoCoMo sponsored	F978	63864
	DoCoMo point	F979	63865
	A Charge	F97A	63866
	Free	F97B	63867
	ID	F97C	63868
	Password	F97D	63869
	Return	F97E	63870
	Clear	F980	63872
	Search (compare)	F981	63873
	New	F982	63874
	Location information	F983	63875
	Toll-free	F984	63876
	Sharp dial	F985	63877

Continued

Symbol	Name	S-JIS Hex Code	S-JIS Decimal Code
ⓑ	Moba Q	F986	63878
①	1	F987	63879
②	2	F988	63880
③	3	F989	63881
④	4	F98A	63882
⑤	5	F98B	63883
⑥	6	F98C	63884
⑦	7	F98D	63885
⑧	8	F98E	63886
⑨	9	F98F	63887
⓪	0	F990	63888
OK	Okay	F9B0	63920
♥	Black heart	F991	63889
💓	Quivering heart	F992	63890
💔	Broken heart	F993	63891
💕	Two hearts	F994	63892
😐	Strict face	F995	63893
😠	Angry face	F996	63894
😞	Disappointed face	F997	63895
😢	Sad face	F998	63896
😖	Indecisive face	F999	63897
↗	Good (getting better)	F99A	63898
♪	Music note	F99B	63899
♨	Hot springs	F99C	63900
◇	Cute	F99D	63901
💋	Kiss mark	F99E	63902
✨	Glittering (new)	F99F	63903

Symbol	Name	S-JIS Hex Code	S-JIS Decimal Code
	A bright idea	F9A0	63904
	Bile rising	F9A1	63905
	Punch	F9A2	63906
	Bomb	F9A3	63907
	Mood	F9A4	63908
	Bad (getting worse)	F9A5	63909
zzz	Sleepy	F9A6	63910
!	Exclamation	F9A7	63911
!?	Exclamation and question	F9A8	63912
!!	Double exclamation	F9A9	63913
	Crash	F9AA	63914
	Flying sweat	F9AB	63915
	Dripping with sweat	F9AC	63916
=3	Dashing	F9AD	63917
	Sound wave 1	F9AE	63918
	Sound wave 2	F9AF	63919

Index

Symbols & Numbers

3G
  bandwidth and, 36
  content and, 35
  defined, 34
  DoCoMo development of, 73–74
  handset use, 34
  investment, 75
  killer app, 74
  licenses, 75
  network goes online, 27
  rollout, 12, 27
  undertaking, 74
501i series handsets. *See also* handsets
  D501i, 5
  illustrated, 4
  N501i, 7
502i series handsets, 8, 12, 26
503i series handsets. *See also* handsets
  emulators and, 118
  Fujitsu, 17
  Mitsubishi, 18
  NEC, 21
  Panasonic, 20
  Sony, 16

A

`<a></a>` tag
  `accesskey` attribute, 126, 128–129
  attributes, 126
  `cti` attribute, 126, 127
  `href` attribute, 126, 127
  in hyperlink dialing, 127–128
  in keypad links, 128–129
  `name` attribute, 126
  `utn` attribute, 126
access keys
  adding, 173–177
  indication, 185
accesskeys
  availability, 169
  indicators of, 175
Active Server Pages (ASP), 111
address book
  account setup management, 192–195
  `address.java`, 257–262
  `address.sql` file, 210
  code testing, 209–210
  creating, 177–210
  data structure, 179–181
  entering/finding addresses, 195–199
  index.php entry page, 181–182
  interface, building, 183–187
  problem troubleshooting, 199–200
  project definition, 178
  searching/editing entries, 200–209
  sign up/logon page, 183–187
  tables, 179–181
  user authentication, 187–192
  user input, 195–199
  user interaction flow, 181–183
`address.java`, 257–262
`address.php`, 246
`addressappli.php` script
  adding, 241
  function, 241
  listing, 239–241
  variables sent to, 241
administration sites, 373
advertising, 91
airline information sites, 343
alignment
  line break, 132–133
  text, 131, 220–222
America Online (AOL), 58
AND operator, 204

457

animations, creating, 230–232
Another You Adventures sites, 360
antialiasing, avoiding, 177
Apache, 314
application descriptor file (ADF), 245
applications. *See also* i-Appli
    `address.java`, 257–262
    address, obtaining, 247
    back-end, building, 239–241
    components, 252–257
    downloading, 247
    interface, building, 242–248
    main method, building, 241–242
    mistakes, 248
    network use in, 248–252
    orientation, 238
    parameters, 238
    programming, 237–264
    running, flow of, 247
    softkeys, 252–257
    standalone programming, 248
    starting screen, 243
    testing, 263
    what you want it to do, 237–238
ARPU (Average Revenue Per User), 82
AT&T Wireless
    DoCoMo investment, 38, 99
    PocketNet service, 99
au
    EZ Web menu, 307
    growth prospects, 304
    memory, 19
    skins, 86–87
audience
    business, 83–84
    DoCoMo recognition, 30–31
    i-mode fit to, 84–87
    i-mode success and, 81
    young people, 81–82
audio. *See also* sounds
    files, 233
    playing back, 233

`AudioPresenter` class
    constructor, 276, 386
    defined, 385
    fields, 386
    methods, 387
    minimum specified functions, 386
    playback with, 233
authentication
    billing system, 182
    user, 187–192

B

Bandai Co., LTD. Channel sites, 355–356
banking, i-mode, 62
banks. *See also* Mobile Banking
    city, 320
    as content providers, 54
    regional, 321–324
`<base>` tag, 129
baseball/soccer sites, 364
batteries, 83
`<blink>` tag, 129
`<blockquote>` tag, 130
Bluetooth
    defined, 79
    devices, 79
    home server, 79
    mobile phones, 79, 80
    spread of, 80
    success, 80
B-MAX (Business-Mobile Access Exchange), 63
`<body>` tag, 131
bookmarking sites, 154–155
books/CDs/games sites, 345–346
Boolean values, 182
`<br>` tag
    `clear` attribute, 131
    defined, 131
    in lists, 165–166
    placing, 174

browsers
    defaults, 46
    NetFront, 49
    running midlets in, 51–52
    XHTML Basic-compliant, 50
business
    i-mode-based solution
        implementation, 83–84
    restaurant chain use, 83
    salesperson use, 83–84
Button class
    constructors, 388
    defined, 226, 388
    inheritance, 387–388
    methods, 388
    use in panel, 226–227

C

Canvas class. *See also* low-level API
    constructor, 389
    defined, 215, 389
    inheritance, 388–389
    methods, 389
    for precision positioning, 217
capacity, DoCoMo problems, 75
car sites, 346–347
cascading style sheets (CSS), 46
case construct, 271
case studies
    Index, 295–306
    Nikkei, 307–312
    Walkerplus.com, 291–294
CD-ROM
    address.sql file, 210
    Adobe GoLive, 315
    Appache, 314
    Chaku Melo Convertor, 276, 315
    config.inc file, 190
    eimode.ttf font, 120
    emulators, 118
    Forte for Java Community
        Edition, 315

i-Jade emulator, 277, 315
i-mode: A Primer, 315
i-topics.php, 175–176
Java 2 SDK, 213
login.html, 184–185
login.php, 187–188, 191–192
with Microsoft Windows, 314
Music MasterWorks Tryout Version,
    315
MySQL, 314
NJ Win, 315
PHP, 314
phpMyAdmin, 315
programs, 313, 314–315
source code examples, 313, 314
symbol font, 118
system requirements, 313–314
topic.php, 173–174
troubleshooting, 315–316
UltraEdit32, 213, 315
utnlogin.html, 186
<center> tag
    defined, 133
    use code, 133
    use results, 134
central office servers, 67
CGI programming, 111
Chaku Melo Convertor
    on CD-ROM, 315
    for converting MIDI files, 161, 276
    defined, 159, 315
    window, 159
changeUserInfo function, 207–209
character encoding, 150
character sites, 352–353
CHECK_BOX list box, 225
cHTML (Compact HTML). *See also* tags
    (cHTML)
    coding limits, 106
    coding small with, 106–107
    defined, 8, 45

continued

cHTML *(continued)*
   development, 47
   editors for, 119–120
   emulators for, 118–119
   file formats, 123
   graphic display, 108
   HTML function comparison, 46–47
   HTML versus, 50–51
   modifying existing HTML documents and, 163–177
   in PHP scripts, 196
   PHS phone viewing, 83
   programming in, 163–210
   tables and, 164
   tags, 45, 124–156, 131, 439–448
   tools, 146
   WML versus, 59
circuit-switching, 64
classes
   AudioPresenter, 233, 385–387
   Button, 226–227, 387–388
   Canvas, 215, 217, 388–389
   Component, 390–391
   ConnectionException, 391–393
   Dialog, 393–396
   Display, 397–399
   Font, 400–401
   Frame, 401–402
   Graphics, 216, 217, 403–405
   Iapplication, 216, 409–410
   Image, 410–411
   ImageLabel, 229–230, 411–412
   importing, 216
   Label, 413–414
   ListBox, 223, 223–227, 224–226, 415–417
   main, 266–267
   MainCanvas, 216, 219
   MediaManager, 419–420
   Panel, 220–236, 422–424
   PhoneSystem, 424–425
   ShortTimer, 425–426
   TextBox, 222–223, 426–428
   Ticker, 227–228, 428–429
   Timer, 429–431
   UIException, 431–433
   UnsupportedOperationException, 433
   URLDecoder, 434
   URLEncoder, 434–435
   VisualPresenter, 230–232, 435–436
C-MAX (Contents-Mobile Access Exchange), 63
coding
   with editors, 119–120
   methods, 251
coding small
   cHTML, 106–107
   graphics sizes, 107
   joy of, 105
   memory/storage, 105–106
   phone hardware, 106
color chart, 156
color screens affordability, 86
colors
   256, 107
   16, 107
   choice of, 177
   codes, 168
   coding small and, 107
   fixed, 169
   font, 136
   game, 265
   gradient, eliminating, 107
   hex codes, 156
   names, 156
   primary, 168
   setting, 268
Component class
   constructor, 390
   defined, 390

inheritance, 390
  methods, 390–391
component listeners
  declaring, 242
  defined, 252
ComponentAction() method
  automating, 254
  components, 253–254
  defined, 252
  listing, 252–253
ComponentListener interface, 391
components, 252
computerchik, 299
config.inc files
  browser checking, 191
  CD-ROM, 190
  defined, 190
  illustrated, 200–201
  IP checking, 191
Connected Limited Device Configuration
    (CLDC), 211
ConnectionException class
  defined, 392
  fields, 392
  inheritance, 391–392
  methods, 393
consumers, 33
content
  3G networks and, 35
  commercial, 96
  community, 96
  depth, 95
  dynamic, hosting, 112–117
  entertainment, 54–55
  essential points, 100–101
  freshness, 95
  moral standards, 93–95
  requirements, 95–98
  tools and utility, 55
  usefulness, 95

content proposals
  creating, 100
  elements, 100
  hints, 100–101
content providers
  banks as, 54
  content proposals, 100–101
  customer/technical support levels, 101
  hobbyist, 102
  official list of, 92
  relationship with DoCoMo, 92
  revenue model, 90, 91–92
  self-regulation, 97
cookies, 112–113
copyright notices, 170
cosmetics/health sites, 348
Could Be Dwango sites, 357
credit card sites, 342
Credit Card/Securities/Insurance. *See also*
    i-mode menu
  credit cards, 342
  insurance, 342–343
  stocks/securities, 341–342
credit union sites. *See also* Mobile
    Banking
  Chugoku, 332
  Hokkaido, 325
  Hokuriku, 330
  Kansai district, 330–331
  Kanto district, Koshinetsu, 326–328
  Kyushu, 333
  Shikoku, 332
  Tohoku, 325–326
  Tokai, 328–329
crypt method, 194
customer-centered approach, 31
customizations
  design and, 87
  drive for, 87

continued

customizations *(continued)*
   ringing tones, 85, 86
   sound, 87
   startup screens, 86
   teens and, 87

D

databases
   MySQL, 178
   passwords in, 192
   SQL, 112–117, 121, 179
   supercharged, 209
   viewed in phpMyAdmin, 116
`date` function, 246
`<dd>` tag, 134–135
DDI Pocket's P-mail DX service, 297
definition lists
   creating, 134
   illustrated, 135
   tags, 134
DES encryption, 194
developers. *See* content providers
dialing
   pauses, 127
   via hyperlinks, 127–128
dialog boxes
   adding, 233–236
   appearance of, 234
   `DIALOG_YESNO`, 234
   `DIALOG_YESNOCANCEL`, 235–236
   error, 234
   info, 234
   in panels, 234
   types of, 233
   warning, 234
`Dialog` class
   constructor, 396
   defined, 393
   exceptions, 393–394
   fields, 395–396
   icons displayed, 395
   inheritance, 393
   maximum message length, 394
   maximum title length, 394
   methods, 396
   selection display, 395
`DIALOG_YESNOCANCEL` dialog box, 235–236
Dictionaries/Tools. *See also* i-mode menu
   dictionaries/handy information, 374
   package tracking, 375
`<dir>` tag, 134
directory lists, 134
`Display` class
   constructor, 398
   defined, 397
   fields, 398–399
   inheritance, 397
   methods, 399
dithering, 107
`<div>` tag
   `align` attribute, 135
   defined, 135
   `<p>` tag within, 136
   use code, 135–136
   use results, 136
`<dl>` tag, 134–135
D-MAX (Database-Mobile Access Exchange), 63
DoCoMo
   3G services, 12, 73
   approach to business, 73
   ARPU (Average Revenue Per User), 82
   AT&T ownership, 99
   capacity problems, 75
   cellular phone subscribers, 24, 25
   chronology, 24–27
   creation, 23–24
   customer service, 97
   defined, 3
   developer relationship with, 92
   digital service, 24, 25

downfall, 37
failures, 31–33
feature introduction, 6
FOMA site, 74
future, 33–39
global market expansion, 37–39
handset maker relationships, 11
handsets, 4, 5, 18
as i-mode architect, 3
inclusion standard, 92–98
infrastructure, updating, 34–36
investors and, 73
key players, 27–29
leveraging user base, 30–31
market, knowing, 36–37
micropayment system, 20
Microsoft analogy, 23
mobile phone purchase, 25
Mova, 24, 29
m-stage initiative, 11
name meaning, 23
NTT and, 77
official sites, 14
Packet Data Communications, 25
pagers, 30
path toward i-mode, 29–33
Pocketboard, 30
revenue model, 14, 17, 89–92
as spin-off, 23–24
stakes in non-Japanese mobile carriers, 38
strategies and growth, 33–39
subscriber billing, 14
subscriber growth, 36
successes, 23, 31, 33
value-added pager services, 25
DoJa
rendering of `Ticker` object, 228
software developer kit, 17
DoPa service, 60, 65, 66

`do-while` loops, 217
Dreamweaver, 106, 117
drop shadows, removing, 167
`<dt>` tag, 134–135
DTDs (Document Type Definitions), 49

E

editors
coding with, 119–120
Dreamweaver, 106, 117
GoLive, 106, 117, 119
WAP Profit, 119–120
WYSIWYG, 117
eimode font, 120
eimode Web site
in full-sized format, 163
pages, breaking up, 169–171
PHP script, 171–173
topics page, 173–174
e-mail
blocking, 9
handsets for, 9–10
i-mode use for, 82, 84–85
in Japan, 85
mass, 9
opt-in, 10
sending, 10
sites, 375–376
size limit, 10
subscription services, 10
emoji symbol characters
`accesskey` availability and, 169
decimal codes, 449–455
decoration as, 126
defined, 124
hex codes, 449–455
memory and, 126
typical use of, 124–125
use illustration, 170
use of, 125–126
employment sites, 346, 373

emulators
    503i-series handsets and, 118
    CD-ROM, 118
    finding, 118–119
    framework, 121
    i-Jade, 18, 277, 315
    KToolbar, 214
    problems, 118
encryption. *See also* security
    `crypt` method, 194
    DES, 194
    password, 192, 194
    with salt, 194
English menu. *See also* i-mode menu
    Database, 383
    Entertainment, 382
    News/Information, 382
    Others, 383
    Ringing Tones, 383
Enoki, Keiichi, 27
entertainment content, 54–55
error dialog box, 234
Europe 3G licenses, 75
`EventListener` interface, 399
events. *See also* games
    finishing, 271
    key pressed, 269
    key released, 269
    media, 269
    processing, 269–275
    separating, 269–270
    series of, 271
    testing, 271–272
    timer expired, 269

F

failures. *See also* DoCoMo
    business-oriented product, 32
    car navigation integration, 32–33
    non-communication-centered
        product, 32

fashion sites, 348
file formats (cHTML), 123
fixed line
    L-mode, 77–78
    phones, 71–73
    subscriber decline, 76
FM sites, 370
`FocusManager`, 399
FOMA. *See also* W-CDMA
    certification refusal, 27
    compatible applications, 74
    data transfer charge, 35
    defined, 34
    download rate, 35
    handsets, 53
    upload rate, 35
`<font>` tag, 136
`Font` class
    constructor, 401
    defined, 400
    fields, 400
    inheritance, 400
    methods, 401
`for` loops, 204, 217, 268
`<form>` tag, 137
Forte for Java Community Edition, 315
fortune telling sites, 362–363
`Frame` class
    constructor, 402
    defined, 402
    fields, 402
    inheritance, 401–402
    methods, 402
Fujitsu
    F502i handset, 26
    F503i handset, 17
functions (PHP)
    `changeUserInfo`, 207–209
    `checkSessionUser`, 203
    `date`, 246
    declaration, 200

defined, 196
forcing into, 199–200
`getnames`, 203–204
`include` file holding, 202
in `library.inc` file, 203, 207
`nextMenu`, 203, 204
`printaddress`, 205–206
`searchform`, 205
`start_session`, 199
storing, 202
`textwrap`, 209, 241
`userform`, 207–209
using, 196

G

Game Taito sites, 360
games
    attractive, 281
    colors, 265
    creating, 265–287
    development, 18
    different screens, 280
    as DoCoMo success, 33
    drawing on screen, 266–269
    event processing, 269–275
    feedback, 281
    first loaded, 278
    i-Appli code listing, 282–287
    in i-Appli menu, 18
    illustrated, 278, 279, 280
    improvements, planning, 280–281
    industry, 21
    main class declaration, 266–267
    play flowchart, 266
    pressing Start button, 279
    project creation, 265–277
    questions, 280–281
    rules, 281
    sounds, 265, 275–277
    squares, 266
    `state`, 275
    testing, 281
    watching, in action, 277–280
    x and y coordinates, 267–269
Games/Fortune Telling. *See also* i-mode
    menu
    Another You Adventures, 360
    Bandai Co., LTD. Channel, 355–356
    Could Be Dwango, 357
    fortune telling, 362–363
    Game Taito, 360
    Konami Net, 357–358
    Namco Station, 356–357
    other games, 360–361
    Sakuma's Suguroku, 358
    Sega Mode, 359
    Tomy-Web, 356
    Webby Hudson, 358–359
`getnames` function, 203–204
GIF images
    file properties, 123
    inserting, 147–149
global market expansion
    DoCoMo, 37–39
    DoCoMo investment in, 38
    failure, 37
    i-mode and, 39
    reasons for, 37–38
Global Packet Radio Service. *See* GPRS
God of Love service. *See also* Index
    advice column, 302
    cost, 300
    defined, 300
    distribution, 296
    growth, 302
    illustrated, 300
    launch, 295
    menu, 301–302
    selling, 300–301

continued

God of Love service (*continued*)
    star signs, 301
    subscribers, 91, 300, 303
    target audience, 301
    tarot readings, 301
golf sites, 365
GoLive, 106, 117, 119, 315
gourmet information sites, 349
GPRS
    defined, 34
    use of, 38
graphics
    cHTML page display, 108
    color choices in, 177
    downloading, when needed, 230
    fist screen, 183
    half-page, 183
    jaggies, 108
    load time, 183
    locks, turning off, 219
    in midlet size limit, 230
    Nikkei use of, 310
    screen size, 108–110
    sizing, 109
    trimming by color, 107
    trimming, to size, 167–169
`Graphics` class
    constructor, 404
    defined, 403
    `drawString` method, 216
    fields, 403
    inheritance, 403
    methods, 281, 404–405
    x and y coordinates and, 217

H

`<h1>...<h4>` tags
    defined, 144
    use code, 144
    use recommendation, 144
    use results, 145

handset maker sites, 355
handsets
    3G, 34
    501i series, 4, 5, 7
    502i series, 8, 12
    503i series, 12, 16, 17, 18, 20, 21
    battery, 83
    color, 5
    cost, 71
    design, 6
    for e-mailing, 9–10
    exploring, 4–12
    features, 11
    FOMA, 53
    generations, 5
    guidelines, 4
    for Internet browsing, 8–9
    introduction of, 6
    Japanese, 7
    manufacturers, 4
    market, 6–7
    names, 4
    NEC, 6, 7
    options, 5
    for phoning, 11–12
    prices, 6
    sales, 6
    size, 5, 6, 7
    testing code for, 209–210
Hanpake, 52
hardware
    i-mode server, 61–63
    network architecture, 60–61
    packet-switched standard and, 57–60
    PDC-P standard, 64–67
`HAVING` operator, 204
`<head>` tag, 145
`Hello.java`, 215
Hello World
    `Hello.java`, 215
    using high-level API, 220–236
    using low-level API, 214–220

Index

high-level API. *See also* Hello World
    animation creation, 230–232
    audio files, 233
    AudioPresenter class, 233
    Button object, 226–227
    defined, 220
    dialog boxes, 233–236
    image labels, 229–230
    ImageLabel object, 229–230
    list box choices, 223–226
    ListBox object, 223–226
    Panel class, 220–237
    scrolling message creation, 227–229
    text alignment/display, 220–222
    text display, 222–223
    Ticker object, 227–229
    VisualPresenter class, 230–232
Hokkaido sites, 376–377
Hokuriku sites, 378
horizontal rules, 145–146, 169
horoscope signs, 124–125
hotel/lodging travel sites, 344
`<hr>` tag
    align attribute, 145
    defined, 145
    size attribute, 145
    use code, 146
    use results, 146
    width attribute, 145
`<html>` tag, 146
HTML. *See also* markup languages
    cHTML function comparison, 46–47
    cHTML versus, 50–51
    i-mode and, 8, 9
    in PHP scripts, 196
    roots, 47
    tags, 439–448
HTML document modification
    access keys, 173–177
    graphics trimming, 167–169
    lessons learned, 177
    tag removal, 165–166
    text-wrap options, 169–173
HTTP servers, 44
HttpConnection interface
    defined, 405
    fields, 405–407
    methods, 407–408
    super interfaces, 405
Hungry Minds Customer Service, 316
hyperlinks, dialing via, 127–128

I

i Playstation, 382
i-Appli
    address.java, 257–262
    address, obtaining, 247
    application registration, 18
    back end, building, 239–241
    components, 252–257
    creating, 214
    defined, 15
    downloading, 247
    instability, 19
    interface, building, 242–248
    J2ME basis, 211
    "killer apps" and, 20
    main method, building, 241–242
    memory limitation, 19
    mistakes, 248
    network use in, 248–252
    news agent product, 310–311
    official sites, 17
    orientation, 238
    parameters, 238
    phone release, 27
    PHP with, 112
    programming, 237–264
    revenue model, 17
    running, flow of, 247
    softkeys, 252–257

continued

i-Appli *(continued)*
    stability, 245
    standalone programming, 248
    standard development, 18
    starting screen, 243
    testing, 263
    what you want it to do, 237–238
i-Appli game. *See also* games
    attractive, 281
    code listing, 282–287
    colors, 265
    creating, 265–287
    development, 18
    different screens, 280
    drawing on screen, 266–269
    event processing, 269–275
    feedback, 281
    first loaded, 278
    illustrated, 278, 279, 280
    improvements, planning, 280–281
    main class declaration, 266–267
    play flowchart, 266
    pressing Start button, 279
    project creation, 265–277
    questions, 280–281
    rules, 281
    sounds, 265, 275–277
    squares, 266
    state, 275
    testing, 281
    watching, in action, 277–280
    x and y coordinates, 267–269
i-Appli menu
    applications, 15–16
    games, 18
Iapplication class
    constructor, 409
    defined, 409
    extensions of, 216
    inheritance, 409
    methods, 409–410

if clauses, 183, 206, 275
i-Jade emulator, 18, 277, 315
Illustrator, exporting images from, 168
ImaDoko service, 183
Image class
    constructor, 410
    defined, 410
    inheritance, 410
    methods, 411
image labels, 229–230
ImageLabel class. *See also* Panel class
    constructors, 412
    defined, 229, 411
    inheritance, 411
    methods, 412
    model-specific issues, 411
    placing images with, 230
    using in a panel, 229
ImageReady, 107
I-MAX (Interface-Mobile Access Exchange), 63
i-melodies, 157, 161
<img> tag
    align attribute, 147, 148
    alt attribute, 147
    defined, 147
    height attribute, 109, 147, 148–149
    hspace attribute, 147
    src attribute, 147
    use code, 147–149
    use results, 148, 149
    vspace attribute, 147
    width attribute, 109, 147, 148–149
i-mimic, 118, 183
i-mode
    architect, 3
    audience, 81–89
    average monthly bill, 13
    content availability, 14–15
    defined, 3–4
    global market expansion and, 39

Index

handsets, 4–12
hardware, 57–67
path toward, 29–33
release, 26
services, 13–21
software, 41–55
sound formats, 157–162
subscriber base, 6, 26
success, 14
technologies, 3
WAP versus, 59
i-mode browsers
   defaults, 46
   NetFront, 49
   running midlets in, 51–52
   XHTML Basic-compliant, 50
The i-mode Incident, 27
i-mode menu
   content proposals, 100–101
   content requirements, 95–98
   Credit Card/Securities/Insurance, 341–343
   defined, 92
   Dictionaries/Tools, 374–375
   E-mail, 375–376
   English menu, 382–383
   Games/Fortune Telling, 355–363
   i Playstation, 382
   i-Navi Link, 381
   inclusion standard, 92–98
   items, 317–318
   Mobile Banking, 320–341
   News/Weather/Information, 318–320
   Regional, 376–381
   Ringing Tones/Screen Pictures, 350–355
   Shopping/Living, 345–350
   site principles of morality, 93–95
   Sports and Entertainment, 364–372
   Town Information/Administration, 373–374
   Travel/Transportation/Maps, 343–344
i-mode networks
   architecture, 60–61
   banking, 62
   behavior, 60
   information flow, 62
   structure illustration, 61
i-mode servers
   B-MAX, 63
   C-MAX, 63
   defined, 62
   D-MAX, 63
   functions, 62
   illustrated, 62
   I-MAX, 63
   M-MAX, 63
   N-MAX, 63
   PDC-P network with, 66
   U-MAX, 63
   W-MAX, 63
i-mode sites
   official, 14, 317–383
   unofficial, 97–98
The i-mode Strategy, 27
i-mode Tool, 120
IMT-2000 standard. *See also* 3G
   defined, 34
   Japan adoption of, 7
   network rollout, 27
i-Navi Link, 381
`include` files, 190, 202
inclusion standard. *See also* i-mode menu
   argument, 98–99
   content requirements, 95–98
   defined, 92
   DoCoMo rights, 92–93
   moral standards, 93–95
   self-regulation, 97
   user complaints and, 97
   violations, 97

Index. *See also* case studies
    accounts receivable, 303
    advantages, 305
    carrier relationship with, 298
    Colosseum Saikyo Densetsu, 296
    company history, 295–296
    defined, 91, 295
    DoCoMo business model move, 298–300
    exclusivity, 305
    focus, 295
    future, 304–306
    God of Love service, 91, 295, 296, 300–304
    growth, 302–303
    growth prospects, 304–305
    lessons learned, 306
    leveraging for the future, 306
    Mail Anime, 296
    mobile commerce, 305
    mobile content, 296–298
    mobile-networked toys, 306
    new revenue growth, 305
    number of sites, 303
    Ochiai, 297, 298, 299, 300, 304
    Odekake Denwa-cho, 295
    Ogawa, 297, 299, 300
    "people" bias, 304
    Private Homepage, 295
    profit-oriented growth, 305
    site additions, 303
    stock options, 304
    strategy planning, 297
    target user, 297
    teams, 303, 304
    VCR control and, 306
    Watanabe, 297, 299, 300
    workforce, 304, 306
    Yoji Baba, 295, 297, 299
`index.php`, 181–182
info dialog box, 234
Init methods, 243
`<input>` tag
    `accesskey` attribute, 138
    `checked` attribute, 138
    defined, 137
    `istyle` attribute, 138
    `maxlength` attribute, 138
    `size` attribute, 137
    `type` attribute, 137, 138
    use results, 140
    `value` attribute, 137
insurance sites, 342–343
`INT` type, 181
`Interactable` interface, 412
interface
    building, 183–187, 242–248
    component location, 245
    declaring elements of, 243–244
Internet browsing
    handsets for, 8–9
    i-mode compatibility, 8
Internet service providers (ISPs), 65
IP numbers, 181, 195
`i-topics.php`, 175–176

J

J2ME
    classes, 211
    Connected Limited Device Configuration (CLDC), 211
    as i-Appli basis, 211
    KVM, 211
    porting programs from, 211
JA. *See also* Mobile Banking
    JA Aichi Prefecture, 335–336
    JA Hiroshima Prefecture, 338–339
    JA Shizuoka Prefecture, 337
    JA Tokyo Metropolis, 338
    JA Wakayama Prefecture, 340
    National Labor Union Credit Union, 341

Net Bank, 341
jam file, 245, 246
Japan
   e-mail in, 85
   Internet usage, 76
   mobile phone use, 85
jar file, 248
Java 2 SDK, 213
Java API
   AudioPresenter, 385–387
   Button, 387–388
   Canvas, 388–389
   Component, 390–391
   ComponentListener, 391
   ConnectionException, 391–393
   Dialog, 393–396
   Display, 397–399
   EventListener, 399
   FocusManager, 399
   Font, 400–401
   Frame, 401–402
   Graphics, 403–405
   HttpConnection, 405–408
   Iapplication, 409–410
   Image, 410–411
   ImageLabel, 411–412
   Interactable, 412
   KeyListener, 413
   Label, 413–414
   LayoutManager, 414–415
   ListBox, 415–417
   MediaData, 417
   MediaImage, 417–418
   MediaListener, 418–419
   MediaManager, 419–420
   MediaPresenter, 420–421
   MediaResource, 421
   MediaSound, 422
   Panel, 422–424
   PhoneSystem, 424–425
   ShortTimer, 425–426
   SoftKeyListener, 426
   TextBox, 426–428
   Ticker, 428–429
   TimeKeeper, 429
   Timer, 429–431
   TimerListener, 431
   UIException, 431–433
   UnsupportedOperationException, 433
   URLDecoder, 434
   URLEncoder, 434–435
   VisualPresenter, 435–437
Java For Dummies, 236
Java midlets
   canvas form, 215
   conversion utilities, 52
   defined, 15, 19
   downloading to scratchpad, 264
   programming, 19
   running, in browser, 51–52
   size limit, 230
Java Server Pages (JSP), 111, 112
J-JIS Japanese encoding, 121
JPEG files, 123
J-Phone
   growth prospects, 304
   J-Sky mobile Internet service, 299, 307
   memory, 19
   Mobile Markup Language (MML), 161

K

Kansai and Chugoku districts sites, 379–380
karaoke sites, 350–352
"keitai," 86
Keitai-Font, 120
key pressed events. *See also* events
   addressing, 271
   code for, 269–270
   as event type, 269

continued

key pressed events *(continued)*
    in `processEvent()` method, 273–274
    in scripts, 273
    using, 271
key released events. *See also* events
    as event type, 269
    using, 271
`KeyListener` interface, 413
killer apps
    i-Appli and, 20
    search for, 73–76
Konami Net sites, 357–358
KPN, 38
KPT, 37
KToolbar
    bug, 214
    Compile button, 214
    compiler, 276
    emulator, 214, 277
    New button, 214
    package creation, 214
    window, 213
KVM (Kilobyte Virtual Machine)
    ADF file, 52
    conversion, 52
    defined, 51
    Unicode, 52
Kyushu sites, 381

L

`Label` class
    constructors, 414
    defined, 413
    fields, 414
    inheritance, 413
    methods, 414
labels
    image, 229–230
    text alignment/display in, 220–222
`LayoutManager` interface, 414–415

lessons learned
    HTML document modification, 177
    Index case study, 306
    Nikkei case study, 312
    Walkerplus.com case study, 294
`library.inc` file
    defined, 202
    functions in, 203, 207
life information sites, 349
Light Transport Protocol (LTP), 44
line breaks, 132–133, 169
links, keypad, 128–129
list boxes
    `CHECK_BOX`, 225
    choices, 223–226
    `MULTIPLE_SELECT`, 225
    `NUMBERED_LIST`, 225
    `RADIO_BUTTON`, 226
    `SINGLE_SELECT`, 226
    types of, 224
`ListBox` class
    constructors, 416
    defined, 415
    fields, 416
    inheritance, 415
    list box types and, 224–226
    methods, 416–417
    use example, 223
lists
    `<br>` tags in, 165–166
    definition, 134, 135
    directory, 134
    numbered, 166
    ordered, 151–152
    reducing/re-ordering, 166
    types of, 415
    unordered, 155
L-mode
    browsers, 77
    content, 77

content development, 78
content market, 78
defined, 77
goal, 77
information package, 78
target market, 78
local broadcaster sites, 368–369
`login.php`, 184–185, 187–188, 191–192
logins
    from outside i-mode, 189
    testing, 189–190
    utn-enabled phone, 189
`lookup.php`, 202–203
loops
    `do-while`, 217
    to fill screen, 217–220
    `for`, 204, 217, 268
    low-level API, 217–220
    to move text around, 217–218
    moving text with, 217–218
    `while`, 217, 251, 275
low-level API. *See also* Hello World
    `Canvas` class, 215, 217
    canvas form, 215
    defined, 214
    `Graphics` class, 216, 217
    `Iapplication` class, 216
    loops, 217–220
    `MainCanvas` class, 216, 219
    package creation, 215–217

M

Macromedia Fireworks, 107
magazine sites, 370–371
Mail Anime, 296
mail melodies, 161
`MainCanvas` class, 216, 219
map sites, 344

market
    global, expanding to, 37–39
    handset, 6–7
    knowing, 36–37
    L-mode, 78
    network gaming, 73
markup languages. *See also* cHTML (Compact HTML)
    defined, 46
    HTML, 46–47, 196
    i-mode versus WAP, 59
    MML, 161–162
    SGML, 47, 48
    WML, 9, 14, 50
    XHTML, 46, 47–50
    XML, 47, 48–49
`<marquee>` tag
    `behavior` attribute, 149
    defined, 149
    demonstration, 150
    `direction` attribute, 149
    `loop` attribute, 149
martial arts sites, 365
matches, displaying, 202
Matsunaga, Mari, 27–29, 31
Media Access Control (MAC) layer, 44
media events, 269
`MediaData` interface, 417
`MediaImage` interface, 417–418
`MediaListener` interface, 418–419
`MediaManager` class
    constructor, 420
    defined, 419
    inheritance, 419
    methods, 420
`MediaPresenter` interface
    defined, 420
    methods, 421
`MediaResource` interface, 421
`MediaSound` interface, 276, 422
Melody Format. *See* MFi

memory
    au, 19
    coding limits, 105–106
    coding small and, 105–106
    emoji and, 126
    i-Appli limitation, 19
    J-Phone, 19
    ringing tones in, 86
    startup screen in, 86
`<menu>` tag, 150
messaging, 33
`<meta>` tag, 150
methods
    AudioPresenter class, 387
    Button class, 388
    Canvas class, 389
    coding, 251
    Component class, 390–391
    ComponentAction(), 252–254
    Dialog class, 396
    Display class, 399
    drawString, 216
    Font class, 401
    Frame class, 402
    GET, 140, 248, 250
    getArgs(), 249
    getnames(), 249, 250–251, 256
    getSound(), 276
    getSourceURL(), 248
    Graphics class, 281, 404–405
    HEAD, 248, 250
    HttpConnection interface, 407–408
    Iapplication class, 409–410
    Image class, 411
    ImageLabel class, 412
    Interactable interface, 412
    KeyListener interface, 413
    Label class, 414
    ListBox class, 416–417
    MediaManager class, 420
    MediaPresenter interface, 421
    MediaResource interface, 421
    MediaSound interface, 422
    Panel class, 423–424
    PhoneSystem class, 424–425
    POST, 140, 248, 249, 250
    printIn(), 263
    processEvents, 269
    readLine, 251
    setCurrent(), 254
    ShortTimer class, 425–426
    softKeyPressed(), 252, 254–256
    squarePosition(), 268–269
    Textbox class, 226, 427–428
    Ticker class, 429
    TimeKeeper interface, 429
    Timer class, 430–431
    TimerListener interface, 431
    UIException class, 432–433
    URLDecoder class, 434
    URLEncoder class, 435
    VisualPresenter class, 437
MFi. *See also* sound formats
    chips, 157–158
    converting to, 159, 233
    defined, 157
    files, 157
    files, creating, 158–161
    files, uploading, 160–161
    i-melodies, 157, 161
    mail melodies, 161
    MIDI versus, 157
    as playback format, 157
MID (Mobile Interfaced Device), 51
MIDI. *See also* sound formats
    compiling, 276
    controllers, 158
    converting, 233, 276
    extension, 276
    input methods, 158
    MFi versus, 158

Index 475

saving as, 159
sound board functions, 277
midlets
    canvas form, 215
    conversion utilities, 52
    defined, 15, 19
    downloading to scratchpad, 264
    programming, 19
    running, in browser, 51–52
    size limit, 230
Ministry of Posts and
    Telecommunications (MPT), 75, 77
MLD format. *See* MFi
M-MAX (Mail-Mobile Access
    Exchange), 63
Mobile Banking. *See also* i-mode menu
    city banks, 320
    JA, 335–341
    national credit unions, 325–333
    regional banks, 321–324
    regional/local banks, 324
    trust associations, 333–335
Mobile Information Device Profile (MIDP)
    API, 211, 212
    classes, 211
    defined, 211
    framework, 211
    use of, 212
Mobile Markup Language (MML), 161
mobile network development
    David versus Goliath struggle, 76–77
    direction of, 71–80
    fixed line revolt and, 71–73
    killer apps and, 73–76
mobile phones. *See also* handsets
    area code, 71
    attitude towards, 72
    Bluetooth-enabled, 79, 80
    cost, 71
    Japanese use, 72, 85
    personalizing, 86

PHS, 83
pre-paid, 72
purchase, 25
serial number, 113
service cost, 72
skins, 86–87
subscribers, 24, 25
text-messaging capabilities, 82
use growth, 72
Mojo, 20
moral standards
    grace, 93–94
    healthy education, 94–95
    sensible, 93
    social ethics, 94
motorsport sites, 365
Mova, 24, 29
movie sites, 366
m-stage
    defined, 73
    initiative, 11
MULTIPLE_SELECT list box, 225
music information sites, 366
Music Markup Language (MML).
    See also sound formats
    converting from, 161
    defined, 161
    resources, 162
Music MasterWorks, 158, 315
My Walker. *See also* Walkerplus.com
    defined, 292
    subscriptions, 292
MySQL
    defined, 314
    PHP use with, 111, 112, 209
    speed, 209
MySQL database
    connecting to, 239
    downloading data from, 178
    uploading data to, 178

N

Namco Station sites, 356–357
national credit unions, 325–333
Natsuno, Takeshi, 28–29
NEC
    N501i handset, 7
    N503i handset, 21
    PC98, 76
NetFront browser, 49, 78
network gaming market, 73
News/Weather/Information menu. *See also* i-mode menu
    general news/weather, 318
    local newspapers/reports, 319
    overseas news, 320
`nextMenu` function, 203, 204
Nikkei. *See also* case studies
    advertising, 309
    audience, 309
    clout, 309
    defined, 307
    graphics use, 310
    home page, 308
    i-Appli news agent, 310–311
    i-mode service, 308, 311
    Internet presence, 307
    lessons learned, 312
    leveraging, 311
    market-geared content, 310–311
    news downloading, 264
    newspapers, 307
    Nikkei Index, 307
    Nikkei Net, 311
    online assets management, 311
    premium service, 310
    presence and profitability, 307–310
    revenue, 308
    Telecom 21, 307
    video and, 311
NJ Star, 120
NJWin
    on CD-ROM, 315
    defined, 212, 315
    installation, 212
N-MAX (Name-Mobile Access Exchange), 63
Nokia NM502i handset, 8, 26
Notepad, 213
NTT. *See also* DoCoMo
    DoCoMo ownership, 77
    fixed-line subscribers, 76
    L-mode, 77–78
    mobile communications business, 24
`NUMBERED_LIST` list box, 225

O

`<object>` tag, 150–151
Ochiai, 297, 298, 299, 300, 304
official i-mode sites. *See also* i-mode menu; i-mode sites
    administration, 373
    airline information, 343
    Another You Adventures, 360
    Bandai Co., LTD. Channel, 355–356
    baseball/soccer, 364
    books/CDs/games, 345–346
    cars, 346–347
    characters, 352–353
    city banks, 320
    cosmetics/health, 348
    Could Be Dwango, 357
    credit cards, 342
    dictionaries/handy information, 374
    employment, 346, 373
    English menu, 382–383
    fashion, 348
    FM, 370
    fortune telling, 362–363
    Game Taito, 360
    general news/weather, 318
    golf, 365

gourmet information, 349
handset makers, 355
Hokkaido, 376–377
Hokuriku, 378
hotel/lodging/travel, 344
i Playstation, 382
i-Navi Link, 381
insurance, 342–343
JA, 335–341
Kansai and Chugoku districts, 379–380
Konami Net, 357–358
Kyushu, 381
life information, 349
list of, 317–383
local broadcasters, 368–369
local newspapers/reports, 319
magazines, 370–371
maps, 344
martial arts, 365
motorsport, 365
movies, 366
music information, 366
Namco Station, 356–357
national credit unions, 325–333
number of, 14
other games, 360–362
outdoor, 364
overseas news, 320
package tracking, 375
pets, 348
prize/lottery/horse racing, 367
public entertainment, 371–372
railroad service information timetable, 373
real estate rental information, 347
recipes, 349–350
regional banks, 321–324
regional/local banks, 324
rent-a-car/traffic information, 344
ringing tones/karaoke, 350–352
Sakuma's Suguroku, 358
Sega Mode, 359
Shikoku, 380
stocks/securities, 341–342
study/qualification, 347
television, 367
television program information, 370
tennis, 365
tickets, 345
Tohoku, 377
Tokai, 377–378
Tomy-Web, 356
town information, 373
train guidance, 343
trust associations, 333–335
variety, 371
visual, 354–355
Webby Hudson, 358–359
official information provider status, 101
Ogawa, 297, 299, 300
`<ol>` tag
    defined, 151
    `start` attribute, 151
    `type` attribute, 151
    use code, 151
    use results, 152
online address book. *See* address book
`<option>` tag
    `selected` attribute, 141
    use code, 141
    use results, 142
    `value` attribute, 140
OR operator, 204
ordered lists, 151–152
outdoor sites, 364

P

`<p>` tag, 152
package tracking sites, 375
packages, creating, 214, 215–217

Packet Data Communications, 25
packet networks
   billing, 13
   cost of, 13
   i-mode use of, 13
   PDC-P, 60-62, 64-67, 110
   WAP services versus, 14
packet-switching
   advantages, 65
   defined, 64
pagers, 30
pages, breaking, 169-173
Panasonic P503i handset, 20
`Panel` class
   `Button` objects, 226-227
   classes used with, 222
   constructor, 423
   defined, 220, 422
   `ImageLabel` object, 229-230
   inheritance, 422
   `ListBox` object, 223-226
   methods, 423-424
   simple display using, 221
   `Ticker` object, 227-229
panels
   declaring, 242
   dialog boxes in, 234
   initializing elements in, 243
   for message display, 220
   scrolling object, 228
   `Textbox` class in, 222
   using, 241
paradigms, 264
partnership proposals, preparing, 100-102
part-time job sites, 346, 373
passwords
   encrypting, 192, 194
   fields, 138
   HTTP transmission, 195
   `NOT NULL`, 180
   transmitting, 194-195
PDC-P networks
   architecture, 60-61
   behavior, 60
   i-mode delivery over, 65, 110
   i-mode servers with, 66
   information flow, 62
   PGW, 65
   PPM, 65
   routing, 65
   structure illustration, 61, 65
   technology, 64
   uses, 64
Perl
   PHP advantages over, 116
   server-side, 111-112
personalization, 86
pet sites, 348
`PhoneSystem` class
   defined, 424
   fields, 424
   inheritance, 424
   methods, 424-425
PHP
   advantages, 116
   Boolean values, 182
   on CD-ROM, 314
   functions, 196
   with i-Appli, 112
   MySQL combination, 111, 112, 209
   scripting features, 116
   server-side, 111-112
   sessions capability, 189
   XHTML and, 50
PHP scripts
   `address.php`, 246
   `addressappli.php`, 239-241
   cHTML in, 196
   for eimode's Web site, 171-173
   HTML in, 196
   `index.php`, 181-182

`i-topics.php`, 175–176
`login.php`, 184–185, 187–188, 191–192
`lookup.php`, 202–203
`signup.php`, 193–194
`topic.php`, 173–174
URLs to connect to, 248
`user.php`, 207
`users.php`, 115–116
phpMyAdmin
  on CD-ROM, 315
  database viewed in, 116
  defined, 114, 315
PHS phone, 83
`<plaintext>` tag
  defined, 152
  HTTP server use of, 153
  use code, 152–153
  use results, 153
PlayStation, 74
"pocket bell" pagers, 296
Pocketboard, 30
Point to Point Protocol (PPP), 44
`<pre>` tag, 154
preformatted text, 154
`print` command, 198
`printaddress` function, 205–206
Private Homepage, 295
prize/lottery/horse racing sites, 367
`processEvent` method
  defined, 269
  `KEY_RELEASED_EVENT` case in, 273–274
  with switch to separate events, 269–270
programming
  CGI, 111
  in cHTML, 163–210
  i-Applis, 237–264
  stand-alone, 248
public entertainment sites, 371–372

pull-down menus
  code, 141
  defining, 140
  illustrated, 142

R

`RADIO_BUTTON` list box, 226
railroad service information timetable site, 373
real estate rental information sites, 347
real-time operating system (RTOS). *See also* software
  adding software on top of, 42
  defined, 41
  different, 42
  functions, 42
recipe sites, 349–350
Regional. *See also* i-mode menu
  Hokkaido, 376–377
  Hokuriku, 378
  Kansai and Chugoku districts, 379–380
  Kyushu, 381
  Shikoku, 380
  Tohoku, 377
  Tokai, 377–378
rent-a-car/traffic information sites, 344
revenue model
  advertising, 91
  content provider, 90, 91–92
  defined, 14
  exploring, 89–92
  i-Appli and, 17
  responsibilities/benefits, 91
  Walker-i, 292
ringing tones
  downloading, 85, 161
  formats, 161
  in memory, 86
  sites, 350–352
  young people use of, 86

Ringing Tones/Screen Pictures. *See also* i-mode menu
  characters, 352–353
  handset marker sites, 355
  ringing tones/karaoke, 350–352
  visual, 354–355

S

Sakuma's Suguroku sites, 358
scratchpad, downloading to, 264
scripting languages, 111–112
`searchform` function, 205
security
  gateway, 187
  password, 192, 194–195
  sessions and, 199
Sega Mode sites, 359
`<select>` tag, 140
serial number
  checking, 181
  as identifier, 113
  security, assuming, 179
  `utn`, 187
servers
  central office, 67
  HTTP, 44
  i-mode, 62–63, 66
server-side model, 111, 121
services
  categories, 14–15
  exploring, 13–21
  i-Appli, 15–21
  subscriptions, 90
sessions
  continuing, 199
  ID, checking, 195
  ID, hijacking, 199
  limitation, 196
  PHP capability, 189
  security and, 199
  starting, 199
  variables, passing, 196

SGML, 47, 48
Shikoku sites, 380
Shopping/Living. *See also* i-mode menu
  books/CDs/games, 345–346
  cars, 346–347
  cosmetics/health, 348
  employment, 346
  fashion, 348
  gourmet information, 349
  life information, 349
  pets, 348
  real estate rental information, 347
  recipes, 349–350
  study/qualification, 347
  tickets, 345
short message service (SMS)
  connections, 31
  i-mode e-mail as extension, 9
  legacy of, 9
  popularity, 31
  uses, 31
`ShortTimer` class
  constructor, 425
  defined, 425
  inheritance, 425
  methods, 425–426
`signup.php`, 193–194
`SINGLE_SELECT` list box, 226
S-JIS encoding, 52, 53, 449–455
"skins," 86–87
softkey listeners
  defined, 252
  events, 256
`SoftKeyListener` interface, 426
`softKeyPressed()` method
  defined, 252
  functions, 255
softkeys
  declaring, 242
  defined, 252
  input, 256
  pressing of, 255

Index

software
    architecture, 41–45
    CSE, 44
    function differences, 42
    HTML browser layer, 44
    layers, 41, 43–44
    MAC layer, 44
    RTOS, 41–43
    structure, 41
    system illustration, 43
    window manager layer, 44
Sony
    502wm handset, 73
    au phone, 86–87
    m-stage initiative and, 11
    PlayStation, 74
    SO502i handset, 12, 26
    SO503i handset, 16
sound formats
    MFi, 157–161
    MIDI, 157
    MML, 161–162
sounds
    game, 265
    loading, 267, 275–276
    varying, 281
speed
    connection, 110
    in eye of beholder, 111
    MySQL, 209
Sports and Entertainment. *See also*
    i-mode menu
    baseball/soccer, 364
    FM, 370
    golf, 365
    local broadcasters, 368–369
    magazines, 370–371
    martial arts, 365
    motorsport, 365
    movies, 366

    music information, 366
    outdoor, 364
    prize/lottery/horse racing, 367
    public entertainment, 371–372
    television, 367
    television program information, 370
    tennis, 365
    variety, 371
SQL databases
    creating, 113
    for hosting dynamic content, 112–117
    using, 121
    viewed in phpMyAdmin, 116
SQL tables, creating, 179–180
`squarePosition()` method
    defined, 268
    listing, 268–269
    starting coordinates, 269
startup screens
    downloading, 86
    in memory, 86
    purchase choice and, 87
stocks/securities sites, 341–342
study/qualification sites, 347
surrogate testers, 119
symbol font, 118
system requirements (CD-ROM), 313–314

T

tables
    in address book example, 179–181
    cHTML and, 164
    cutting out, 165
    defined, 179
    SQL, 179–180
    user, 179–180
tags (cHTML)
    <a>, 126–129
    available, 439–448
    <base>, 129

continued

tags *(continued)*
    `<blink>`, 129
    `<blockquote>`, 130
    `<body>`, 131
    `<br>`, 131–133
    `<center>`, 133–134
    `<dd>`, 134–135
    defined, 45
    `<dir>`, 134
    `<div>`, 135–136
    `<dl>`, 135
    `<dt>`, 134–135
    `<font>`, 136
    `<form>`, 137
    `<h1>`...`<h4>`, 144–145
    `<head>`, 145
    `<hr>`, 145–146
    `<html>`, 146
    `<img>`, 147–149
    `<input>`, 137–140
    `<marquee>`, 149–150
    `<menu>`, 150
    `<meta>`, 150
    `<object>`, 150–151
    `<ol>`, 151–152
    `<option>`, 140–142
    `<p>`, 152
    `<plaintext>`, 152–153
    `<pre>`, 154
    `<select>`, 140
    `<textarea>`, 142–144
    `<title>`, 154–155
    `<ul>`, 155
    using, 46
TCP/IP, 44
Telecom 21, 307
television program information sites, 370
television sites, 367
tennis sites, 365
testers, 119
testing
    code for handsets, 209–210
    events, 271–272
    games, 281
    i-Applis, 263
    logins, 189–190
text
    alignment, 131
    alignment/display, in label, 220–222
    body definition, 131
    centering, 133–134
    color, 136
    displaying, 222–223
    formatting, 154
    formatting, cutting out, 173
    moving, across page, 149–150
    moving, with loops, 217–218
    preformatted, 154
    quoting within, 130
    view during graphics load, 183
    wraps, 131–133
text editors, 213
`<textarea>` tag
    `cols` attribute, 143
    defined, 143
    `istyle` attribute, 143
    maximum characters, 143
    `name` attribute, 142
    `rows` attribute, 142–143
    use code, 143
    use results, 144
    within `<form>` tags, 142
`TextBox` class
    constructor, 427
    defined, 222, 427
    fields, 223, 427
    inheritance, 426–427
    methods, 226, 427–428
    in the panel, 222
    simple example, 222
`textwrap` function, 209, 241

text-wrap options, setting, 169–173
`textwrap` script, 177
"thumb tribe," 10
`Ticker` class. *See also* `Panel` class
   constructors, 428
   defined, 227, 428
   displaying scrolling message, 228
   DoJa emulator rendering, 228
   inheritance, 428
   methods, 429
ticket sites, 345
`TimeKeeper` interface, 429
`Timer` class
   constructor, 430
   defined, 430
   inheritance, 429–430
   methods, 430–431
timer expired events, 269
`TimerListener` interface, 431
`<title>` tag, 154–155
Tohoku sites, 377
Tokai sites, 377–378
Tomy-Web sites, 356
tools and utility content, 55
`topic.php`, 173–174
Town Information/Administration. *See also* i-mode menu
   administration, 373
   employment, 373
   railroad service information timetable, 373
   sport, 374
   town information, 373
train guidance sites, 343
transparency support, 168
Travel/Transportation/Maps. *See also* i-mode menu
   airline information, 343
   hotel/lodging/travel, 344
   maps, 344
   rent-a-car/traffic information, 344
   train guidance, 343

troubleshooting
   CD-ROM, 315–316
   online address book, 199–200
trust association sites. *See also* Mobile Banking
   Chugoku, 335
   Hokkaido, 333
   Hokuriku, 334
   Kansai district, 334
   Kanto district, Koshinetsu, 334
   Kyushu, 335
   postal savings, 335
   Tohoku, 333
   Tokai, 334
Tsutaya Online, 82

U

`UIException` class
   constructors, 432
   defined, 431
   fields, 432
   inheritance, 431
   methods, 432–433
`<ul>` tag, 155
UltraEdit32, 213, 315
U-MAX (User-Mobile Access Exchange), 63
UMTS
   defined, 34
   standard adoption, 38
   W-CDMA networks, 39
Unicode, 52
U.S.
   3G licenses, 75
   Internet usage, 76
Universal Mobile Telecommunications System. *See* UMTS
unofficial sites. *See also* i-mode sites
   dating sites, 97
   defined, 97
   profitability of, 98

unordered lists, 155
UnsupportedOperationException class, 433
URLDecoder class, 434
URLEncoder class
   constructor, 435
   defined, 435
   inheritance, 434
   methods, 435
user.php, 115–116, 207
user tables, creating, 179–180
userform function, 207–209
users
   authenticating, 187–192
   registration code, 114–115
   sign-up screen, 115
users.php, 115–116
utn attribute, 113, 179, 181
   in <a> tags, 187
   for handset identification, 210
   serial number, 187
utn-enabled handsets
   graphics for, 183
   login test, 189
   menu and sign up page, 186
   serial number and, 190
   users logged in with, 185
utnlogin.html, 186

V

variables
   declaring, 246
   sent to addressappli.php, 241
   session, passing, 196
variety sites, 371
visual sites, 354–355
VisualPresenter class
   constructor, 436
   defined, 230, 435
   fields, 436
   inheritance, 435
   methods, 437

   for showing animated gif, 231, 232
   with softkey, 232
voice network, billing, 13

W

Walker-i site
   copyright, 292
   earnings, 292
   offerings, 294
   revenue model, 292
   start page, 291
   turning revenue from, 293
Walkerplus.com. *See also* case studies
   company history, 291–293
   D2C agent, 292
   defined, 291
   lessons learned, 294
   My Walker, 292
   operating costs, 293
   revenue model, 292
   strategy analysis, 293–294
   Walker magazines, 292, 293
   Web site, 293
WAP (Wireless Application Protocol)
   billing, 14
   circuit-switched networks, 14
   defined, 9, 58
   failure, 21
   Forum, 9, 58
   i-mode versus, 59
   "priority" network competition with, 57–60
   Profit editor, 119–120
   slow uptake, 59
   WML language, 9, 59
warning dialog box, 234
Watanabe, 297, 299, 300
W-CDMA. *See also* FOMA
   defined, 34
   DoCoMo investment in, 36
   UMTS, 39
Webby Hudson sites, 358–359

while loops, 217, 251, 275
window managing layer, 44
wireless content providers, 89
W-MAX (Web-Mobile Access Exchange), 63
WML (Wireless Markup Language). *See also* markup languages; WAP (Wireless Application Protocol)
    cHTML versus, 59
    defined, 9
    development for, 50
    limited use of, 14
WYSIWYG tools, 46, 106–107, 117

X

x and y coordinates
    declarations, 267
    setting, 268–269
    starting location, 269
    variables, 267

XHTML. *See also* markup languages
    Basic, 49, 50
    defined, 46, 47
    DTDs, 49
    flexibility, 51
    functionality missing from, 51
    modules, 49
    PHP and, 50
    tags, 439–448
    version 1.1, 49
XML. *See also* markup languages
    defined, 47, 48
    document, 48–49
    flexibility, 51
    tags and, 49

Y

Yoji Baba, 295, 297, 299

Hungry Minds, Inc.
End-User License Agreement

READ THIS. You should carefully read these terms and conditions before opening the software packet(s) included with this book ("Book"). This is a license agreement ("Agreement") between you and Hungry Minds, Inc. ("HMI"). By opening the accompanying software packet(s), you acknowledge that you have read and accept the following terms and conditions. If you do not agree and do not want to be bound by such terms and conditions, promptly return the Book and the unopened software packet(s) to the place you obtained them for a full refund.

1. **License Grant.** HMI grants to you (either an individual or entity) a nonexclusive license to use one copy of the enclosed software program(s) (collectively, the "Software") solely for your own personal or business purposes on a single computer (whether a standard computer or a workstation component of a multi-user network). The Software is in use on a computer when it is loaded into temporary memory (RAM) or installed into permanent memory (hard disk, CD-ROM, or other storage device). HMI reserves all rights not expressly granted herein.

2. **Ownership.** HMI is the owner of all right, title, and interest, including copyright, in and to the compilation of the Software recorded on the disk(s) or CD-ROM ("Software Media"). Copyright to the individual programs recorded on the Software Media is owned by the author or other authorized copyright owner of each program. Ownership of the Software and all proprietary rights relating thereto remain with HMI and its licensers.

3. **Restrictions On Use and Transfer.**

    (a) You may only (i) make one copy of the Software for backup or archival purposes, or (ii) transfer the Software to a single hard disk, provided that you keep the original for backup or archival purposes. You may not (i) rent or lease the Software, (ii) copy or reproduce the Software through a LAN or other network system or through any computer subscriber system or bulletin-board system, or (iii) modify, adapt, or create derivative works based on the Software.

    (b) You may not reverse engineer, decompile, or disassemble the Software. You may transfer the Software and user documentation on a permanent basis, provided that the transferee agrees to accept the terms and conditions of this Agreement and you retain no copies. If the Software is an update or has been updated, any transfer must include the most recent update and all prior versions.

4. **Restrictions on Use of Individual Programs.** You must follow the individual requirements and restrictions detailed for each individual program in the "What's on the CD-ROM?" appendix of this Book. These limitations are also contained in the individual license agreements recorded on the Software Media. These limitations may include a requirement that after using the program for a specified period of time, the user must pay a registration fee or discontinue use. By opening the Software packet(s), you will be agreeing to abide by the licenses and restrictions for these individual programs that are detailed in the "What's on the CD-ROM?" appendix and on the Software Media. None of the material on this Software Media or listed in this Book may ever be redistributed, in original or modified form, for commercial purposes.

5. **Limited Warranty.**

    (a) HMI warrants that the Software and Software Media are free from defects in materials and workmanship under normal use for a period of sixty (60) days from the date of purchase of this Book. If HMI receives notification within the warranty period of defects in materials or workmanship, HMI will replace the defective Software Media.

    (b) HMI AND THE AUTHOR OF THE BOOK DISCLAIM ALL OTHER WARRANTIES, EXPRESS OR IMPLIED, INCLUDING WITHOUT LIMITATION IMPLIED WARRANTIES OF MERCHANTABILITY AND FITNESS FOR A PARTICULAR PURPOSE, WITH RESPECT TO THE SOFTWARE, THE PROGRAMS, THE SOURCE CODE CONTAINED THEREIN, AND/OR THE TECHNIQUES DESCRIBED IN THIS BOOK. HMI DOES NOT WARRANT THAT THE FUNCTIONS CONTAINED IN THE SOFTWARE WILL MEET YOUR REQUIREMENTS OR THAT THE OPERATION OF THE SOFTWARE WILL BE ERROR FREE.

    (c) This limited warranty gives you specific legal rights, and you may have other rights that vary from jurisdiction to jurisdiction.

6. **Remedies.**

    (a) HMI's entire liability and your exclusive remedy for defects in materials and workmanship shall be limited to replacement of the Software Media, which may be returned to HMI with a copy of your receipt at the following address: Software Media Fulfillment Department, Attn.: *i-mode: A Primer*, Hungry Minds, Inc., 10475 Crosspoint Blvd., Indianapolis, IN 46256, or call 1-800-762-2974. Please allow four to six weeks for delivery. This Limited Warranty is void if failure of the Software Media has resulted from accident, abuse, or misapplication. Any replacement Software Media will be warranted for the remainder of the original warranty period or thirty (30) days, whichever is longer.

(b) In no event shall HMI or the author be liable for any damages whatsoever (including without limitation damages for loss of business profits, business interruption, loss of business information, or any other pecuniary loss) arising from the use of or inability to use the Book or the Software, even if HMI has been advised of the possibility of such damages.

(c) Because some jurisdictions do not allow the exclusion or limitation of liability for consequential or incidental damages, the above limitation or exclusion may not apply to you.

7. **U.S. Government Restricted Rights.** Use, duplication, or disclosure of the Software for or on behalf of the United States of America, its agencies and/or instrumentalities (the "U.S. Government") is subject to restrictions as stated in paragraph (c)(1)(ii) of the Rights in Technical Data and Computer Software clause of DFARS 252.227-7013, or subparagraphs (c)(1) and (2) of the Commercial Computer Software-Restricted Rights clause at FAR 52.227-19, and in similar clauses in the NASA FAR supplement, as applicable.

8. **General.** This Agreement constitutes the entire understanding of the parties and revokes and supersedes all prior agreements, oral or written, between them and may not be modified or amended except in a writing signed by both parties hereto that specifically refers to this Agreement. This Agreement shall take precedence over any other documents that may be in conflict herewith. If any one or more provisions contained in this Agreement is held by any court or tribunal to be invalid, illegal, or otherwise unenforceable, each and every other provision shall remain in full force and effect.

GNU GENERAL PUBLIC LICENSE

Version 2, June 1991
Copyright (c) 1989, 1991 Free Software Foundation, Inc.
59 Temple Place, Suite 330, Boston, MA 02111-1307, USA
Everyone is permitted to copy and distribute verbatim copies of this license document, but changing it is not allowed.

Preamble

The licenses for most software are designed to take away your freedom to share and change it. By contrast, the GNU General Public License is intended to guarantee your freedom to share and change free software — to make sure the software is free for all its users. This General Public License applies to most of the Free Software Foundation's software and to any other program whose authors commit to using it. (Some other Free Software Foundation software is covered by the GNU Library General Public License instead.) You can apply it to your programs, too.

When we speak of free software, we are referring to freedom, not price. Our General Public Licenses are designed to make sure that you have the freedom to distribute copies of free software (and charge for this service if you wish), that you receive source code or can get it if you want it, that you can change the software or use pieces of it in new free programs; and that you know you can do these things.

To protect your rights, we need to make restrictions that forbid anyone to deny you these rights or to ask you to surrender the rights. These restrictions translate to certain responsibilities for you if you distribute copies of the software, or if you modify it.

For example, if you distribute copies of such a program, whether gratis or for a fee, you must give the recipients all the rights that you have. You must make sure that they, too, receive or can get the source code. And you must show them these terms so they know their rights.

We protect your rights with two steps: (1) copyright the software, and (2) offer you this license which gives you legal permission to copy, distribute and/or modify the software.

Also, for each author's protection and ours, we want to make certain that everyone understands that there is no warranty for this free software. If the software is modified by someone else and passed on, we want its recipients to know that what they have is not the original, so that any problems introduced by others will not reflect on the original authors' reputations.

Finally, any free program is threatened constantly by software patents. We wish to avoid the danger that redistributors of a free program will individually obtain patent licenses, in effect making the program proprietary. To prevent this, we have made it clear that any patent must be licensed for everyone's free use or not licensed at all.

The precise terms and conditions for copying, distribution and modification follow.

TERMS AND CONDITIONS FOR COPYING, DISTRIBUTION, AND MODIFICATION

0. This License applies to any program or other work which contains a notice placed by the copyright holder saying it may be distributed under the terms of this General Public License. The "Program", below, refers to any such program or work, and a "work based on the Program" means either the Program or any derivative work under copyright law: that is to say, a work containing the Program or a portion of it, either verbatim or with modifications and/or translated into another language. (Hereinafter, translation is included without limitation in the term "modification".) Each licensee is addressed as "you".

   Activities other than copying, distribution and modification are not covered by this License; they are outside its scope. The act of running the Program is not restricted, and the output from the Program is covered only if its contents constitute a work based on the Program (independent of having been made by running the Program). Whether that is true depends on what the Program does.

1. You may copy and distribute verbatim copies of the Program's source code as you receive it, in any medium, provided that you conspicuously and appropriately publish on each copy an appropriate copyright notice and disclaimer of warranty; keep intact all the notices that refer to this License and to the absence of any warranty; and give any other recipients of the Program a copy of this License along with the Program.

   You may charge a fee for the physical act of transferring a copy, and you may at your option offer warranty protection in exchange for a fee.

2. You may modify your copy or copies of the Program or any portion of it, thus forming a work based on the Program, and copy and distribute such modifications or work under the terms of Section 1 above, provided that you also meet all of these conditions:

   a) You must cause the modified files to carry prominent notices stating that you changed the files and the date of any change.

   b) You must cause any work that you distribute or publish, that in whole or in part contains or is derived from the Program or any part thereof, to be licensed as a whole at no charge to all third parties under the terms of this License.

   c) If the modified program normally reads commands interactively when run, you must cause it, when started running for such interactive use in the most ordinary way, to print or display an announcement including an appropriate copyright notice and a notice that there is no warranty (or else, saying that you provide a warranty) and that users may redistribute the program under these conditions, and telling the user how to view a copy of this License. (Exception: if the Program itself is interactive but does not normally print such an announcement, your work based on the Program is not required to print an announcement.)

These requirements apply to the modified work as a whole. If identifiable sections of that work are not derived from the Program, and can be reasonably considered independent and separate works in themselves, then this License, and its terms, do not apply to those sections when you distribute them as separate works. But when you distribute the same sections as part of a whole which is a work based on the Program, the distribution of the whole must be on the terms of this License, whose permissions for other licensees extend to the entire whole, and thus to each and every part regardless of who wrote it.

Thus, it is not the intent of this section to claim rights or contest your rights to work written entirely by you; rather, the intent is to exercise the right to control the distribution of derivative or collective works based on the Program.

In addition, mere aggregation of another work not based on the Program with the Program (or with a work based on the Program) on a volume of a storage or distribution medium does not bring the other work under the scope of this License.

3. You may copy and distribute the Program (or a work based on it, under Section 2) in object code or executable form under the terms of Sections 1 and 2 above provided that you also do one of the following:

   a) Accompany it with the complete corresponding machine-readable source code, which must be distributed under the terms of Sections 1 and 2 above on a medium customarily used for software interchange; or,

   b) Accompany it with a written offer, valid for at least three years, to give any third party, for a charge no more than your cost of physically performing source distribution, a complete machine-readable copy of the corresponding source code, to be distributed under the terms of Sections 1 and 2 above on a medium customarily used for software interchange; or,

   c) Accompany it with the information you received as to the offer to distribute corresponding source code. (This alternative is allowed only for noncommercial distribution and only if you received the program in object code or executable form with such an offer, in accord with Subsection b above.)

The source code for a work means the preferred form of the work for making modifications to it. For an executable work, complete source code means all the source code for all modules it contains, plus any associated interface definition files, plus the scripts used to control compilation and installation of the executable. However, as a special exception, the source code distributed need not include anything that is normally distributed (in either source or binary form) with the major components (compiler, kernel, and so on) of the operating system on which the executable runs, unless that component itself accompanies the executable.

If distribution of executable or object code is made by offering access to copy from a designated place, then offering equivalent access to copy the source code from the same place counts as distribution of the source code, even though third parties are not compelled to copy the source along with the object code.

4. You may not copy, modify, sublicense, or distribute the Program except as expressly provided under this License. Any attempt otherwise to copy, modify, sublicense or distribute the Program is void, and will automatically terminate your rights under this License. However, parties who have received copies, or rights, from you under this License will not have their licenses terminated so long as such parties remain in full compliance.

5. You are not required to accept this License, since you have not signed it. However, nothing else grants you permission to modify or distribute the Program or its derivative works. These actions are prohibited by law if you do not accept this License. Therefore, by modifying or distributing the Program (or any work based on the Program), you indicate your acceptance of this License to do so, and all its terms and conditions for copying, distributing or modifying the Program or works based on it.

6. Each time you redistribute the Program (or any work based on the Program), the recipient automatically receives a license from the original licensor to copy, distribute or modify the Program subject to these terms and conditions. You may not impose any further restrictions on the recipients' exercise of the rights granted herein. You are not responsible for enforcing compliance by third parties to this License.

7. If, as a consequence of a court judgment or allegation of patent infringement or for any other reason (not limited to patent issues), conditions are imposed on you (whether by court order, agreement or otherwise) that contradict the conditions of this License, they do not excuse you from the conditions of this License. If you cannot distribute so as to satisfy simultaneously your obligations under this License and any other pertinent obligations, then as a consequence you may not distribute the Program at all. For example, if a patent license would not permit royalty-free redistribution of the Program by all those who receive copies directly or indirectly

through you, then the only way you could satisfy both it and this License would be to refrain entirely from distribution of the Program.

If any portion of this section is held invalid or unenforceable under any particular circumstance, the balance of the section is intended to apply and the section as a whole is intended to apply in other circumstances.

It is not the purpose of this section to induce you to infringe any patents or other property right claims or to contest validity of any such claims; this section has the sole purpose of protecting the integrity of the free software distribution system, which is implemented by public license practices. Many people have made generous contributions to the wide range of software distributed through that system in reliance on consistent application of that system; it is up to the author/donor to decide if he or she is willing to distribute software through any other system and a licensee cannot impose that choice.

This section is intended to make thoroughly clear what is believed to be a consequence of the rest of this License.

8. If the distribution and/or use of the Program is restricted in certain countries either by patents or by copyrighted interfaces, the original copyright holder who places the Program under this License may add an explicit geographical distribution limitation excluding those countries, so that distribution is permitted only in or among countries not thus excluded. In such case, this License incorporates the limitation as if written in the body of this License.

9. The Free Software Foundation may publish revised and/or new versions of the General Public License from time to time. Such new versions will be similar in spirit to the present version, but may differ in detail to address new problems or concerns.

Each version is given a distinguishing version number. If the Program specifies a version number of this License which applies to it and "any later version", you have the option of following the terms and conditions either of that version or of any later version published by the Free Software Foundation. If the Program does not specify a version number of this License, you may choose any version ever published by the Free Software Foundation.

10. If you wish to incorporate parts of the Program into other free programs whose distribution conditions are different, write to the author to ask for permission. For software which is copyrighted by the Free Software Foundation, write to the Free Software Foundation; we sometimes make exceptions for this. Our decision will be guided by the two goals of preserving the free status of all derivatives of our free software and of promoting the sharing and reuse of software generally.

NO WARRANTY

11. BECAUSE THE PROGRAM IS LICENSED FREE OF CHARGE, THERE IS NO WARRANTY FOR THE PROGRAM, TO THE EXTENT PERMITTED BY APPLICABLE LAW. EXCEPT WHEN OTHERWISE STATED IN WRITING THE COPYRIGHT HOLDERS AND/OR OTHER PARTIES PROVIDE THE PROGRAM "AS IS" WITHOUT WARRANTY OF ANY KIND, EITHER EXPRESSED OR IMPLIED, INCLUDING, BUT NOT LIMITED TO, THE IMPLIED WARRANTIES OF MERCHANTABILITY AND FITNESS FOR A PARTICULAR PURPOSE. THE ENTIRE RISK AS TO THE QUALITY AND PERFORMANCE OF THE PROGRAM IS WITH YOU. SHOULD THE PROGRAM PROVE DEFECTIVE, YOU ASSUME THE COST OF ALL NECESSARY SERVICING, REPAIR OR CORRECTION.

12. IN NO EVENT UNLESS REQUIRED BY APPLICABLE LAW OR AGREED TO IN WRITING WILL ANY COPYRIGHT HOLDER, OR ANY OTHER PARTY WHO MAY MODIFY AND/OR REDISTRIBUTE THE PROGRAM AS PERMITTED ABOVE, BE LIABLE TO YOU FOR DAMAGES, INCLUDING ANY GENERAL, SPECIAL, INCIDENTAL OR CONSEQUENTIAL DAMAGES ARISING OUT OF THE USE OR INABILITY TO USE THE PROGRAM (INCLUDING BUT NOT LIMITED TO LOSS OF DATA OR DATA BEING RENDERED INACCURATE OR LOSSES SUSTAINED BY YOU OR THIRD PARTIES OR A FAILURE OF THE PROGRAM TO OPERATE WITH ANY OTHER PROGRAMS), EVEN IF SUCH HOLDER OR OTHER PARTY HAS BEEN ADVISED OF THE POSSIBILITY OF SUCH DAMAGES.

LICENSE AGREEMENT:

Forte for Java, release 3.0, Community Edition, English

To obtain Forte for Java, release 3.0, Community Edition, English, you must agree to the software license below.

Sun Microsystems, Inc. Binary Code License Agreement

READ THE TERMS OF THIS AGREEMENT AND ANY PROVIDED SUPPLEMENTAL LICENSE TERMS (COLLECTIVELY "AGREEMENT") CAREFULLY BEFORE OPENING THE SOFTWARE MEDIA PACKAGE. BY OPENING THE SOFTWARE MEDIA PACKAGE, YOU AGREE TO THE TERMS OF THIS AGREEMENT. IF YOU ARE ACCESSING THE SOFTWARE ELECTRONICALLY, INDICATE YOUR ACCEPTANCE OF THESE TERMS BY SELECTING THE "ACCEPT" BUTTON AT THE END OF THIS AGREEMENT. IF YOU DO NOT AGREE TO ALL THESE TERMS, PROMPTLY RETURN THE UNUSED SOFTWARE TO YOUR PLACE OF PURCHASE FOR A REFUND OR, IF THE SOFTWARE IS ACCESSED ELECTRONICALLY, SELECT THE "DECLINE" BUTTON AT THE END OF THIS AGREEMENT.

1. <u>LICENSE TO USE</u>. Sun grants you a non-exclusive and non-transferable license for the internal use only of the accompanying software and documentation and any error corrections provided by Sun (collectively "Software"), by the number of users and the class of computer hardware for which the corresponding fee has been paid.

2. <u>RESTRICTIONS</u>. Software is confidential and copyrighted. Title to Software and all associated intellectual property rights is retained by Sun and/or its licensors. Except as specifically authorized in any Supplemental License Terms, you may not make copies of Software, other than a single copy of Software for archival purposes. Unless enforcement is prohibited by applicable law, you may not modify, decompile, or reverse engineer Software. You acknowledge that Software is not designed, licensed or intended for use in the design, construction, operation or maintenance of any nuclear facility. Sun disclaims any express or implied warranty of fitness for such uses. No right, title or interest in or to any trademark, service mark, logo or trade name of Sun or its licensors is granted under this Agreement.

3. <u>LIMITED WARRANTY</u>. Sun warrants to you that for a period of ninety (90) days from the date of purchase, as evidenced by a copy of the receipt, the media on which Software is furnished (if any) will be free of defects in materials and workmanship under normal use. Except for the foregoing, Software is provided "AS IS". Your exclusive remedy and Sun's entire liability under this limited warranty will be at Sun's option to replace Software media or refund the fee paid for Software.

4. **DISCLAIMER OF WARRANTY**. UNLESS SPECIFIED IN THIS AGREEMENT, ALL EXPRESS OR IMPLIED CONDITIONS, REPRESENTATIONS AND WARRANTIES, INCLUDING ANY IMPLIED WARRANTY OF MERCHANTABILITY, FITNESS FOR A PARTICULAR PURPOSE OR NON-INFRINGEMENT ARE DISCLAIMED, EXCEPT TO THE EXTENT THAT THESE DISCLAIMERS ARE HELD TO BE LEGALLY INVALID.

5. **LIMITATION OF LIABILITY**. TO THE EXTENT NOT PROHIBITED BY LAW, IN NO EVENT WILL SUN OR ITS LICENSORS BE LIABLE FOR ANY LOST REVENUE, PROFIT OR DATA, OR FOR SPECIAL, INDIRECT, CONSEQUENTIAL, INCIDENTAL OR PUNITIVE DAMAGES, HOWEVER CAUSED REGARDLESS OF THE THEORY OF LIABILITY, ARISING OUT OF OR RELATED TO THE USE OF OR INABILITY TO USE SOFTWARE, EVEN IF SUN HAS BEEN ADVISED OF THE POSSIBILITY OF SUCH DAMAGES.

   In no event will Sun's liability to you, whether in contract, tort (including negligence), or otherwise, exceed the amount paid by you for Software under this Agreement. The foregoing limitations will apply even if the above stated warranty fails of its essential purpose.

6. **Termination**. This Agreement is effective until terminated. You may terminate this Agreement at any time by destroying all copies of Software. This Agreement will terminate immediately without notice from Sun if you fail to comply with any provision of this Agreement. Upon Termination, you must destroy all copies of Software.

7. **Export Regulations**. All Software and technical data delivered under this Agreement are subject to US export control laws and may be subject to export or import regulations in other countries. You agree to comply strictly with all such laws and regulations and acknowledge that you have the responsibility to obtain such licenses to export, re-export, or import as may be required after delivery to you.

8. **U.S. Government Restricted Rights**. If Software is being acquired by or on behalf of the U.S. Government or by a U.S. Government prime contractor or subcontractor (at any tier), then the Government's rights in Software and accompanying documentation will be only as set forth in this Agreement; this is in accordance with 48 CFR 227.7201 through 227.7202-4 (for Department of Defense (DOD) acquisitions) and with 48 CFR 2.101 and 12.212 (for non-DOD acquisitions).

9. **Governing Law**. Any action related to this Agreement will be governed by California law and controlling U.S. federal law. No choice of law rules of any jurisdiction will apply.

10. **Severability**. If any provision of this Agreement is held to be unenforceable, this Agreement will remain in effect with the provision omitted, unless omission would frustrate the intent of the parties, in which case this Agreement will immediately terminate.

11. **Integration**. This Agreement is the entire agreement between you and Sun relating to its subject matter. It supersedes all prior or contemporaneous oral or written communications, proposals, representations and warranties and prevails over any conflicting or additional terms of any quote, order, acknowledgment, or other communication between the parties relating to its subject matter during the term of this Agreement. No modification of this Agreement will be binding, unless in writing and signed by an authorized representative of each party.

FORTE(TM) FOR JAVA(TM), RELEASE 3.0, COMMUNITY EDITION SUPPLEMENTAL LICENSE TERMS

These supplemental license terms ("Supplemental Terms") add to or modify the terms of the Binary Code License Agreement (collectively, the "Agreement"). Capitalized terms not defined in these Supplemental Terms shall have the same meanings ascribed to them in the Agreement. These Supplemental Terms shall supersede any inconsistent or conflicting terms in the Agreement, or in any license contained within the Software.

1. **Software Internal Use and Development License Grant**. Subject to the terms and conditions of this Agreement, including, but not limited to Section 4 (Java(TM) Technology Restrictions) of these Supplemental Terms, Sun grants you a non-exclusive, non-transferable, limited license to reproduce internally and use internally the binary form of the Software complete and unmodified for the sole purpose of designing, developing and testing your Java applets and applications intended to run on the Java platform ("Programs").

2. **License to Distribute Software**. Subject to the terms and conditions of this Agreement, including, but not limited to Section 4 (Java (TM) Technology Restrictions) of these Supplemental Terms, Sun grants you a non-exclusive, non-transferable, limited license to reproduce and distribute the Software in binary code form only, provided that (i) you distribute the Software complete and unmodified and only bundled as part of, and for the sole purpose of running, your Programs, (ii) the Programs add significant and primary functionality to the Software, (iii) you do not distribute additional software intended to replace any component(s) of the Software, (iv) for a particular version of the Java platform, any executable output generated by a compiler that is contained in the Software must (a) only be compiled from source code that conforms to the corresponding version of the OEM Java Language Specification; (b) be in the class file format defined by the corresponding version of the OEM Java Virtual Machine Specification; and (c) execute properly on a reference runtime, as specified by Sun, associated with such version of the Java platform, (v) you do not remove or alter any proprietary legends or notices contained

in the Software, (v) you only distribute the Software subject to a license agreement that protects Sun's interests consistent with the terms contained in this Agreement, and (vi) you agree to defend and indemnify Sun and its licensors from and against any damages, costs, liabilities, settlement amounts and/or expenses (including attorneys' fees) incurred in connection with any claim, lawsuit or action by any third party that arises or results from the use or distribution of any and all Programs and/or Software.

3. **License to Distribute Redistributables**. Subject to the terms and conditions of this Agreement, including but not limited to Section 4 (Java Technology Restrictions) of these Supplemental Terms, Sun grants you a non-exclusive, non-transferable, limited license to reproduce and distribute the binary form of those files specifically identified as redistributable in the Software "RELEASE NOTES" file ("Redistributables") provided that: (i) you distribute the Redistributables complete and unmodified (unless otherwise specified in the applicable RELEASE NOTES file), and only bundled as part of Programs, (ii) you do not distribute additional software intended to supersede any component(s) of the Redistributables, (iii) you do not remove or alter any proprietary legends or notices contained in or on the Redistributables, (iv) for a particular version of the Java platform, any executable output generated by a compiler that is contained in the Software must (a) only be compiled from source code that conforms to the corresponding version of the OEM Java Language Specification; (b) be in the class file format defined by the corresponding version of the OEM Java Virtual Machine Specification; and (c) execute properly on a reference runtime, as specified by Sun, associated with such version of the Java platform, (v) you only distribute the Redistributables pursuant to a license agreement that protects Sun's interests consistent with the terms contained in the Agreement, and (v) you agree to defend and indemnify Sun and its licensors from and against any damages, costs, liabilities, settlement amounts and/or expenses (including attorneys' fees) incurred in connection with any claim, lawsuit or action by any third party that arises or results from the use or distribution of any and all Programs and/or Software.

4. **Java Technology Restrictions**. You may not modify the Java Platform Interface ("JPI", identified as classes contained within the "java" package or any subpackages of the "java" package), by creating additional classes within the JPI or otherwise causing the addition to or modification of the classes in the JPI. In the event that you create an additional class and associated API(s) which (i) extends the functionality of the Java platform, and (ii) is exposed to third party software developers for the purpose of developing additional software which invokes such additional API, you must promptly publish broadly an accurate specification for such API for free use by all developers. You may not create, or authorize your licensees to create, additional classes, interfaces, or subpackages that are in any

way identified as "java", "javax", "sun" or similar convention as specified by Sun in any naming convention designation.

5. **Java Runtime Availability**. Refer to the appropriate version of the Java Runtime Environment binary code license (currently located at http://www.java.sun.com/jdk/index.html) for the availability of runtime code which may be distributed with Java applets and applications.

6. **Trademarks and Logos**. You acknowledge and agree as between you and Sun that Sun owns the SUN, SOLARIS, JAVA, JINI, FORTE, and iPLANET trademarks and all SUN, SOLARIS, JAVA, JINI, FORTE, and iPLANET-related trademarks, service marks, logos and other brand designations ("Sun Marks"), and you agree to comply with the Sun Trademark and Logo Usage Requirements currently located at http://www.sun.com/policies/trademarks. Any use you make of the Sun Marks inures to Sun's benefit.

7. **Source Code**. Software may contain source code that is provided solely for reference purposes pursuant to the terms of this Agreement. Source code may not be redistributed unless expressly provided for in this Agreement.

8. **Termination for Infringement**. Either party may terminate this Agreement immediately should any Software become, or in either party's opinion be likely to become, the subject of a claim of infringement of any intellectual property right.

For inquiries please contact: Sun Microsystems, Inc. 901 San Antonio Road, Palo Alto, California 94303 (LFI#91205/Form ID#011801)

TERMS AND CONDITIONS OF THE LICENSE & EXPORT FOR
JAVA™ 2 SDK, STANDARD EDITION 1.3.

Sun Microsystems, Inc.
Binary Code License Agreement

READ THE TERMS OF THIS AGREEMENT AND ANY PROVIDED SUPPLEMENTAL LICENSE TERMS (COLLECTIVELY "AGREEMENT") CAREFULLY BEFORE OPENING THE SOFTWARE MEDIA PACKAGE. BY OPENING THE SOFTWARE MEDIA PACKAGE, YOU AGREE TO THE TERMS OF THIS AGREEMENT. IF YOU ARE ACCESSING THE SOFTWARE ELECTRONICALLY, INDICATE YOUR ACCEPTANCE OF THESE TERMS BY SELECTING THE "ACCEPT" BUTTON AT THE END OF THIS AGREEMENT. IF YOU DO NOT AGREE TO ALL THESE TERMS, PROMPTLY RETURN THE UNUSED SOFTWARE TO YOUR PLACE OF PURCHASE FOR A REFUND OR, IF THE SOFTWARE IS ACCESSED ELECTRONICALLY, SELECT THE "DECLINE" BUTTON AT THE END OF THIS AGREEMENT.

1. <u>LICENSE TO USE</u>. Sun grants you a non-exclusive and non-transferable license for the internal use only of the accompanying software and documentation and any error corrections provided by Sun (collectively "Software"), by the number of users and the class of computer hardware for which the corresponding fee has been paid.

2. <u>RESTRICTIONS</u>. Software is confidential and copyrighted. Title to Software and all associated intellectual property rights is retained by Sun and/or its licensors. Except as specifically authorized in any Supplemental License Terms, you may not make copies of Software, other than a single copy of Software for archival purposes. Unless enforcement is prohibited by applicable law, you may not modify, decompile, or reverse engineer Software. You acknowledge that Software is not designed, licensed or intended for use in the design, construction, operation or maintenance of any nuclear facility. Sun disclaims any express or implied warranty of fitness for such uses. No right, title or interest in or to any trademark, service mark, logo or trade name of Sun or its licensors is granted under this Agreement.

3. <u>LIMITED WARRANTY</u>. Sun warrants to you that for a period of ninety (90) days from the date of purchase, as evidenced by a copy of the receipt, the media on which Software is furnished (if any) will be free of defects in materials and workmanship under normal use. Except for the foregoing, Software is provided "AS IS". Your exclusive remedy and Sun's entire liability under this limited warranty will be at Sun's option to replace Software media or refund the fee paid for Software.

4. **DISCLAIMER OF WARRANTY**. UNLESS SPECIFIED IN THIS AGREEMENT, ALL EXPRESS OR IMPLIED CONDITIONS, REPRESENTATIONS AND WARRANTIES, INCLUDING ANY IMPLIED WARRANTY OF MERCHANTABILITY, FITNESS FOR A PARTICULAR PURPOSE OR NON-INFRINGEMENT ARE DISCLAIMED, EXCEPT TO THE EXTENT THAT THESE DISCLAIMERS ARE HELD TO BE LEGALLY INVALID.

5. **LIMITATION OF LIABILITY**. TO THE EXTENT NOT PROHIBITED BY LAW, IN NO EVENT WILL SUN OR ITS LICENSORS BE LIABLE FOR ANY LOST REVENUE, PROFIT OR DATA, OR FOR SPECIAL, INDIRECT, CONSEQUENTIAL, INCIDENTAL OR PUNITIVE DAMAGES, HOWEVER CAUSED REGARDLESS OF THE THEORY OF LIABILITY, ARISING OUT OF OR RELATED TO THE USE OF OR INABILITY TO USE SOFTWARE, EVEN IF SUN HAS BEEN ADVISED OF THE POSSIBILITY OF SUCH DAMAGES. In no event will Sun's liability to you, whether in contract, tort (including negligence), or otherwise, exceed the amount paid by you for Software under this Agreement. The foregoing limitations will apply even if the above stated warranty fails of its essential purpose.

6. **Termination**. This Agreement is effective until terminated. You may terminate this Agreement at any time by destroying all copies of Software. This Agreement will terminate immediately without notice from Sun if you fail to comply with any provision of this Agreement. Upon Termination, you must destroy all copies of Software.

7. **Export Regulations**. All Software and technical data delivered under this Agreement are subject to US export control laws and may be subject to export or import regulations in other countries. You agree to comply strictly with all such laws and regulations and acknowledge that you have the responsibility to obtain such licenses to export, re-export, or import as may be required after delivery to you.

8. **U.S. Government Restricted Rights**. If Software is being acquired by or on behalf of the U.S. Government or by a U.S. Government prime contractor or subcontractor (at any tier), then the Government's rights in Software and accompanying documentation will be only as set forth in this Agreement; this is in accordance with 48 CFR 227.7201 through 227.7202-4 (for Department of Defense (DOD) acquisitions) and with 48 CFR 2.101 and 12.212 (for non-DOD acquisitions).

9. **Governing Law**. Any action related to this Agreement will be governed by California law and controlling U.S. federal law. No choice of law rules of any jurisdiction will apply.

10. **Severability**. If any provision of this Agreement is held to be unenforceable, this Agreement will remain in effect with the provision omitted, unless omission would frustrate the intent of the parties, in which case this Agreement will immediately terminate.

11. <u>Integration</u>. This Agreement is the entire agreement between you and Sun relating to its subject matter. It supersedes all prior or contemporaneous oral or written communications, proposals, representations and warranties and prevails over any conflicting or additional terms of any quote, order, acknowledgment, or other communication between the parties relating to its subject matter during the term of this Agreement. No modification of this Agreement will be binding, unless in writing and signed by an authorized representative of each party.

JAVA™ 2 SOFTWARE DEVELOPMENT KIT (J2SDK), STANDARD EDITION, VERSION 1.3.X

SUPPLEMENTAL LICENSE TERMS

These supplemental license terms ("Supplemental Terms") add to or modify the terms of the Binary Code License Agreement (collectively, the "Agreement"). Capitalized terms not defined in these Supplemental Terms shall have the same meanings ascribed to them in the Agreement. These Supplemental Terms shall supersede any inconsistent or conflicting terms in the Agreement, or in any license contained within the Software.

1. <u>Software Internal Use and Development License Grant</u>. Subject to the terms and conditions of this Agreement, including, but not limited to Section 4 (Java™ Technology Restrictions) of these Supplemental Terms, Sun grants you a non-exclusive, non-transferable, limited license to reproduce internally and use internally the binary form of the Software complete and unmodified for the sole purpose of designing, developing and testing your Java applets and applications intended to run on the Java platform ("Programs").

2. <u>License to Distribute Software</u>. Subject to the terms and conditions of this Agreement, including, but not limited to Section 4 (Java ™ Technology Restrictions) of these Supplemental Terms, Sun grants you a non-exclusive, non-transferable, limited license to reproduce and distribute the Software in binary code form only, provided that (i) you distribute the Software complete and unmodified and only bundled as part of, and for the sole purpose of running, your Programs, (ii) the Programs add significant and primary functionality to the Software, (iii) you do not distribute additional software intended to replace any component(s) of the Software, (iv) you do not remove or alter any proprietary legends or notices contained in the Software, (v) you only distribute the Software subject to a license agreement that protects Sun's interests consistent with the terms contained in this Agreement, and (vi) you agree to defend and indemnify Sun and its licensors from and against any damages, costs, liabilities, settlement amounts

and/or expenses (including attorneys' fees) incurred in connection with any claim, lawsuit or action by any third party that arises or results from the use or distribution of any and all Programs and/or Software.

3. <u>License to Distribute Redistributables</u>. Subject to the terms and conditions of this Agreement, including but not limited to Section 4 (Java Technology Restrictions) of these Supplemental Terms, Sun grants you a non-exclusive, non-transferable, limited license to reproduce and distribute the binary form of those files specifically identified as redistributable in the Software "README" file ("Redistributables") provided that: (i) you distribute the Redistributables complete and unmodified (unless otherwise specified in the applicable README file), and only bundled as part of Programs, (ii) you do not distribute additional software intended to supersede any component(s) of the Redistributables, (iii) you do not remove or alter any proprietary legends or notices contained in or on the Redistributables, (iv) you only distribute the Redistributables pursuant to a license agreement that protects Sun's interests consistent with the terms contained in the Agreement, and (v) you agree to defend and indemnify Sun and its licensors from and against any damages, costs, liabilities, settlement amounts and/or expenses (including attorneys' fees) incurred in connection with any claim, lawsuit or action by any third party that arises or results from the use or distribution of any and all Programs and/or Software.

4. <u>Java Technology Restrictions</u>. You may not modify the Java Platform Interface ("JPI", identified as classes contained within the "java" package or any subpackages of the "java" package), by creating additional classes within the JPI or otherwise causing the addition to or modification of the classes in the JPI. In the event that you create an additional class and associated API(s) which (i) extends the functionality of the Java platform, and (ii) is exposed to third party software developers for the purpose of developing additional software which invokes such additional API, you must promptly publish broadly an accurate specification for such API for free use by all developers. You may not create, or authorize your licensees to create, additional classes, interfaces, or subpackages that are in any way identified as "java", "javax", "sun" or similar convention as specified by Sun in any naming convention designation.

5. <u>Trademarks and Logos</u>. You acknowledge and agree as between you and Sun that Sun owns the SUN, SOLARIS, JAVA, JINI, FORTE, and iPLANET trademarks and all SUN, SOLARIS, JAVA, JINI, FORTE, and iPLANET-related trademarks, service marks, logos and other brand designations ("Sun Marks"), and you agree to comply with the Sun Trademark and Logo Usage Requirements currently located at http://www.sun.com/policies/trademarks. Any use you make of the Sun Marks inures to Sun's benefit.

6. **Source Code**. Software may contain source code that is provided solely for reference purposes pursuant to the terms of this Agreement. Source code may not be redistributed unless expressly provided for in this Agreement.

7. **Termination for Infringement**. Either party may terminate this Agreement immediately should any Software become, or in either party's opinion be likely to become, the subject of a claim of infringement of any intellectual property right.

For inquiries please contact: Sun Microsystems, Inc. 901 San Antonio Road, Palo Alto, California 94303 (LFI#90955/Form ID#011801)

Professional Mindware™

Master today's cutting-edge technologies with M&T Books™

As an IT professional, you know you can count on M&T Books for authoritative coverage of today's hottest topics. From ASP+ to XML, just turn to M&T Books for the answers you need.

Written by top IT professionals, M&T Books delivers the tools you need to get the job done, whether you're a programmer, a Web developer, or a network administrator.

Open Source: The Unauthorized White Papers
408 pp • 0-7645-4660-0 • $19.99 U.S. • $29.99 Can.

XHTML™: Moving Toward XML
456 pp • 0-7645-4709-7 • $29.99 U.S. • 44.99 Can.

Cisco® IP Routing Handbook
552 pp • 0-7645-4695-3 • 29.99 U.S. • $44.99 Can.

Cross-Platform Perl, 2nd Edition
648 pp • 0-7645-4729-1 • 39.99 U.S. • $59.99 Can.

Linux® Rapid Application Development
648 pp • 0-7645-4740-2 • $39.99 U.S. • $59.99 Can.

XHTML In Plain English
750 pp • 0-7645-4743-7 • $19.99 U.S. • $29.99 Can.

XML In Plain English, 2nd Edition
750 pp • 0-7645-4744-5 • $19.99 U.S. • $29.99 Can.

The SuSE™ Linux® Server
600 pp • 0-7645-4765-8 • $39.99 U.S. • $59.99 Can.

Red Hat® Linux® Server, 2nd Edition
816 pp • 0-7645-4786-0 • $39.99 U.S. • $59.99 Can.

Java™ In Plain English, 3rd Edition
750 pp • 0-7645-3539-0 • $19.99 U.S. • $29.99 Can.

The Samba Book *(Available Spring '01)*
550 pp • 0-7645-4773-9 • $39.99 U.S. • $59.99 Can.

Managing Linux® Clusters *(Available Spring '01)*
xx pp • 0-7645-4763-1 • $24.99 U.S. • $37.99 Can.

MySQL™/PHP Database Applications
(Available Spring '01)
504 pp • 0-7645-3537-4 • $39.99 U.S. • $59.99 Can.

Available wherever the very best technology books are sold.
For more information, visit us at www.mandtbooks.com

©2002 Hungry Minds, Inc. All rights resrved. M&TBooks, the M&T Books logo and Professional Mindware are trademarks of Hungry Minds. All other trademarks are property of their respective owners.